Genetic Differentiation and Dispersal in Plants

NATO ASI Series

Advanced Science Institutes Series

A series presenting the results of activities sponsored by the NATO Science Committee, which aims at the dissemination of advanced scientific and technological knowledge, with a view to strengthening links between scientific communities.

The Series is published by an international board of publishers in conjunction with the NATO Scientific Affairs Division

A Life Sciences **B Physics**	Plenum Publishing Corporation London and New York
C Mathematical and Physical Sciences	D. Reidel Publishing Company Dordrecht, Boston and Lancaster
D Behavioural and Social Sciences **E Applied Sciences**	Martinus Nijhoff Publishers Boston, The Hague, Dordrecht and Lancaster
F Computer and Systems Sciences **G Ecological Sciences**	Springer-Verlag Berlin Heidelberg New York Tokyo

Series G: Ecological Sciences No. 5

Genetic Differentiation and Dispersal in Plants

Edited by

P. Jacquard, G. Heim and J. Antonovics

CNRS, Centre L. Emberger
Route de Mende, B.P. 5051
F-34033 Montpellier

Springer-Verlag Berlin Heidelberg New York Tokyo
Published in cooperation with NATO Scientific Affairs Division

Proceedings of the NATO Advanced Research Workshop on Population Biology of Plants held at Montpellier, May 21–25, 1984

ISBN 3-540-15977-0 Springer-Verlag Berlin Heidelberg New York Tokyo
ISBN 0-387-15977-0 Springer-Verlag New York Heidelberg Berlin Tokyo

Library of Congress Cataloging in Publication Data. NATO Advanced Research Workshop on Population Biology of Plants (1984 : Montpellier, France) Genetic differentiation and dispersal in plants. (NATO ASI series. Series G, Ecological sciences ; no. 5) "Proceedings of the NATO Advanced Research Workshop on Population Biology of Plants held at Montpellier, May 21–25, 1984"–T.p. verso. Bibliography: p. Includes index. 1. Plant genetics–Congresses. 2. Botany–Variation–Congresses. 3. Plant populations–Congresses. 4. Seeds–Dispersal–Congresses. 5. Plants–Migration–Congresses. I. Jacquard, P. II. Heim, G. III. Antonovics, J. IV. North Atlantic Treaty Organization. Scientific Affairs Division. V. Title. VI. Series. QK981.N35 1984 581.1'5 85-25004
ISBN 0-387-15977-0 (U.S.)

Printed in Germany

Printing: Beltz Offsetdruck, Hemsbach; Bookbinding: J. Schäffer OHG, Grünstadt
2131/3140-5 4 3 2 1 0

CONTENTS

CONTRIBUTORS

ANTONOVICS, J. : Department of Botany, Duke University, Durham, N.C. 27706, U.S.A.

ARDOUIN, P. : Laboratoire de Phytochimie, Biologie Végétale, Université Lyon I, F-69622 Villeurbanne Cedex, France.

BARRIERE, C. : Laboratoire de Botanique, Université Bordeaux I, F-33405 Talence Cedex, France.

BERGER, A. : Biologie des Populations et des Peuplements, Centre L. Emberger, C.N.R.S., F-34033 Montpellier Cedex, France.

BERTHAUD, J. : Génétique et Physiologie du Développement des Plantes, C.N.R.S., F-91190 Gif-sur-Yvette, France.

BOS, M. : Department of Genetics, University of Groningen, Kerklaan 30, 9751 NN Haren (Gn), The Netherlands.

BOUAB, N. : Biologie des Populations et des Peuplements, Centre L. Emberger, BP 5051, F-34033 Montpellier Cedex, France.

BRIANE, J.P. : Laboratoire Associé 121 (CNRS), Université Paris XI, F-91405 Orsay Cedex, France.

CUGUEN, J. : Biologie des Populations et des Peuplements, Centre L. Emberger, C.N.R.S., BP 5051, F-34033 Montpellier Cedex, France.

DOMMEE, B. : Biologie des Populations et des Peuplements, Centre L. Emberger, C.N.R.S., BP 5051, F-34033 Montpellier Cedex, France.

ELLSTRAND, N.C. :Department of Botany and Plant Sciences, University of California, Riverside, California, U.S.A.

ESCARRE, J. : Laboratoire Associé 121 (CNRS), Université Paris Xl, F-91405 Orsay Cedex, France.

FIASSO, J.L. : Laboratoire de Phytochimie, Biologie Végétale, Université Lyon I, F-69622 Villeurbanne Cedex, France.

GASQUEZ, J. : Laboratoire de Malherbologie, I.N.R.A., BV 1540, F- 21034 Dijon Cedex, France.

GLIDDON, C. : School of Plant Biology, University College of North Wales, Bangor, Gwynedd LL57 2UW, United Kingdom.

GOUYON, P.H. : Biologie des Populations et des Peuplements, Centre L. Emberger, C.N.R.S., BP 5051, F-34033 Montpellier Cedex, France.

GRAVELAND, J. : Institute of Ecological Research, Duinzoom 20a, 3233 Eg Oostvoorne, The Netherlands.

HARMENS, H. : Department of Genetics, University of Groningen, Kerklaan 30, 9751 NN Haren (Gn), The Netherlands.

HAYWARD, M.D. : Welsh Plant Breeding Station, University College of Wales, Aberystwyth, United Kingdom.

HOUSSARD, C. : Laboratoire Associé 121 (C.N.R.S.), Université Paris XI, F-91405 Orsay Cedex, France.

HUBAC, J.M. : Biologie Végétale, Université Paris-Sud, Centre d'Orsay, F-91405 Orsay Cedex, France.

IBRAHIM, M. : Biologie des Populations et des Peuplements, Centre L. Emberger, C.N.R.S., B P 5051, F-34033 Montpellier Cedex, France.

JACQUARD, P. : Biologie des Populations et des Peuplements, Centre L. Emberger, C.N.R.S., B.P. 5051, F-34033 Montpellier Cedex, France.

JAY, M. : Laboratoire de Phytochimie, Biologie Végétale, Université Lyon I, F-69622 Villeurbanne Cedex, France.

JOLY, H. : Génétique et Physiologie du Développement des Plantes, C.N.R.S., F-91190 Gif-sur-Yvette, France.

KELLEY, S.E. : Department of Botany, Duke University, Durham, N.C. 27706, USA.

KJELLBERG, F. : Biologie des Populations et des Peuplements, Centre L. Emberger, C.N.R.S., BP 5051, F-34033 Montpellier Cedex, France.

KUIPER, D. : Department of Plant Physiology, University of Groningen, P.O. Box 14, 9750 AA Haren (Gn), The Netherlands.

LEFEBVRE, C. : Laboratoire d'Ecologie Végétale et de Génétique, Université Libre de Bruxelles, 1850 Chaussée de Wavre, 1160 Brussels, Belgium.

LUMARET, R. : Biologie des Populations et des Peuplements, Centre L. Emberger, C.N.R.S., BP 5051, F-34033 Montpellier Cedex, France.

MATHEZ J. : Laboratoire de Systématique et Ecologie Méditerranéennes, Institut de Botanique, 163 rue Auguste Broussonet, F-34000 Montpellier, France.

MAZZONI, C. : Biologie des Populations et des Peuplements, Centre L. Emberger, C.N.R.S., BP 5051, F-34033 Montpellier Cedex, France.

MORISHIMA, H. : National Institute of Genetics, Misima, 411 Japan.

NGUYEN VAN, E. : Génétique et Physiologie du Développement des Plantes, C.N.R.S., F-91180 Gif-sur-Yvette, France.

N'TSIBA, F. : Biologie des Populations et des Peuplements, Centre L. Emberger C.N.R.S., BP 5051, F-34033 Montpellier Cedex, France.

OLIVIERI, I. : Biologie des Populations et des Peuplements, Centre L. Emberger, C.N.R.S., BP 5051, F-34033 Montpellier Cedex, France.

PERNES, J. : Génétique et Physiologie du Développement des Plantes, C.N.R.S., F-91190 Gif-sur-Yvette, France.

RAYMOND, M. : Biologie des populations et des peuplements, Centre L. Emberger, C.N.R.S., BP 5051, F-34033 Montpellier Cedex, France.

ROACH, D.A. : Department of Botany, Duke University, Durham, N.C. 27706, U.S.A.

ROSS, M.D. : Department of Forest Genetics and Forest Tree Breeding, university of Göttingen, D-3400 Göttingen - Weende, West Germany.

SACKSCHEWSKY, M.R. : Department of Botany, Washington State University, Pullman, W.A 99164-4320, U.S.A.

SALEEM, M. : School of Plant Biology, University College of North Wales, Gwynedd LL 57 2UW, United Kingdom.

SARR, A. : Génétique et Physiologie du Développement des Plantes C.N.R.S. F-91190 Gif-sur-Yvette, France.

SECOND, G. : Biologie des Populations et des Peuplements, Centre L. Emberger, C.N.R.S., B.P. 5051, F-34033 Montpellier Cedex, France.

STEEN, R. : Departement of Genetics, University of Groningen Kerklaan 30, 9751 NN Haren (Gn), The Netherlands.

THIEBAUT, B. : Biologie des Populations et des Peuplements, Centre L. Emberger, C.N.R.S., BP 5051, F-34033 Montpellier Cedex, France.

VALDEYRON, G. : Biologie des populations et des Peuplements, Centre L. Emberger, C.N.R.S., BP 5051, F-34033 Montpellier Cedex, France.

VALERO, M. : Laboratoire de Génétique Ecologique et Biologie des Populations Végétales, U.S.T.L., F-59655 Villeneuve d'Asq Cedex, France.

VALIZADEH, M. : Biologie des Populations et des peuplements, Centre L. Emberger, C.N.R.S., BP 5051, F-34033 Montpellier-Cedex, France.

VAN DAMME, J.MM.: Institute of Ecological Research, Duinzoom 20a, 3233 EG Oostvoorne, The Netherlands.

VAN DIJK, H. : Department of Genetics, Kerklaan 30, 9751 NN Haren (Gn), The Netherlands.

WACQUANT, J.P. :Biologie des Populations et des Peuplements, Centre L. Emberger, C.N.R.SX., BP 5051, F-34033 Montpellier Cedex, France.

WILLIAMS, G.Z. :Department of Botany, Washington State University, Pullman, WA 99164-4320, U.S.A.

XENA de ENRECH, N. : Laboratoire de Systématique et Ecologie Méditerranéennes, Institut de Botanique, 163 rue Auguste Broussonnet, F-34000 Montpellier, France.

PREFACE

This volume is based on a workshop on "Population Biology of Plants : The Interfaces (Genetics, Physiology, Demography, Biogeography)", with a specific profile on "Diversification of Plant Populations in relation to Modes of Reproduction and Dispersal : Genetic and Physiological Mechanisms", held in Port-Camargue, France, from May 21-25, 1984. This workshop was initiated by the "Unit of Population and Community Biology", in Montpellier, and sponsored by the NATO Scientific Affairs Division (ARW grant 876/83) and by the CNRS (Table ronde).

All populations are subjected to environmental "screening". Given a genetic diversity whose expression can be modified by a degree of demographic and individual plasticity (at the morphological and physiological levels), they present a structure related to their environment. Ideally populations should be studied simultaneously from the point of view of the population geneticist, the physiologist and the demographer . These specific approaches only become fully meaning full in the "light of Evolution".

Among the evolutionary forces that quantitatively act on the frequencies, objects of interest of workers specialising in Population Biology are selection and drift.

An other main object of the study must be dispersal. But, its playing extent - relative to the other forces - in the adjustment to the environment is not fully recognized.

One might envisage a critical dispersal distance beyond which the continued distribution of propagules would tend to the existence of only one taxon (except for some specific situations).

What evidence do we have to corroborate such a hypothesis or in what extent this statement and its reciprocal are true ? Do the modes of reproduction (vegetative or sexual, autogamous or allogamous) significantly modify the dispersal ?

Since some years, the group of Plant Population Biology, from CNRS, in Montpellier, has been working these questions. It has seemed to this group that its own working environment might be a propitious venue for discussion at the level of world specialist, in Botany. The non participation of zoologists, or only in a passive way, was one of the disappointment of the meeting. In the future it would be essential, to develop comparisons of the two types of organisms, or for a clarification of the important role of animals in the migration of plants.

Nevertheless, the papers presented at the meeting as well as the formal and non formal discussions contribute to a review of results concerned with diversification at two levels :

1. gene and molecular level ;

2. phenotypic and fitness level ;

and with dispersal at also two levels :

1. gene level, as gene flow ;

2. individual level :

The papers are revised versions of those presented at Port-Camargue. They were reviewed by : J. Antonovics, A. Berger, P.-H. Gouyon, J. Harper, M. Hayward, G. Heim, P. Jacquard, J. Pernès, L. Thaler, K.M. Urbanska, G. Valdeyron, Ph. Vernet and J.-M. Woldendorp. We wish to thank all the reviewers for their help.

Special thanks are due to Olga Caggia for assistance in organizing the meeting and Elisabeth Ruiz for editorial assistance.

P. JACQUARD

Montpellier
Mars 1985

Genetic Differentiation: Variation in Single Genes or Molecular Variation

Chemical Diversification Within The Dactylis glomerata L. Polyploid Complex (Gramineae).

P. Ardouin, J.L. Fiasson, M. JAY, R. Lumaret* and J.M. Hubac**

Laboratoire de Phytochimie, Biologie Végétale, Université Lyon I, F 69622 Villeurbanne Cedex

ABSTRACT

The complex of Cocksfoot (Dactylis glomerata) consists of diploids and tetraploids. Diploid taxa are morphologically and physiologically differentiated ; they are geographically separated and occur in specialized habitats. Tetraploid plants, less differentiated and specialized, occur in numerous and various sites ; they represent the complex over most of its distribution area. The present paper deals with polyphenolic compounds as chemical markers of genetical diversity between and within natural populations ; multivariate analysis of HPLC profiles from more than 200 representatives of di- and tetraploid populations gives an original view of the diversification trends in this species. A finer study of several stations with a sympatric occurence of both cytotypes allows to appreciate at the micro-local scale the range of chemical diversification following the polyploidisation.

INTRODUCTION

The complex of Dactylis glomerata comprises two chromosomic types : diploids (2n = 14) and cytologically autotetraploids (2n = 48). The diploids are morphologically and physiologically differentiated. Each of 14 subspecies has a limited

---

(*) Biologie des Populations et des Peuplements, Centre L. Emberger, C.N.R.S., F 34033 Montpellier Cedex
(**) Biologie Végétale, Université Paris-Sud, Centre d'Orsay, F 91405 Orsay Cedex

NATO ASI Series, Vol. G5
Genetic Differentiation and Dispersal in Plants
Edited by P. Jacquard et al.

distribution (Fig. 1) and a specialized habitat, e.g. Dactylis glomerata ssp. aschersoniana which occurs in most of Europe but mainly grows in undisturbed forest patches. The tetraploids assigned to three subspecies appear less differentiated and less specialized : ssp. glomerata occurs in temperate eurasian countries, ssp. hispanica and marina have a circum-mediterranean distribution. Except for the latter taxon, restricted to the seashore, tetraploids are capable of colonizing a greater range of habitats than diploids, and have a much larger and continuous distribution (Borrill 1977).

The aim of the present work was to study the organization within the D. glomerata complex with help of polyphenolic markers distribution. Two parallel approaches were attempted. The first dealt with 214 individuals originating from 40 natural diploid and tetraploid populations, the sampling being representative of the complex as a whole. The second approach was focused on various sites cohabited by diploid and tetraploid populations. Those stations were particularily suited for appreciating the potential of metabolic and chemical diversification and the incidence of gene flow between and within cytotypes.

Flavonoids (polyphenolic compounds belonging to the secondary metabolism) have been chosen as chemical markers : recent data demonstrate that they are very useful in studies dealing with Population Biology of Plants (Jay & Gorenflot 1980 ; Nicholls & Bohm 1982 ; Gorenflot & al. 1983 ; Jay & al. 1984, 1985 ; Ardouin & al. 1985 ; Ismaili & Jay 1985). Flavonoids are considered as allelochemics (Harborne 1979 ; Harborne & Ingham 1979) and their accumulation can be related to a plant's answer to environmental conditions. Futhermore, they offer a wide range of structural diversity, based on various hydroxytions, methylations and glycosylations, that allows a large range of metabolic choices. This diversity, organized with reference to biosynthetic branchings and sequences, could be helpful when a dynamic interpretation of the populational structure is considered.

MATERIAL AND METHODS

Plant Material

The plants were grown from seeds collected in various natural habitats and cultivated during 2-5 years at the Experimental Garden of C.E.P.E., Montpellier, France. The material was taken from young autumnal sprouts as a general precaution against physiological variation in flavonoid pattern.

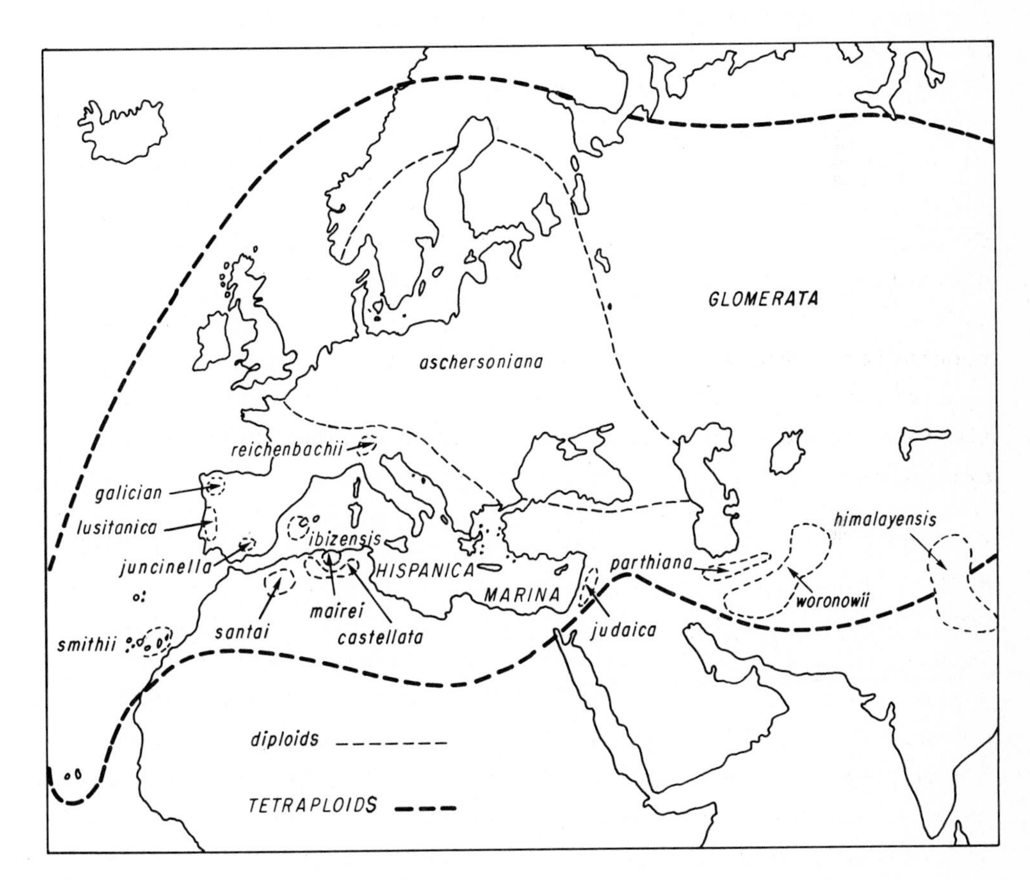

Figure 1 : The primary distribution of D. glomerata

The origin of seed collections has been indicated in detail in previous papers (Lumaret 1981, 1984 a & b) : Dactylis glomerata ssp. aschersoniana Greabner : Landskrana - Sweden, Saint Nabor - France ; ss. castellata Borrill : Le Kef - Tunisia,

Bouira - Algeria, Gued Chiffa - Algeria ; ssp. "Galician" Borrill : Convite - Spain ; ssp. glomerata Hayek : La Clusaz - France, Puy de Dôme - France, Loch Ness - Scotland ; ssp. himalayensis Domin : Gangi Tehri - India ; ssp. hispanica Roth Olmi Capella - Corsica, Sierra Nevada - Spain, Bono - Sardinia Mas de Ricome - France ; ssp. ibizensis Stebb. & Zoh. : Ibiza-Balearic Isles ; ssp. juncinella Bory : Sierra Nevada - Spain ; ssp. judaica Stebb. & Zoh. : Safad - Israel, Mont Kenaan - Lebanon ; ssp. lusitanica Stebb. & Zoh. : Serra de Cintra - Portugal ; ssp. mairei Stebb. & Zoh. : Kerrata gorges - Algeria ; ssp. marina Borrill : Foz d'Arelho - Portugal, Nazare - Portugal, Cape St Vincento - Protugal, Sanguinaires Isles - France ; ssp. parthiana Park. & Borr. : Mont Elbruz - Iran ; ssp. reichenbachi Stebb. & Zoh. : Dolomites - Italy ; ssp. santai Stebb. & Zoh. : Relizane - Algeria ; ssp. smithii Link.: La Grovata - Canary Islands, Barraco de Ruiz - Canary Islands ; ssp. woronowii Ovezinn : Katal Yekchian - Iran.

Phytochemical Analysis

Flavonoid analysis was performed on leaf material. Air-dried leaves were extracted by an alcoholic mixture ; after concentrating under reduced pressure, the residue was dissolved in boiling water (discarding lipophilic contaminants) and shaken against n-Butanol (where the flavonoids, but not the more hydrophilic contaminants, migrate). Butanolic epiphase was evaporated and dry residue taken up in a small volume of methanol.

The methanolic solution was subsequently analysed by High Pressure Liquid Chromatography procedure (C_{18} column, elution by a gradient of CH_3CN in diluted acetic acid). The components of the flavonoid pattern were located by mean of the retention time of eluted peaks ; the quantitative estimation was got through the peak heights. Structural identification was achieved by cochromatography with authentic samples.

The flavonoid analysis from leaf extracts of D. glomerata s.l. revealed 43 chromatographic peaks, each representing at least one flavonoid compound ; table 1 gives some examples of the main classes characteristics for this species.

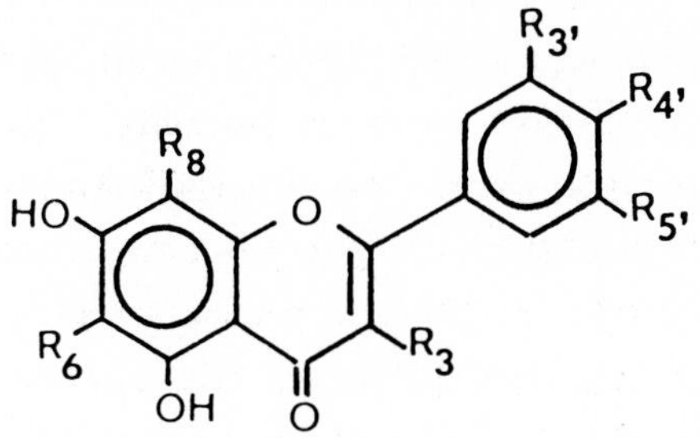

C-glycosyl flavones	R3	R3'	R4'	R5'	R6	R8
isoorientin	H	OH	OH	H	gluc.	H
orientin	H	OH	OH	H	H	gluc.
lucenin 3	H	OH	OH	H	gluc.	xyl.
isovitexin	H	H	OH	H	gluc.	H
schaftoside	H	H	OH	H	gluc.	arab.
O-glycosyl flavones						
4'-O-glucosyl tricin	H	OMe	O-gl.	OMe	H	H
O-glycosyl flavonols						
3-O-glucosyl isorhamnetin	O-gl.	OMe	OH	H	H	H

Table 1 : Major representatives of the main classes of flavonoids in the leaves of Dactylis glomerata L.

Multivariate Analysis

The data from chromatographic patterns were submitted to the Factorial Analysis of Correspondences.

RESULTS

The ordination of the plant collection by F.A.C. of the flavonic content is presented in Fig. 2 with respect to the first two axes which were accounting for 33 % of the total variation of the matrix.

Diversification in the diploid Populations (Fig. 2)

Variability within subspecies : some subspecies have been sampled from one natural population, the others from two or three. Generally, the range of the sampling was related to the size of the geographical area. Notwithstanding the number of samples the chemical diversity at the population level had the same order of magnitude as at the subspecies level. In contrast, the chemical diversity between the 14 diploid taxa was significantly greater.

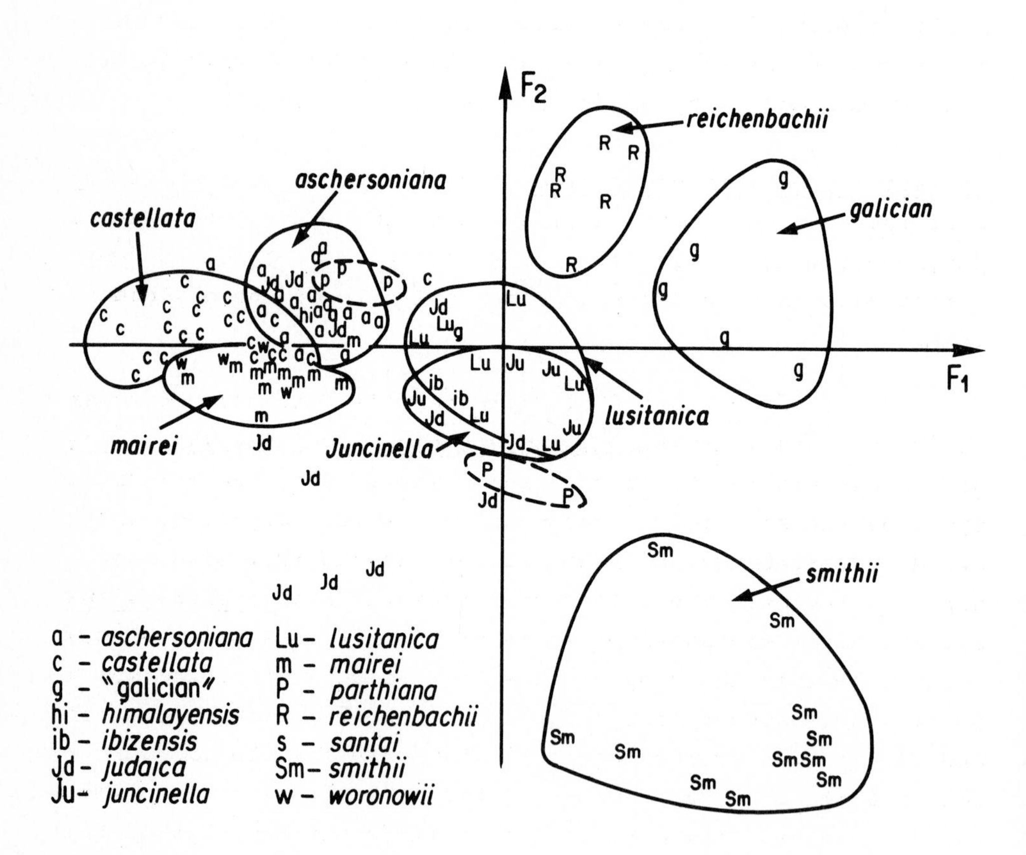

Figure 2 : Ordination obtained by F.A.C. of H.P.L.C. flavonoid data concerning 120 individuals of 20 diploid populations of D. glomerata. Axes 1 and 2 account for 33% of the total variation.

Among the particular cases, ssp. parthiana displays two differential phenolic pathways, with emphasis on either mono-C-glycosyl- or di-C-glycosyl-luteolins ; such a differentiation has not appeared so marked in morphological or electrophoretical data (Borrill 1977 ; Lumaret 1981).

The flavonoid variation of ssp. judaica is of special interest. While all individuals from two populations (Lebanon and Israel) were well defined by an -as yet unknown- flavonoid pattern, some of them managed a biosynthetic pathway resembling the chemical profile of temperate cocksfoot. This heterogeneity could be related to the evolutionary lag between temperate electrophoretical pattern (Lumaret 1981, 1984) and mediterranean morphology (Borrill 1977) shown by this taxon.

Variability between subspecies : the polyphenolic results support the previous individualization of the subspecies based on geographical, morphological and electrophoretical data. They demonstrate once more the genetic isolation of the diploid taxa.

The ordination obtained from phytochemical data puts together the representatives of ssp. aschersoniana and ssp. himalayensis ; this similarity of strategies was already known from enzymology (Lumaret 1981, 1984). As far as the morphology is concerned, Stebbins and Zohary (1959) argued that plants of ssp. himalayensis are taller than those of ssp. aschersoniana and have more elongated panicles with numerous smaller glomerules ; however the same authors pointed out that there was a considerable amount of overlapping in morphological characters and that many specimens could not be correctly placed, were the localities of collection unknown. This leads to question the taxonomical viewpoint of Systematists who distinguish the ssp. himalayensis apparently on the single basis of geographical origin.

In the same way, the flavonoid pattern also fails to distinguish ssp. castellata from ssp. santai. Both taxa show almost the same geographical distribution and the same electrophoretical profile but they were recognized by distinct phytodermological characters viz. stomatal density, size of epidermal and guard cells. Such differences have to be used cautiously, but they reflect differential selective pressures associated with the divergent ecological preferences of the two taxa.

More generally, the diploid part of the D. glomerata complex is gathered in two large clusters : - one of them in the upper part of the figure embraces the temperate taxa, ssp. aschersoniana, himalayensis, parthiana, reichenbachi, "galician" and lusitanica ; - the other on the lower part of the figure is constituted by the mediterranean and subtropical taxa. This distribution into two groups against the first two axes of the ordination, resulting from the characteristic biosynthetic capabilities, is in perfect agreement with the conclusions drawn from morphology (Borrill 1977) and enzymology (Lumaret 1981, 1984).

The organization of the upper group is supported by the importance of C-glycosyl flavones on one hand (eurasian populations ssp. aschersoniana, himalayensis, parthiana) and O-glycosyl flavones on the other hand ("latin" populations : ssp. reichenbachii, "galician", lusitanica). A more complex metabolic situation appears in the lower group, four chemotypes being characterized : 1) an original mono-6-C-glycosyl flavone based chemotype expressed in subtropical ssp. smithii ; 2) a C-glycosyl flavone chemotype based on luteolin derivatives for ssp. castellata and santai ; 3) a peculiar, not yet elucidated, chemotype for ssp. judaica ; 4) a very simple phenolic pattern mainly based on di- and mono-C-glycosides of luteolin for ssp. mairei and woronovii. These very different chemotypes emphasize the diversity of the metabolic answers instigated by the selective pressures operating with in the mediterranean basin.

Diversification in the tetraploid Populations (Fig. 3)

In the tetraploid part of the complex, the concept of population is much more accurate than in the diploid part. However the chemical intrapopulational variation remains more important than variation between populations of a given taxon. This observation can be verified as well in ssp. glomerata as in ssp. hispanica, and is again in full agreement with previous observations (Borrill 1977 ; Lumaret 1981, 1984).

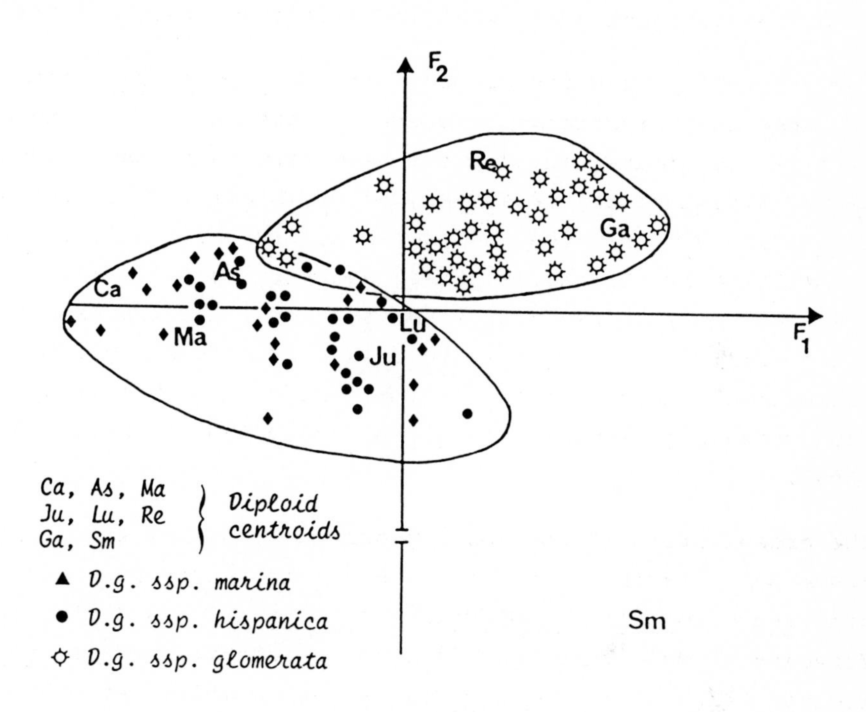

Figure 3 : Ordination obtained by F.A.C. of H.P.L.C. flavonoid data concerning 94 individuals of 12 tetraploid populations of D. glomerata. Axes 1 and 2 account for 33% of the total variation.

Another important trait is the almost complete demarcation between temperate and mediterranean populations as far as the polyphenolic strategy is concerned ; from the biochemical view point it is an original result against the electrophoretical data. Moreover, a good metabolic similarity is observed between a given tetraploid taxon and the various diploid subspecies within the same climatic / geographical group. This encompassing, by the tetraploid's variation, of the characters of the diverse diploids occurring in its distribution area, has been considered by Borrill (1977) and by Lumaret (1981,

1984) as either a phylogenetic relationship or an adaptative convergence.

Particular study of the sites cohabited by Di- and Tetraploids

The phytochemical traits have also been investigated in particular sites where diploids and tetraploids occur together. One of the three stations studied is that of ssp. mairei in the Kerrata Gorges (Algeria), where 14 tetraploids and 15 diploids were collected for a phytochemical approach (Fig. 4).

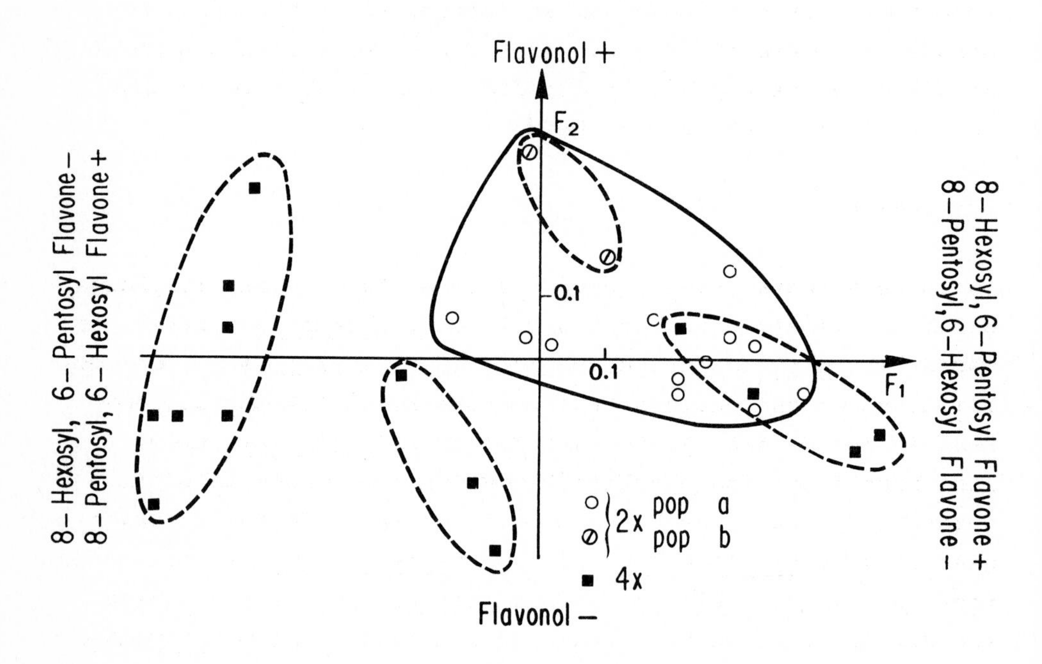

Figure 4 : Ordination obtained by F.A.C. of H.P.L.C. flavonoid data about 29 sympatric individuals of D. glomerata ssp. mairei (Kerrata Gorges, Algeria) ; axes 1 and 2 account for 64% of the total variation.

It is obvious that the doubling ofploidy level allowed new genetic combinations, reflected by new equilibria of polyphenolic metabolism. The diploid individuals are characterized by the presence of flavonol and a good balance between hexosyl transferase activities concerning respectively the C-6 and C-8 positions on the flavone skeleton. The tetraploids (probably auto-) initiated three new phenolic equilibria : 1) giving privilege to hexosyl-transferase activity on C-6 ; 2) with emphasis on C-8 hexosylation ; 3) preserving the balance of hexosylation but loosing the ability to synthetize the flavonol skeleton. Such a large chemical diversification suggests that the phenolic variation of the tetraploid Cocksfoot as a whole might have resulted firstly from regulative alterations at the time of the polyploidisation. The distance separating the extreme tetraploids from the syntopic diploids can be interpretated as the result of an introgression from another, parapatric, tetraploid e.g. in the case of Kerrata Gorges a gene flow between the autotetraploid ssp. *mairei* and another tetraploid growing around the Gorges.

CONCLUSION

The present work gives a new complement to previous morphological and isozymic approaches of the *D. glomerata* complex. As far as the *diploids* are concerned, a relatively good individualization of the various taxa was observed. Of special interest is the original and heterogenous ssp. *judaica*, some individuals showing characteristics of "temperate" phenolic metabolism. The study revealed similarity in flavonic strategies of ssp. *himalayensis* and ssp. *aschersoniana* as well as in those of ssp. *castellata* and ssp. *santai*. On the other hand, two differential pathways have been found in ssp. *parthiana*. In the *tetraploids* a significative disjunction is observed according to the climatic groups : temperate ssp. *glomerata* was quite distinct from mediterranean ssp. *hispanica* and ssp. *marina*, each tetraploid showing similarities with diploids of the same area. In particular cases of *sympatric occurrence* of diploids and tetraploids, metabolic modifications induced by autopolyploidisation and possibly followed by further diversification due to gene flow from another local tetraploid, were clearly present.

Thus, secondary metabolism seems to be very hepful in studies on the genetic variability in the D. glomerata. The intrinsic characteristics of such markers give a new light on the organisation of the polyploid complex.

REFERENCES

Ardouin P, Jay M, Lumaret R(1985)Etude d'une situation de sympatrie entre diploides et tétraploides de Dactylis glomerata sur la base du polymorphisme enzymatique et phénolique. Can. J. Bot. in press

Borrill M (1977) Evolution and genetic resources in Cocksfoot Ann. Report Welsh Plant Breed. Station : 90-209

Gorenflot R, Hubac JM, Jay M, Lalande P (1983) Geographic distribution, polyploidy and pattern of flavonoids in Phragmites australis (Cav.) Trin.ex Steud. In : Felsenstein J (ed) Numerical Taxonomy. Springer Berlin, p 474-478

Harborne JB (1979) Flavonoid Pigments. In : Rosenthal GA, Janzen DH (eds) Herbivores, the interactions with secondary metabolits. Acad. Press London, p. 619-655

Harborne JB, Ingham JL (1979) Biochemical aspects of the Co-evolution of Higher Plants with their fungal parasites. In : Harborne JB (ed) Biochemical aspects of plants and animals co-evolution. Acad. Press London, p. 342-406

Ismaili A, Jay M (1985) Polyphenolic pattern in Arrhenatherum elatius. Phytochemistry in press

Jay M, Gorenflot R (1980) Premières observations relatives à la variation de l'expression du métabolisme flavonique chez le Phragmites australis (Cav.) Trin. ex Steud. Rev. Gen. Bot Fr. 8:261-274

Jay M, Ismaili A, Khalfallah N, Lacoste A (1985) Polymorphisme flavonique comme moyen de mesure des stratégies adaptatives et de la hiérarchie phylogénétique au sein du genre Arrhenatherum. Taxon in press

Jay M, Plenet D, Ardouin P, Lumaret R, Jacquard P (1984) Flavonoid variation in seven tetraploid populations of Dactylis glomerata L. Biochem. Syst. & Ecol. 12:193-198

Lumaret R (1981) Structure génétique d'un complexe polyploide *Dactylis glomerata* L. Thesis Montpellier France.

Lumaret R (1984a) The role of polyploidy in the adaptative significance of polymorphism at the GOT 1 locus in *Dactylis glomerata* complex. Heredity 52:153-169

Lumaret R (1984b) Further contribution to the knowledge of evolution in *Dactylis glomerata* (polyploid complex) from enzymatic polymorphism studied in several diploid and tetraploid subspecies. In : Whyte RO (ed) The paleoenvironment of East Asia. Hong-Kong Univ. Publ., p.

Nicholls KW, Bohm BA (1982) Quantitative flavonoid variation in *Lupinus sericeus*. Biochem. Syst. & Ecol. 10:225-231

Stebbins GL, Zohary D (1959) Cytogenetics and Evolutionary studies in the genus *Dactylis*. 1-morphology and interrelationships of the diploid subspecies. Univ. California Publ. Bot. 31:1-40

Enzymatic Variability of Beechstands (*Fagus sylvatica L.*) on three Scales in Europe : Evolutionary Mechanisms

J. CUGUEN*, B. THIEBAUT***, F. NTSIBA* and G. BARRIERE**

* Biologie des Populations et des Peuplements, Centre L. Emberger, C.N.R.S., B.P. 5051, 34033 Montpellier CEDEX, France.

ABSTRACT

Enzymatic polymorphism of beech is examined by electrophoresis with two loci of peroxidases (Px1, Px2).

Today, there are many studies concerning the enzymatic polymorphism of grassy and bushy plants with "short generation" because in this case, it is easier to reveal genetic variation with experimentation. It is the contrary in the case of trees with "long generation" (beech become adult about 40-50 years in the best) which is why they are less often studied.

In the case of beech, in order to overcome this difficulty, we have chosen to analyse beechstands in as different environments as possible, firstly on a large European scale in the whole beech area, secondly on a regional scale in the southern region and finally, thirdly, on a local scale in six beechstands of the Aigoual-mountain (Cevennes).

On the two first scales, allelic frequencies are not randomly distributed, on the contrary they constitute geographical clines.

** Laboratoire de Botanique, Université de Bordeaux I, Bordeaux, France.

*** Laboratoire de Systématique et d'Ecologie méditerranéennes, Université des Sciences et Techniques du Languedoc, Institut de Botanique, 34000 Montpellier, France.

NATO ASI Series, Vol. G5
Genetic Differentiation and Dispersal in Plants
Edited by P. Jacquard et al.

Some of these clines are related to climatic variation; whereas the other clines do not seem to be related to environmental conditions. In the first case, the role of the natural selection is strongly suggested by the biochemical variations which were always observed with the same ecological modification, in distant regions and on different scales. But in the second case, genic migration, stochastic forces and the history of the beechstands can explain allelic variation.

These studies give us a momentary image of the beechstand's allelic structures. But it is obvious that today's variability which has been gradually established by means of evolutionary mechanisms may continue to evolve.

On a local scale we can study these mechanisms. From one generation to the next, significant differences appear. In and between 15 years old regenerations significant differences also appear. These examples show that today's allelic structures are in evolution. These results are now going to lead us to consider the local scale spatial and temporal genetic structuration of beechstands and the mating and dispersion system of the beech.

INTRODUCTION

Genetic variability in natural populations is due to a series of factors whose actions are antagonistic, homogenizing or diversifying. Allozymes have already been used to describe genetic differentiations between plants which are close together, have a continuous geographic distribution and even for cases of long-lived trees (GRANT & MITTON 1977, BERGMANN 1978, LUNDKVIST 1979, MITTON et al. 1980, LINHART et al. 1981).

In order to study the enzymatic polymorphism of beech, we decided to start with the resources of the field -we are not denying the importance of experimentation- by analysing the enzymatic polymorphism of beechstands located in as different environments as possible. This investigation was carried out firstly on a large European scale in the whole beech area, secondly on a regional scale in the southern regions and thirdly, on the scale of a local forest in six beechstands on the Aigoual mountain (Cévennes-France). We describe some of the alloenzymatic variability of beechstands and attempt to clarify the evolutionary mechanisms which are responsible.

The beech is considered a climax species in most parts of Europe (fig. 1) where it grows in very different environments. Particularly in the southern part of its area, beechstands develop in very contrasted climates: atlantic and mediterranean, plains and mountains. The beech is a monoecious tree, with male and female flowers, which doesn't mate before the tree is at least 40 or 50 years old. Its pollination is anemophilious, and its mating system is essentially allogamous (SCHAFFALITSKY -de-MUCKADELL 1955).

We know a little about the history of the European beech starting from the last glacial epoch. Today's stands of beech apparently regrew from an important source located in the Balkans and from several other secondary sources located around the Mediterranean Sea and in the south-western part of France on the Atlantic coast (BEUG 1967, SERCELJ 1970, JALUT 1974, TRIAT-LAVAL 1978, PONS 1983 ...). The beech seems to have reached its present distribution at different periods for each region: since 5000 or 4000 BP in the southern regions but only since 3000 or 2500 BP in the northern and western plains (DE BEAULIEU & PONS 1979, VERNET 1981). Considering that one beech generation is on a minimum of 60 to 100 years long, the number of generations in these forests is relatively small. Today's beechstands are thus genetically young material. But according to this data however, there may have twice as many generations in the south than in the north.

So we can see how one ecological factor and two particular historical circumtances have distinguished southern beechstands and have encouraged a polymorphism of the beech in these regions: contrasted climates, a plurality of sources and a greater number of generations.

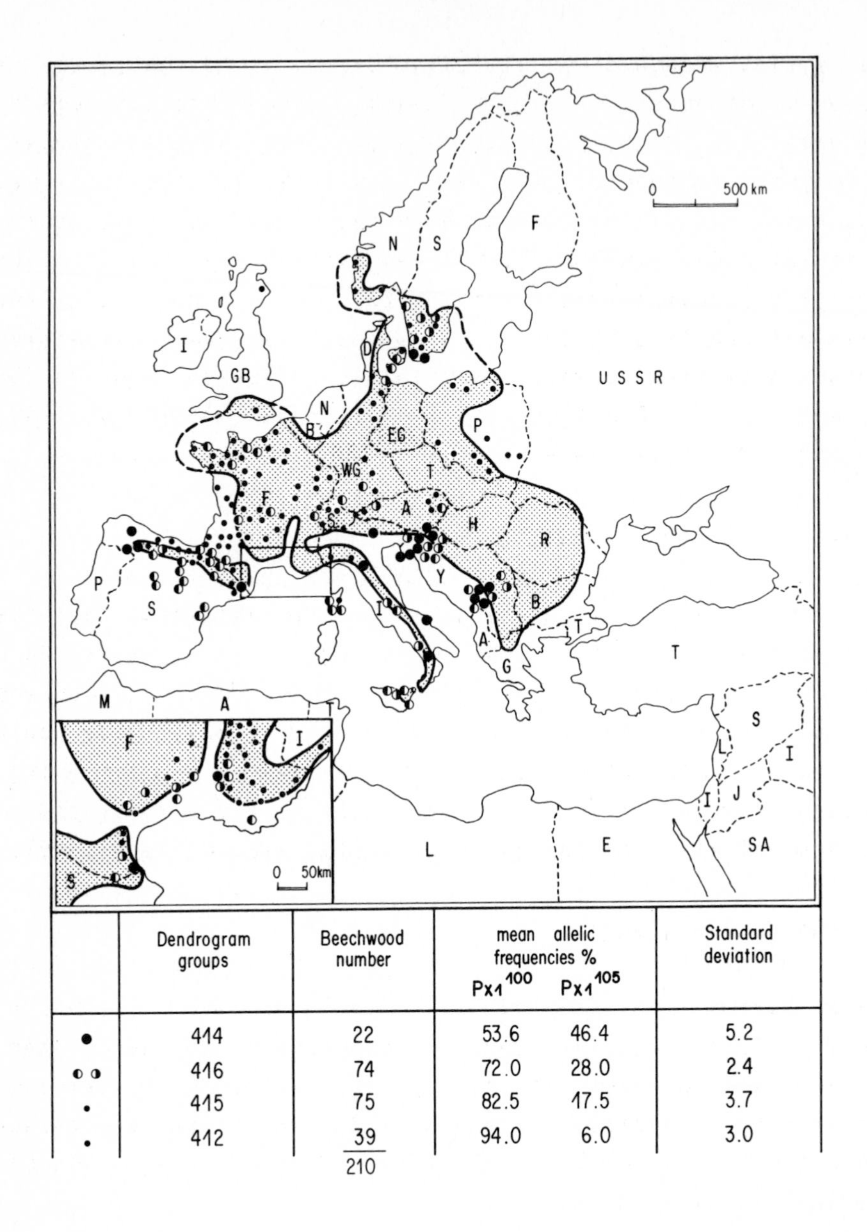

	Dendrogram groups	Beechwood number	mean allelic frequencies % $Px1^{100}$	$Px1^{105}$	Standard deviation
●	414	22	53.6	46.4	5.2
◐ ◑	416	74	72.0	28.0	2.4
•	415	75	82.5	17.5	3.7
·	412	39	94.0	6.0	3.0
		210			

Figure 1 - Beech area (Fagus sylvatica L.) and localisation of studied beechwoods. Peroxidases 1 (Px1): geographical distribution and allelic diversity.

The genetic variability of beech seems quite high. The morphological and sylvan characteristics of the trees vary according to their origins (KRAHL-URBAN 1958, GALOUX 1966, GOHRN 1972, GARELKOVA 1977 and 1983, TEISSIER-du-CROS et al. 1981), but there is not enough genetic isolation for us to infer any clear speciation effects. For the moment, the european beech is considered -and rightly so- to be one single species without any well-defined subspecies (TUTIN et al. 1964). The biochemical characteristics of the beech have been studied more recently, starting with KIM (1979 and 1980) who had already shown variations in enzymatic polymorphism between two origins and at different stages of seedling growth. In Montpellier, we are analysing systematically the enzymatic variability of beechstands in Europe (THIEBAUT 1982, THIEBAUT et al. 1982;, FELBER & THIEBAUT 1982 and 1984, BARRIERE et al. in press).

1 - MATERIAL AND METHODS

The genetic variability of beech is examined by starch electrophoresis analysis, with two genetic markers in the peroxidases: Px1 and Px2. According to the accepted hypothesis of their genetic determinism (THIEBAUT et al. 1982), these would be monomeric enzymes controlled by two polymorphic loci with two codominant alleles (Px1-100 and Px1-105) for the first and three codominant alleles for the second (Px2-13, Px2-26 and Px2-39).

Large and regional scales

210 beechstands were sampled in the whole beech area (fig. 1) at different latitudes, in varied topographic locations and on different types of soil. In each forest, the material analysed was taken from approximately 50 trees of different ages and chosen at random from a stand of 3 to 4 hectares characterized by homogenous ecological conditions.

Local scale

Firstly, two generations per forest were studied in six

Mathematical treatments of the large European and regional scales have been realised by Mr. Ph. BONNERIC (C.I.T.I.M. - U.S.T.L., Montpellier) on the C.N.U.S.C. Computer.

beechstands on the Aigoual mountain, and secondly six natural regenerations with more or less vigorous individuals were studied in one beechstand. In each case, the samples were chosen from 50 individuals per generation or per degree of vigor.

2 - RESULTS

2.1. - Large European scale

In most cases, the mating system seems to be panmitic. Of the 210 beechstands analysed, only 20 forests for Px1 and 8 others for Px2 do not correspond to the expectations of the Hardy-Weinberg equilibrium.

The five alleles of peroxidases are present in almost all of the forests. However, allelic frequencies vary considerably between beechstands, non randomly and according to precise geographic gradients.

Px1 locus: In general, the Px1-100 allele is more frequent (between 41 and 100%) than Px1-105 (between 0 and 59%). A hierarchical classification (fig. 2) reveals four groups of forests according to their allelic diversity. Their diversity varies a lot in the southern and northern beechstands, where it is high or low (fig. 1), while it is generally low in middle European forests. The diversity variations are more apparent, in the south and north, on the latitudinal limits of the beech area where climates are more extremes.

Px2 locus: The Px2-26 allele is the most frequent (between 54 and 96%), Px2-39 less frequent (between 6 and 40%) and Px2-13 even rarer (between 0 and 20%). These allelic frequencies change according to two geographic gradients, the first latitudinal and the second longitudinal.

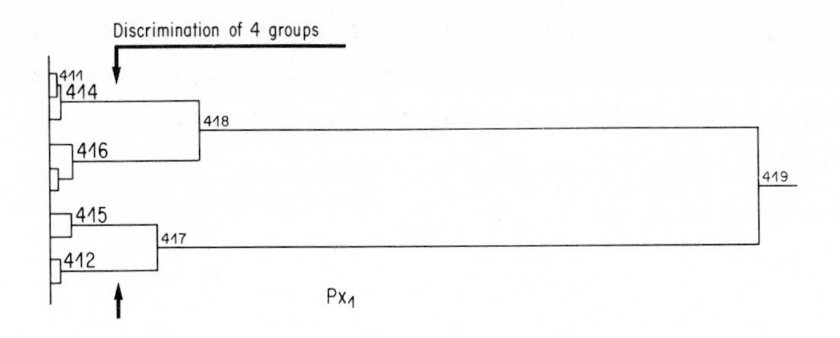

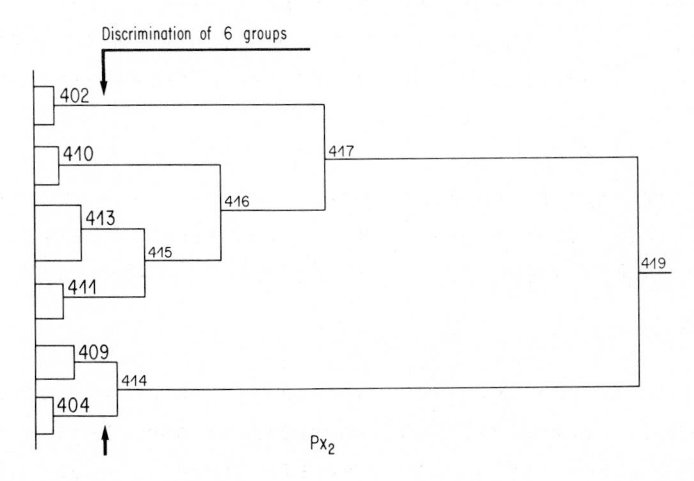

Figure 2 - Dendrograms of peroxidases I (Px1) and peroxidases 2 (Px2): 210 beechwods, chi-square - distances and aggregation according to the mean distance

We can distinguish six groups of forests according to their allelic diversity (fig. 2). It varies more in southern regions (fig. 3), thus on the limit of the beech distribution in harsh climates.

The frequencies of the Px2-13 allele varies from the west to the east (table 1): rare in the west (Cantabrical Pyrenees), it becomes more frequent in the Spanish and French Mediterranean regions, then its frequency decreases towards the east until Yugoslavia, including Corsica and the Italian Peninsula. This last gradient seems to be independent of the general ecological conditions.

We can thus notice that the variations in allelic diversity are simultaneously higher for the two allozymes in the South of France. Are these frequencies randomly distributed or not ?

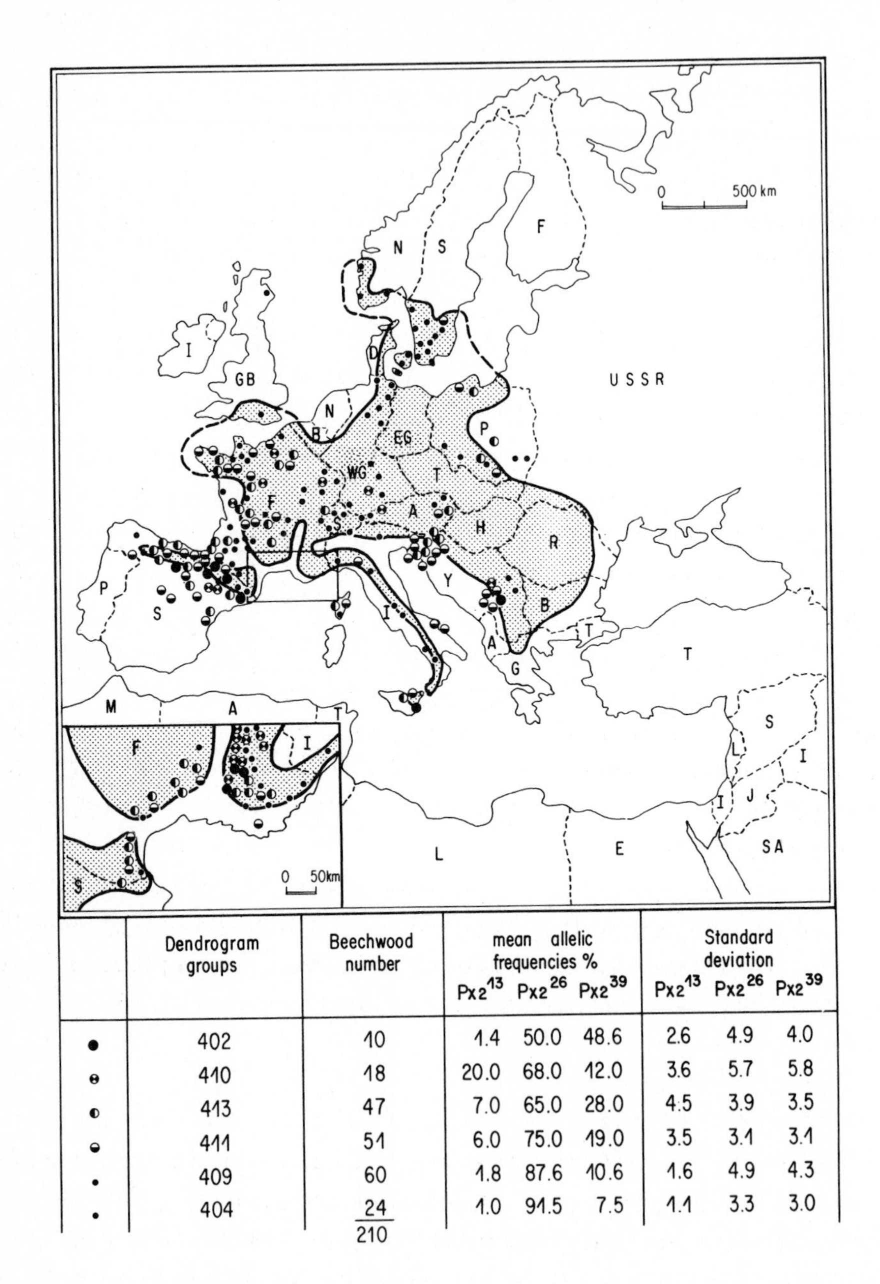

	Dendrogram groups	Beechwood number	mean allelic frequencies %			Standard deviation		
			$Px2^{13}$	$Px2^{26}$	$Px2^{39}$	$Px2^{13}$	$Px2^{26}$	$Px2^{39}$
●	402	10	1.4	50.0	48.6	2.6	4.9	4.0
⊖	410	18	20.0	68.0	12.0	3.6	5.7	5.8
◐	413	47	7.0	65.0	28.0	4.5	3.9	3.5
◒	411	51	6.0	75.0	19.0	3.5	3.1	3.1
•	409	60	1.8	87.6	10.6	1.6	4.9	4.3
•	404	24	1.0	91.5	7.5	1.1	3.3	3.0
		210						

Figure 3 - Peroxidases 2 (Px2) : geographical distribution and allelic diversity of beechwoods.

Table 1: Peroxidases 2 (PX2) - Variations of allelic diversity from west to east in southern europe.

Regions	PX2-13 X	PX2-13 conf.	PX2-13 extreme values	PX2-39 X	PX2-39 conf.	PX2-39 extreme values
Cantabrican Pyrenees 13 beechwoods 1,358 alleles	2.8	0.90	1.90 3.70	22.7	2.27	20.43 24.97
Aquitaine, Atlantic and Central Pyrenees 17 beechwoods 1,776 alleles	8.5	1.32	7.18 9.82	20.9	1.93	18.97 22.83
Western Mediterranean Basin (west of the Rhone river) 38 beechwoods 3,739 alleles	11.8	1.06	10.74 12.86	20.7	1.33	19.37 22.03
Eastern Mediterranean Basin (east of the Rhone river) 53 beechwoods 5,245 alleles	3.9	0.53	3.37 4.43	17.5	1.05	16.45 18.55

Mean allelic frequencies(X), confidence limits at the 5% level (conf.) and extreme values are calculated from absolute allelic frequencies.

2.2. - Regional scale

Near the Mediterranean sea, the beechstands grow in the mountains. The optimal beechstands are located at an average altitude with a cool climate and we can recognize a lower limit where the climate becomes too dry for beech and a upper limit where it becomes both too cold and too dry. 102 beechstands were examined, equally sampled from these three situations.

Px1 locus : In the mediterranean region, allelic frequencies vary significantly according to the location of the forests (fig. 4), and the allelic diversity is generally higher at the two limits of the beech distribution.

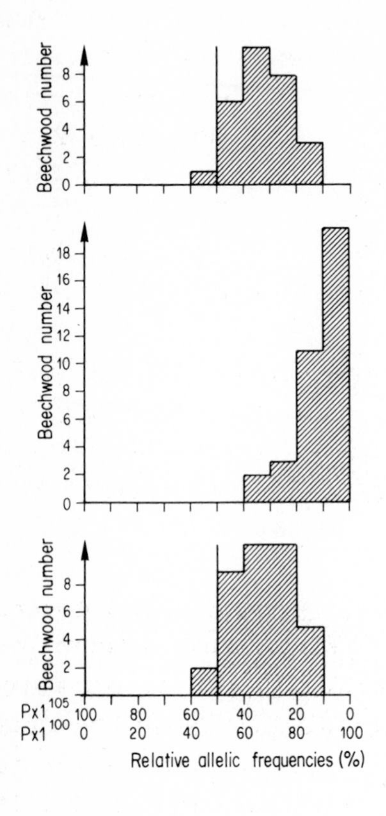

Figure 4 - Peroxidases 1 (Px1) allelic frequencies according to altitude near the mediterranean region.

We also compared beechstands located on the same slope. The differences between neighboring forests are not always significant. However, the ratio of the allelic diversity index (SHANNON 1948) between a forest located at the upper or lower limit and an optimal forest is often greater than one (table 2).

These two series of results make it clear that allelic diversity is significantly higher at the limits where climatic conditions are harsher for the beech.

Table 2: Peroxidases 1 (PX1) - Variations of allelic diversity between forests located on the same mountain slope in the mediterranean region.

Ratios between Shannon's diversities		1	1	Total	X2	P
between an upper limit forest	obs.	27	3	30	19.20	<1%.
and an optimal one	exp.*	15	15			
between a lower limit forest	obs.	15	6	21	3.86	<5%
and an optimal one	exp.*	10.5	10.5			
between an upper limit forest	obs.	25	18	43	1.14	NS
and a lower limit one	exp.*	21.5	21.5			

* according to the hypothesis that there are as many ratios >1 as ones <1.

Px2 locus : Allelic diversity was compared in the same way between beechstands on the same slope. The differences between two neighboring forests located at different altitudes are not always significant. However, the ratio of the allelic diversity index between a forest located at a high altitude and a forest located at a lower altitude is often greater than one (table 3). The diversity thus tends to increase with altitude and a colder climate. In most cases, this tendency is linked with the Px2-39 allelic frequency which is greater at higher altitudes.

2.3. - Local scale : Allelic variation between generations

Since the generalized deforestation during the XVIII th and XIX th centuries, the forest massif of the Aigoual mountain has been replenished for about 150 years. Today's beechstands are thus recent. We examined two generations of trees in six forests less than 15 kms apart at an altitude of 1200 to 1400 meters: the oldest were 150 to 200 years old and the youngest 10 to 40 years old.

Table 3: Peroxidases 2 (PX2) - Variations of allelic diversity between forests located on the same mountain slope in the mediterranean region.

Ratios between Shannon's diversities		1	1	Total	X2	P
between an upper limit forest	obs.	101	37	138	29.65	<1%.
and one located lower	exp.*	69	69			

* according to the hypothesis that there are as many ratios >1 as ones <1.

Px1 locus : From one forest to another, the allelic frequencies vary more among the young than among the old trees (table 4). The variation amplitudes are 29% for the first and only 9% for the latter. The homogeneity tests show that these differences are highly significant for the young trees ($p < 1\%.$) but not significant for the old ones ($p = 0{,}81$).

In each forest, differences appear between the two generations. In four beechstands, the Px1-105 allele is rarer in the young trees. In Peyrebesse its frequency is more of less stable from one generation to the next. At Charbonières in contrast, it increased in the young trees. But these differences are not significant, except for the Lingas forest. On the other hand, if we examine the two generations in all of the beechstands, the frequencies are as a whole significantly different ($p = 0{,}012$).

It is thus clear that on the Aigoual massif from one generation to the next, we observe a diversification of the forests and a decrease in the average allelic diversity.

Px2 locus : From one forest to another, the allelic frequencies of old trees vary more than those of young trees (table 5). The homogeneity tests show that these differences are highly significant for the oldest trees ($p < 1\%.$) but not for the young ones ($p = 0{,}69$).

Table 4: Peroxidases 1 (PX1) - Allelic frequencies(%) between two generations in six beechwoods of Aigoual mountain (Cevennes, France).

Beechwoods	Old			Regeneration			P(X2)a
		alleles			alleles		
	N	100	105	N	100	105	
Plo du Four	48	69	31	51	79	21	.09
Charbonnieres	47	68	32	45	58	42	.15
Aigoual	39	65	35	48	76	24	.12
Cerrereyde	50	61	39	52	72	28	.09
Lingas	42	70	30	50	87	13	<1%
Peyrebesse	41	67	33	36	65	35	.81
variation amplitude of allelic frequencies		9	9		29	29	
total or average	267	66	34	282	73	27	.012

P(X2)a: comparison between 2 generations in each beechwood.

The differences between the two generations are not significant in the forests, except for Lingas. And if we examine all of the beechstands, there is no significant difference between the two generations (p = 0,13).

Contrary to the results obtained with Px1, the allelic structures are more homogeneous for the young trees with Px2, although the average allelic diversity remains quite unchanged.

In one beechstand, the variations observed on a local scale between two generations may be as important as those which were noticed on a regional scale between different forests located on the same slope, especially for Px1.

Table 5: Peroxidases 2 (PX2) - Allelic frequencies(%) between two generations in six beechwoods of Aigoual mountain (Cevennes, France).

Beechwoods	Old				Regeneration				P(X2)b
		alleles				alleles			
	N	13	26	39	N	13	26	39	
Plo au Four	44	10	64	26	52	9	64	27	.90
Charbonnieres	42	7	58	35	31	10	66	24	.18
Aigoual	38	3	83	14	51	6	74	21	.29
Cerrereyde	50	7	65	28	51	8	63	29	.82
Lingas	48	0	56	44	50	8	64	29	.03
Peyrebesse	42	2	70	27	33	3	76	21	.38
variation amplitude of allelic frequencies		10	27	30		7	13	8	
total or average	264	5	66	29	268	7	68	25	.13

P(X2)b: insufficient samples to compute chi-square without grouping(alleles 13 and 26 pooled together).

Several evolutionary mechanisms may be responsible for the changes between two generations in one beechstand. But the diversifying selective action can mask the action of the other mechanisms. In order to clarify the latter, we analysed six natural regenerations and the neighboring adult trees in an environment as homogeneous as possible in order to minimize the diversifying selective power.

24 local scale : Allelic variation in natural regenerations

Under 120 years old trees, the regenerations in the Lingas forest are all 15 years old. These regenerations are dense (200 to 300 individuals per m2), making the interindividual competition high.

This means that there are different vegetative forms and degrees of vigor for plants of the same age, and these differences confer distinct statuses. The two extremes are:

- on one hand, tall individuals, 140 cms tall on an average, with well-developed canopies: they are thus dominant, their viability is high and from now on, the adult trees for the forest will be recruited among them;
- on the other hand, short individuals, 30 cms tall on an average with less developed canopies: they are thus dominated, their mortality rate is high, and their viability is low.

We chose six regeneration plots, no more than 20 meters apart. Samples were taken from each plot from an area of 4 to 6 m2 according to the density of the seedlings, from 50 dominant individuals, 50 dominated individuals and from adult trees located above them.

Px1 locus : The adults in this forest, including the trees located above the regeneration, have a low allelic diversity (table 6).

From one plot to another, the allelic frequencies of the dominant trees vary more than those of the dominated ones. But homogeneity tests show that these variations are not significant in neither cases ($p = 0,16$ for the dominants and $p = 0,74$ for the dominated).

For each regeneration, the differences between dominants and dominated are not significant and, if we examine all the regenerations, the two statuses are not significantly different ($p = 0,92$).

Px2 locus : In contrast to Px1, the adult trees in the forest, including the beechs located above the regenerations, have a high allelic diversity (table 7).

Table 6: Peroxidases 1 (PX1) - Allelic frequencies(%) in six natural regenerations in Lingas forest.

genotypes of adults trees above the regeneration		regeneration						
		dominant			dominated			
		alleles			alleles			P(X2)a
		N	100	105	N	100	105	
2*(100-100)	1	48	91	9	48	86	14	.36
2*(100-100)+(105-105)	2	50	90	10	48	83	17	.17
2*(100-100)	3	47	82	18	47	90	10	.09
2*(100-100)	4	33	80	20	48	89	11	.15
(100-100)	5	48	91	9	46	85	15	.22
2*(100-100)	6	48	86	14	48	87	13	.83
variation amplitude of allelic frequencies			11	11		7	7	
total or average		274	87	13	285	87	13	.92

	N	100	105
adult trees in the beechwood	45	88	12

P(X2)a: comparison between two social statuses in each regeneration.

The allelic frequencies of trees of different status vary from one plot to another in a highly significant way ($p < 1\%$. in both cases).

For the regenerations, the differences between the dominants and the dominated are numerous and significant in two out of six cases. While the frequencies of Px2-13 do not vary a lot, the frequencies of Px2-39 are higher among the dominants. If we

Table 7: Peroxidases 2 (PX2) - Allelic frequencies(%) in six natural regenerations in Lingas forest.

genotypes of adults trees above the regeneration		regeneration								
		dominant				dominated				
			alleles				alleles			P(X2)a
		N	13	26	39	N	13	26	39	
(26-39)+(39-39)	1	47	6	44	50	48	6	58	35	.04
2*(26-26)+(26-39)	2	46	3	54	42	41	9	59	33	.20
2*(26-26)	3	48	2	66	32	47	2	85	13	.0013
2*(26-26)	4	32	5	80	16	46	7	86	8	.11
(13-26)	5	48	23	66	11	46	20	67	13	.74
(26-26)+(39-39)	6	48	5	73	22	48	6	77	17	.36
variation amplitude of allelic frequencies			21	36	39		18	28	27	
total or average		269	7	24	29	276	8	72	20	<1%

	N	13	26	39
adult trees in the beechwood	44	3	74	23

P(X2)b: comparison between two social statuses in each regeneration(alleles 13 and 26 pooled together).

examine all the regenerations, the two statuses are significantly different (p<1%.).

The allelic structure of the regenerations is influenced by the genotypes of adult trees in the forest, and especially by those which are located very near by. In the case of Px1, when the adult allelic diversity is low, it is also low for the regenerations which resemble them. On the contrary, in the case of Px2, when the diversity is high for the adults, it is just as high for the regenerations and the six allelic structures are different. In a homogeneous environment and only a few meters apart, these differences reveal that the neighboring phenomena are more important than the homogenizing migrations.

Sometimes, the contribution of the reproducers located above the regenerations seems obvious. For regeneration number 5 the high frequency of Px2-13 seems connected with the presence of a heterozygote (13 - 26). For regenerations 1 and 2, the high frequency of Px2-39 seems to be due to the presence of carriers of this allele above them. For regeneration 6 however, the proximity of a homozygote (39 - 39) does not seems to have much influence on its allelic structure. Thus all trees do not participate equally in the regeneration and so manifest different mating capacities.

But the allelic structure of the regenerations does not depend only on the trees located directly above them. The presence of Px2-13 in 5 regenerations, although this allele was absent in the trees directly above them, proves that there is certainly some migration. This was confirmed by the presence of adult heterozygotes, carriers of this allele, 15 meters around each of these 5 regenerations.

In each regeneration, Px2-39 is regularly more frequent (there is only one exception) among the dominants than among the dominated. At least two possible explanations may be suggested:

- Since the 39 allele is always present around the regenerations, no matter who the allele giver, this allele may confer a better viability on 15 years old young trees;
- However, since there are no many givers of this allele around the six regenerations which were studied, it is also possible that the best performing seed-trees near by possessed the Px2-39 allele fortuitously, and transmitted it to their descendants. Complementary analyses would be necessary in order to reach a final conclusion. In any case, the future adult trees in the

Lingas forest will be recruited from among the dominants, and the Px2-39 allele will surely be well represented later on in this high altitude forest. This result is compatible with other observations given earlier on a regional scale showing that the frequency of Px2-39 tends to increase with the altitude.

III - DISCUSSION AND CONCLUSION

On a large European scale and on a regional scale, the biochemical results obtained fot Px1 and Px2 reveal the action of homogeneizing forces and even more so of diversifying forces: homogeneization of beechstands, the 5 alleles studied are present in the whole beech area, and there are no qualitative differences from one region to another;
diversification of beechstands, there are however important quantitative differences and the allelic frequencies do not seem to be randomly distributed since they obey an order geographic gradients which is very clear on the European and regional scales. The more pronounced variations in the southern part of the beech area may be due to several ecological and historical factors. Certain gradients, latitudinal, appear at the same time as certain climatic modifications, thus emphasizing the environmental influence on the allelic structure of beechstands. Other gradients, longitudinal, seem independant of their environment, and tend to show that homogeneizing genic migrations have little impact.

The investigation done on a large European scale and on a regional scale gives a momentary image of their allelic structures, but these structures are subject to change. The local scale analysis has clearly shown the modifications which are already taken place and the importance of certain evolutionary mechanisms.

In the context of the reforestation of the Aigoual mountain, the allelic structure evolve from one generation to the next, over intervals of approximately one century. The modifications are particularly clear for Px1: in the new generation, the allelic frequencies vary more from one forest to another but the average allelic diversity is lower. This decrease is occuring while the forest environment is growing back. Today, living conditions in the Aigoual beechstands are closer to the "optimal atmosphere" for beech than they were during the past century when the forest was fragmented and subject to bad climatic conditions. Supposing the allelic structure of the young trees continues until they become adults, a high degree of allelic diversity would apparently correspond to difficult conditions whereas there would be a lower diversity in optimal conditions. In the case of beech, we believe it to be important that the variation observed on a local scale in time correspond with the observations made on the large European and regional scales in space.

In addition, these results strongly suggest that natural selection plays an important part since identical biochemical modifications are repeated in the same ecological circumtances on three different scales, in space and time, and in distant regions. However, this evidence will remain only empirical as long as what influences of selection, migration and genetic drift has not been more precisely evaluated.

By analysis six neighboring regenerations in the same environment, the allelic variations observed in the Lingas forest can no longer be considered a result of a diversifying selective effect, and other evolutionary mechanisms become more visible. In a beechstand characterized by important historical changes in living conditions, the following mechanisms were distinguished:

- the influence of neighbors, little migration and consequently rather limited gene flow in a closed forest;
- different mating capacities among adult trees;
- different viabilities for each status of young beech.

But our data are still insufficient on a local scale for us to estimate the importance of each of these mechanisms which have just begun to be studied for the beech.

If we compare our results with those obtained for other herbaceous or ligneous species, it seems that intrapopulation genetic variations are in general characteristic of vegetal communities (LINHART et al. 1981) and it would be extremely interesting to know if these variations are as important in a stable environment. It is of course hard to define stability for the beech which has such a long life-time because what may seem like minimal perturbations for a human life-time, like felling every 120 or 150 years, for example, can have a strong impact on beech which may live up to 300 years, and thus on the genetic structure of its populations.

REFERENCES

Barriere G, Comps B, Cuguen J, Ntsiba F, Thiebaut B (to be published) La variabilité génétique écologique du hêtre (Fagus Sylvatica L.) en Europe. Etude alloenzymatique : phénomènes évolutifs et isolements génétiques des hêtraies. - Symposium on "Improvement and Sylviculture of Beech", I.U.F.R.O. group P1-10-00 may-june 1984, Hamburg, West Germany.

Beaulieu JL, Pons A (1979) Répertoire analytique des travaux concernant les plantes et la végétation du Parc National des Cévennes. Annales du Parc National des Cévennes, Florac, 1 : 215-226.

Beug HJ (1967) On the forest history of the Dalmatian coast. Rev Paleobot. Palynol. 2 : 271-279.

Bergmann F (1978) The allelic distribution at an acid phosphatase locus in Norway spruce (Picea abies L.) along similar climatic gradients. Theor. Appl. Genet., 52 : 57-64.

Felber F, Thiebaut B (1982) La hetraie méridionale française: structure génétique en relation avec les conditions écologiques. In Struktur und Dynamik von Waldern. Bericht des 25. Internationalen Symposions der Internationalen Vereinigung fur Vegetationskunde, Rinteln 13-16. 4. 1981. Vaduz, Liechtenstein; A.R. Gantner Verlag: 459-473.

Felber F, Thiebaut B (1984) Etude préliminaire sur le polymorphisme enzymatique du hêtre (Fagus sylvatica L.), variabilité génétique de deux systèmes de peroxydases en relation avec les conditions écologiques. Acta Oecologica, Oecologia plantarum 5 : 133-150.

Galoux A (1966) La variabilité génécologique du hêtre commun (Fagus sylvatica L.) en Belgique. Travaux Station de Recherche des Eaux et Forêts, Groenendaal-Hoeilaart, Belgique, série A, 11 : 121 p.

Garelkova Z (1977) Ecological variation of beech in north-western Bulgaria. Gorskostopanska Nauka 14 : 19-31.

Garelkova Z (1980) Photosynthesis and transpiration of beech seedlings from different above sea-level. Gorskostopanska Nauka 20 : 19-27.

Gohrn V (1982) Provenance and progeny trials with european beech Forstl. Forsoebvaes. Dan. 33 : 82-113.

Grant MC, Mitton JB (1977) Genetic differentiation among growth forms of Engelmann Spruce and subalpine fir at tree line. Artic and Alpine Res. 9 : 259-263.

Jalut M (1974) Evolution de la végétation et variations climatiques durant les quinze derniers millénaires dans l'extrémité orientale des Pyrénées. Thèse doct. d'Etat, Univ. Paul Sabatier Toulouse, 181 p.

Kim ZS (1979) Inheritance of leucine aminipeptidase and acid phosphatase isoenzymes in beech (Fagus sylvatica L.). Silvae Genet., 28 : 68-71.

Kim ZS (1980) Veranderung der genetischen struktur von buchenpopulationen durch viabilitatsselektion im keimlingstadium. Forstwiss. Diss. Univ. Gottingen, 96 p.

Krahl-Urban J (1958) Vorlaufige Ergebnisse von Buchen-Provenienzversuchen. Allg. Forst. Jagdztg 129 : 242-251.

Linhart YB, Mitton JB, Sturgeon KB, Davis ML (1981) Genetic variations in space and time in a population of Ponderosa pine. Heredity 46 : 407-426.

Lundkvist K (1979) Allozyme frequency distribution in four swedish populations of Norway spruce (Picea abies K.). 1 Estimations of genetic variation within and among populations, genetic linkage and a mating system parameter. Hereditas 90 : 124-143.

Mitton JB, Sturgeon KB, Davis ML (1981) Genetic differenciation in Ponderosa pine along a steep elevational transact. Silvae Genet. 29 : 100-103.

Pons A (1983) La paléoécologie face aux variations spatiales du bioclimat méditerranéen. Colloque "Bioclimatologie méditerranéenne" - Fondation L. Emberger, Ch Sauvage, Montpellier 18-20 mai 1983. II. 3 : 1-9.

Schaffalitzky-de-Muckadell M (1955) A development stage in Fagus sylvatica L. characterized by abundant flowering. Physiologia Plantarum : 370-373.

Sercelj A (1970) Das Refugial problem die Spatglaziale Vegetation-sentwicklung im Norfeld des sudost-Alpenraumes. Mitt. Ostalp.WPflanzensoziol. Arbeitsgem 10 : 76-78.

Shannon CE (1948) A mathematical theory of communication. Bull. Syst. Tech. Journal 27 : 379-423 et 623-656.

Teissier-du-Cros E, Le Tacon F, Nepveu G, Parde J, Perrin R, Timbal J (1980) Le hêtre I.N.R.A., Département des recherches forestières, Paris, 613 p.

Thiebaut B (1984) Variabilité génétique écologique du Hêtre commun(Fagus sylvatica L.) en Europe, application à la sylviculture. Colloque sciences et industries du bois. Grenoble, septembre 1982. Ministère de la Recherche et de la Technologie, Paris, France: 97-110.

Thiebaut B, Lumaret R, Vernet P (1982) The bud enzymes of beech (Fagus sylvatica L.). Genetic distinction and analysis of polymorphism in several french populations. Silvae Genet., 31 : 51-60.

Triat-Laval (1978) Contribution polleanalytique à l'histoire tardiet post-glaciaire de la végétation de la basse vallée du Rhône. Thèse Doct. d'Etat, Sciences, Univ. Aix-Marseille III, 307 p.

Tutin TG, Heywood VH, Burges NA, Valentine DH, Walters SM, Webb DA(1964) Flora Europaea, vol. 1. Cambridge Univ. Press, 464 p.

Vernet JL (1981) Histoire du hêtre. In : Le hêtre, I.N.R.A., Département des recherches forestières, p 49-57.

Geographic Origins, Genetic Diversity and the Molecular Clock Hypothesis in the *Oryzeae*

G. SECOND
Biologie des Populations et des Peuplements, Centre Louis Emberger, C.N.R.S., B.P. 5051. F-34033 Montpellier Cedex (France). And Institut Français de Recherche Scientifique pour le Développement en Coopération (ORSTOM), 24 rue Bayard. 75008 Paris.

ABSTRACT

The two species groups *Sativa* and *Latifolia* of the genus *Oryza* seem to be genetically independant in their variation. They have both a pan-tropical distribution. A study using isozyme electrophoresis of strains representing most of the area of distribution at 16 to 40 loci, shows a genetic structure in disagreement with that previously established on a morphological basis (Morishima 1969 and others). The isozyme structure shows, in both groups, a maximum divergence between Australian and other taxa, while American taxa are closely related to their Asian counterparts.

The application of the calibration of the electrophoretic clock by Sarich (1977) to the distances found points to a time of divergence for the Australian taxa compatible with geological data of a collision between Australasia and South-East Asia some 15 millions years ago. The same calibration gives estimates compatible with our knowledge on the palaeoenvironment in particular with regards to :

1)- The possibilities of migration between Eurasia and Africa (interrupted earlier for species of rice adapted to forest or inundated plain than for those species adapted to temporary pools in arid savannas)

2)- The emergence of the Himalayan barrier between China and South Asia.

The evolutionary picture which emerges is that of an Eurasian

NATO ASI Series, Vol. G5
Genetic Differentiation and Dispersal in Plants
Edited by P. Jacquard et al.

origin of the genus with migration to Australia and to Africa during the Tertiary period and to America probably during the historical epoch. The morphological and adaptive divergence of these species in America would be a case of rapid adaptation to new ecological niches. Allotetraploid species of rice would be recently evolved species induced by the man's activities.

A way to further test the validity of the molecular clock hypothesis among the Oryzeae tribe is suggested. An explanation of why the opposite view of a direct adaptive significance of isozyme polymorphism has so often been envisaged in plants is put forward.

INTRODUCTION

The use of the diversity of genomes to determine a time scale according to the so-called molecular evolutionary clock hypothesis is an attractive possibility to explore the "fourth dimension".

An example may illustrate in particular the importance of a time scale in the interpretation of a genetic structure: the case of the subspecies differentiation of the common cultivated rice. It is presented in a simplistic frame in Fig. 1 which reflects two main ways of looking at the evolutionary paths leading to an observed differentiation at the subspecies level.

The existence of two subspecies among the common cultivated rice was evidenced in Japan in the 1920's. It appeared later that an intergradation of forms existed between them and their origin was considered in terms of disruptive selection in the course of domestication (Oka, 1974, 1982).

However, it was recently discovered that a large isozyme electrophoretic distance (D of Nei of the order of 0.3 over 40 loci) was also related to this subspecies differentiation. According to the widely accepted view of protein polymorphism evolution and also because no new isozyme was found in the African cultivated rice species compared to its immediate wild ancestor, it was inferred that such an enormous amount of molecular differentiation should need more than a few thousands generations to accumulate.

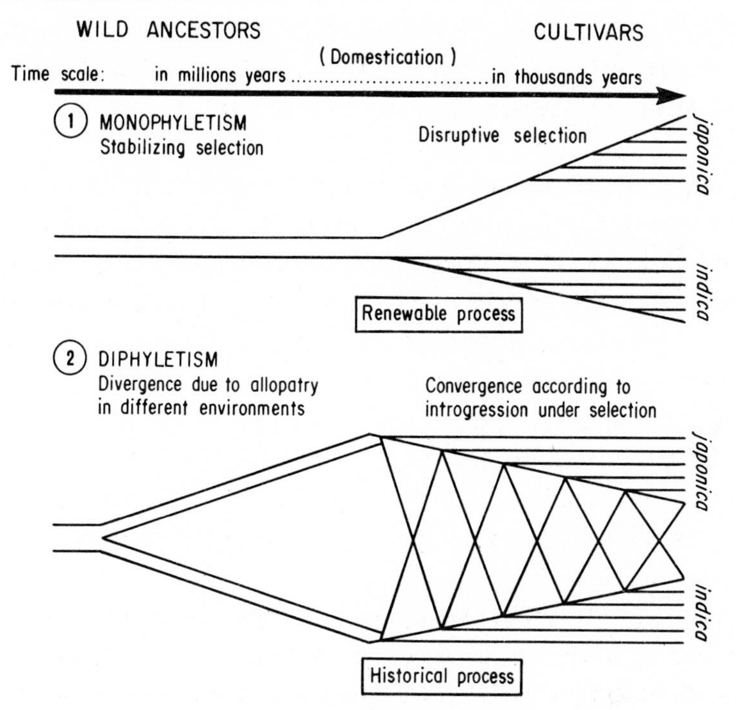

Fig. 1 - Two main hypothesis for the evolution of a subspecific differentiation with intergradation of forms. The case of the indica and japonica subspecies of cultivated rice is presented as an example. Hypothesis 1 is the author's interpretation of the ideas of Oka (1974, 1982). Hypothesis 2 is according to Second (1982).

The latter finding thus cast doubt on the former interpretation and pointed to a differentiation in the wild ancestors, previous to domestication (Second, 1982). This hypothesis was supported by the finding that Chinese wild rice are genetically differentiated from their South Asian counterpart, an information that was not previously available (Second, 1984a).

It is intended to show here that a judicious comparison of the genetically diversified geographical origins in the section Eu-Oryza of genus Oryza allows to obtain estimates of time of divergence which can be put together in a reasonable way with past tectonic and climatic events in the last 20 millions years. Then it is suggested that the tribe Oryzeae is suitable to further test the molecular clock hypothesis in plants against vertebrates possibly in the last 100 millions years. Finally, a way to explain why isozymes so often seem to be associated with environmental factors, particularly in plants, is pointed out.

GEOGRAPHIC ORIGINS: THE TRIBE ORYZEAE AND THE SECTION EU-ORYZA

The tribe Oryzeae is a small tribe considered as primitive among the grasses (Stebbins 1956, Tateoka 1957, Clayton 1975). Its distribution, with generally different genera on different continents, suggests its existence at the time of the fragmentation of Gondwanaland (when grasses might have been quite unobtrusive in the plant cover and thus do not appear in the fossil record until the Eocene epoch).

Among the genus Oryza, two species groups have a pan-tropical distribution and we may thus ask whether the same geographic barriers have acted upon their isozyme differentiation in the same way, independently of their life history characteristics. Their species composition is outlined in Table 1. One group, the Sativa group comprises, as is suggested by its name, all cultivated forms and their closest wild and weedy relatives. They are all diploid and share only one genome (A) as reported in the literature on the basis of chromosome pairing at meiosis in F1 hybrids.

The other, the Latifolia group, includes no domesticated form, but four genomes are recognized: B, C, D and E. The B, C and E genomes are known at the diploid level with their maximum distribution in Africa, South Asia and Australia, respectively. The D genome is not known at the diploid level. It was presumed to correspond to O.officinalis in China (Second, 1984b) although this remains to be proved as its cytogenetic relationship with other genomes has not yet been studied. Two allotetraploid genomes are known in nature: BC in Africa and Asia and CD, mainly in America but probably in Asia also.

Table 1. The species of section Eu-Oryza of genus Oryza

	Area of distribution	Biological type *	Genome **
SATIVA GROUP			
Cultivated species :			
O.sativa (two sub-species indica and japonica)	Asian origin	I	A
O.glaberrima	African origin	A	A
Wild species :			
O.rufipogon (complex species)			
- Oceanian form	Australasia	A or I	A
- Asian form	China, South Asia	A+I+P	A
- American form	Tropical America	I	A
O.longistaminata	Africa	P	A
O.breviligulata	Africa	A	A
LATIFOLIA GROUP (all wild)			
O.punctata(diploid)	Africa	A	B
O.punctata (tetraploid)	Africa	P	BC
O.minuta	South-East Asia	P	BC
O.malampuzahensis	Asia	P	BC
O.officinalis	Asia	P	C+D(?)
O.eichingeri	Africa	P	C
O.latifolia			
O.alta	Tropical	P	CD
O.grandiglumis	America		
O.australiensis	Australia	I	E

* A : annual ; I : intermediate ; P : perennial
** Genomes according to the observation of pairing of chromosomes at meiosis, as reported in the literature. For the full nomenclature of species, readers may consult Tateoka (1963).

GENETIC DIVERSITY: ISOZYME ELECTROPHORETIC STUDIES

Representatives of worldwide collections (except _O.officinalis_ from China) were studied at a few discriminant isozyme loci, and a condensed collection was constitued on the basis of the taxonomical classification, the geographical or ecological origins and the preliminary isozyme survey. The condensed assembly of the strains was further studied by isozyme electrophoresis.

For the _O.sativa_ group, 181 strains were analyzed at 24 isozyme loci, including 15 enzyme systems commonly stained in plant isozyme studies (Second, 1984 b). Only one locus was monomorphic and up to 10 alleles were distinguished at a single locus with a mean of 4.1 allele per locus.

In the _Latifolia_ group, 155 strains were screened at 8 loci and a condensed assembly of 25 strains was studied at 17 loci (Second 1984b). One locus (a different one than in the Sativa group) was also monomorphic. A mean of 3.3 alleles per locus were distinguished.

Nei's standard genetic distances (D) were computed from strain to strain in all possible combinations within groups in order not to make any a priori classification. While the maximum distances found were D = 1.3 in the two groups, the median values were much greater in the comparisons within the _Latifolia_ group than within the Sativa group in accordance with the values of the average calculated heterozygosity (1 - xi , xi = frequency of allele i). The table 2 summarizes these various data.

The data were further analysed through several methods of numerical taxonomy, using the presence or absence of alleles in particular strains or the matrix of distances between strains. They gave congruent results. Fig. 2 and 3 show the clusters of strains assembled on the plane defined by the two first vectors extracted in a Principal Coordinate Analysis of the matrix of Nei's distances calculated for pairs of strains in the _Sativa_ and _Latifolia_ groups, respectively.

In Fig. 2A, which includes the world accessions in the Sativa group, three clusters are separated. One includes the strains from Australia and some of the strains from New Guinea (_O. rufipogon_, Oceanian form); it is the most wide apart. Another one includes only the strains of the African perennial life form (_O.longistaminata_). The third cluster is composite, comprising all Asian and American strains and the other New Guinean strains (_O. rufipogon_, Asian and American forms), the African annual life form (_O. breviligulata_) and all cultivated forms (which all clustered together with O.breviligulata).

TABLE 2. Various data concerning the isozyme electrophoretic analysis of section Eu-Oryza.

	Sativa group		Latifolia group	
	1 *	2	1	2
Number of strains or vegetatively propagated plants studied	181	25	25	25
Number of loci considered	24	15	17	15
Number of alleles per locus	4.1	3.6	3.3	3.3
Average calculated heterozygosity	0.32	0.32	0.47	0.45
Median value of Nei's distance between lines	0.33	0.39	0.49	0.51**
Maximum value of Nei's distance between lines	1.26	1.30	1.45	1.32

* In view of the fact that different numbers of loci and lines were studied in each group, two sets of data are presented for each group : 1) those concerning the entire assembly of individuals and loci studied, 2) a restricted assembly more liable for a comparison between groups with only, for both groups, 25 individuals chosen at random in the various species or geographic forms and the 15 homologous loci studied in common.
**0.95 when only the diploid strains were considered.

It was a surprise to note that the morphologically distinct American strains (Morishima 1969) shared all their frequent alleles with their Asian relatives. On the contrary, O.breviligulata, along with the African cultivated rice O.glaberrima, could be clearly distinguished from others on the basis of one specific allele Pgi-A3 as well as of their short ligule. Also, when only Asian strains were considered, as shown in Fig. 2B, it appeared that some Chinese accessions were clearly distinguishable from their South Asian relatives .

The differentiation between Chinese and South Asian strains was, for a large part, parallel to the differentiation of cultivars in indica and japonica type but the amount of variation was much greater in wild strains than in cultivated ones.

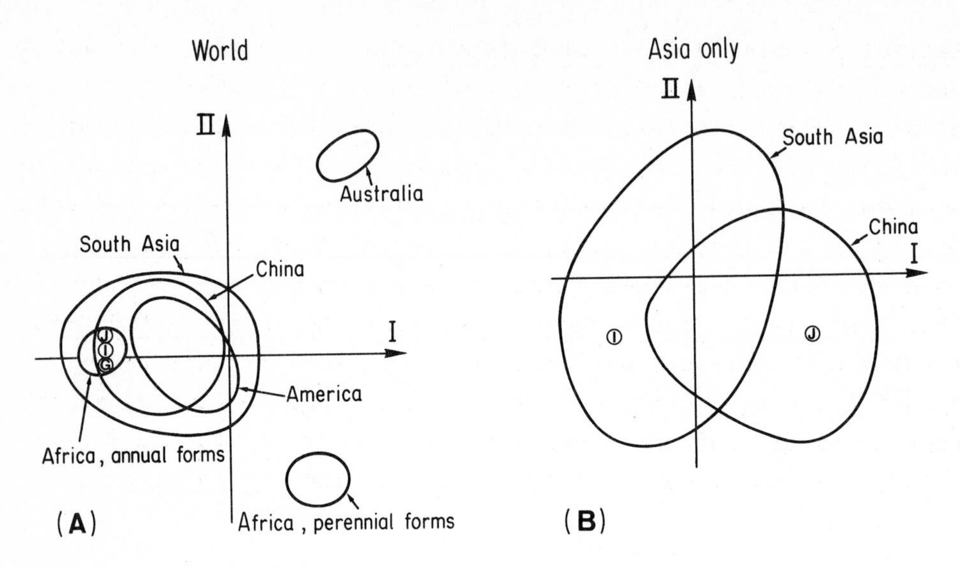

Fig. 2 . The geographic distribution of the spontaneous forms of the Sativa group as they cluster in the first plane of a principal coordinate analysis of the genetic diversity scored at 24 isozyme loci.
A)- 181 strains representative of the whole species group are considered.
B)- 110 strains originating in Asia are only considered.
The three basic isozyme patterns of the indica (I), japonica (J), and glaberrima (G) types of cultivated rice are plotted as surnumerary variables. The taxonomical classification is given in table 1.

Up to 43 isozyme loci involving 15 different enzyme systems were further stained in cultivated rice and in some wild representatives of the Sativa group (Second, 1982 and unpublished data). The genetic structure described above was only reinforced when more loci were considered. In short, four main areas of geographic differentiation appear in the Old World: Africa, South Asia, China and Australasia. The American taxa were distinguishable from the Asian ones on the basis of their genotypic combinations but not of any specific allele. Only within African accessions could the annual and perennial life forms be unambiguously distinguished on an isozyme basis.

In Fig. 3, which shows the distribution of strains of the

Latifolia group, diploid strains originating in Africa, South Asia and Australia were found at the extremes in the distribution while allotetraploid strains, but also some diploid strains belonging to the genome C, were found to be intermediate. With exceptions at only two loci (in the species O.minuta) the intermediate strains with BC and C genomes showed no allele of their own but alleles of both the B and C genomes. A putative D genome isozyme pattern was thus tentatively determined by comparing the isozymes found in strains with CD and C genomes. It showed specific alleles. Assuming, as stated above, the origin of D genome in China, the four areas of geographic differentiation noticed in the Sativa group, are thus remarkably found also in the Latifolia group.

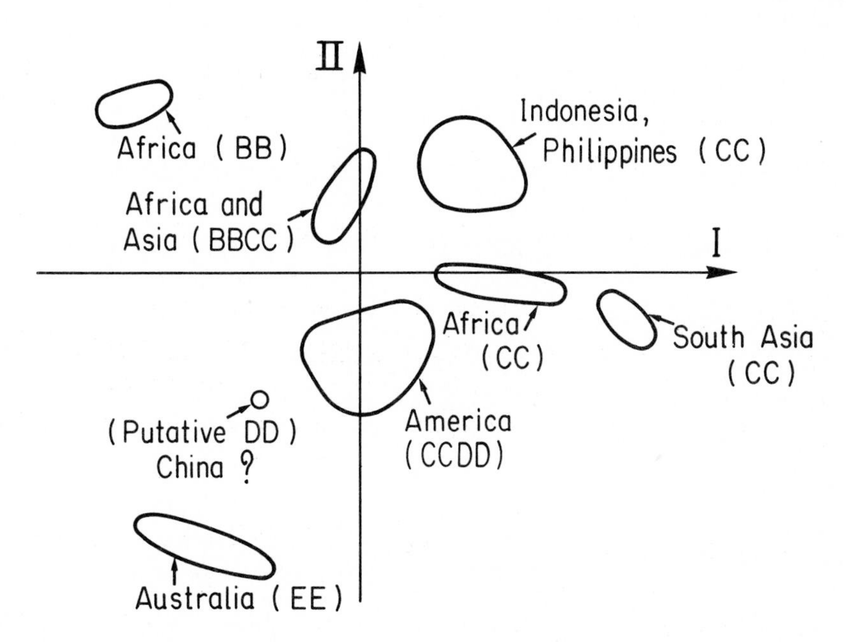

Fig. 3 - A similar study as in Fig. 2 of 26 strains of the Latifolia species group scored at 17 isozyme loci. The taxonomical classification according to the genomes or subgenomes is given in table 1.

THE ELECTROPHORETIC CLOCK AND A SCENARIO OF EVOLUTION

Turning to the interpretation, it appeared that, although apparent convergence in isozyme patterns between geographically isolated strains sometimes could be distinguished (as for example between the annual african O.breviligulata and all cultivated forms), no clear overall association between types of habitat (or life forms) and isozyme patterns could be seen at a continental scale. On the contrary, it appeared that a simple interpretation could be reached in the light of the mutation-random drift hypothesis of molecular evolution (Kimura, 1983); in other words, considering the divergence in isozyme patterns as reflecting a long time of genetic isolation (after the gene flow between the two groups of strains compared was suppressed). The intermediate patterns would then represent recently intermixed genotypes.

Sarich (1977) gave a good understanding of the molecular clock based on electrophoretic data and attempted its calibration in vertebrates on the basis of immunological distances related to paleontological data. His calibration (see the medium curve in Fig. 4 which corresponds to our mixed assembly of loci, Second 1984a and c) points, in our case to the Miocene epoch as the time of divergence to account for the maximum distances found in both the Sativa and Latifolia groups between Australian and non Oceanian strains.

This dating can be put together with the well documented collision of Australasia with South-East Asia in the course of its Northward rafting, some 15 millions years ago. The isolation of some populations of rice in the moving topography of the Malesian archipelago could have taken place at this epoch. These populations could have subsequently migrated up to Australia and remained isolated according to the so-called Wallace line. It seems thus reasonable to calibrate our electrophoretic clock upon this geological event which allows the following further determinations of times of divergence :

- 15 million years also between African and Asian representatives

of the *Latifolia* group.
- 7 million years between the perennial African (*O.longistaminata*) and Asian *O.rufipogon*.
- 2 or 3 million years between the ancestors of the three basic types of cultivated rice (*indica*, *japonica* and *glaberrima*), that is between african *O.breviligulata* and its South Asian and Chinese relatives.

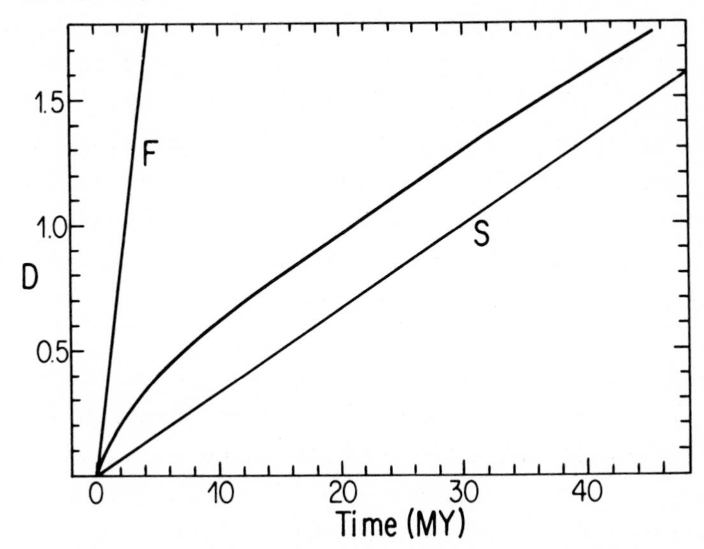

Fig. 4 - The accumulation of electrophoretic genetic distance (D) as a function of time for the rapidly evolving proteins (F), slowly evolving proteins (S), and a mixed system involving 3/4 of slowly evolving, and 1/4 of rapidly evolving proteins (unlabelled line) calculated according to Sarich (1977).

The migration was probably from Asia to Africa because the forest adapted species are found in Asia. With regards to the ecological requirements of the species concerned, these estimates are in agreement with the probable sequence for the establishment of a climatic barrier between Africa and Eurasia: tropical forest disappeared first, then humid savanna, and arid savanna before the establishment of desertic or temperate areas.

Further, the 2 or 3 million years of divergence found between the ancestors of the *indica* and *japonica* types of cultivated rice can be put together with the geologicaly recent rise of the Himalayas which may not have become a barrier for the migration of the *Hipparion* (three-toed horse) fauna (hence of annual wild rice) until the Pliocene epoch.

On the other hand, although they appear as thorougly naturalized, the American forms or species may represent very recent introductions. Arguments coming from different fields further support this finding (Second and Ghesquiere, in preparation).

The above stated apparent convergence of all cultivated rice with *O.breviligulata* could be due to their domestication from an annual life form which may have crossed the African climatic barrier and the Himalayas long after the migration of species adapted to forests or innundated plains was interrupted.

It is the reasonableness of the estimated dates that gives confidence in the use of the electrophoretic clock although the hypothesis of a universal molecular clock valid for plants as well as for animals and irrespective of generation lengths is rather controversial (see however Prager et al., 1976 and Hori and Osawa, 1979 for findings related to the question).

It seems that the validity of such a hypothesis could be further tested in the *Oryzeae* tribe with regards 1) to geological events; 2) to other organisms whose distribution may have been affected in a parallel way by the same tectonic events.

To illustrate the approach proposed, one may consider 1) the possible phylogenetic tree for the genus *Oryza* (Fig. 5), starting some 100 million years ago, that can be taken as a working hypothesis related to geological events (Second 1984c), 2) the example of such groups as frogs which show large molecular diversity in spite of a low morphological and chromosomal polymorphism. In one example, the immunological distances found between representatives of mammals and frogs in Australia and America were comparable and allowed the determination of a time of divergence of 60 - 70 million years, consistent with the study of

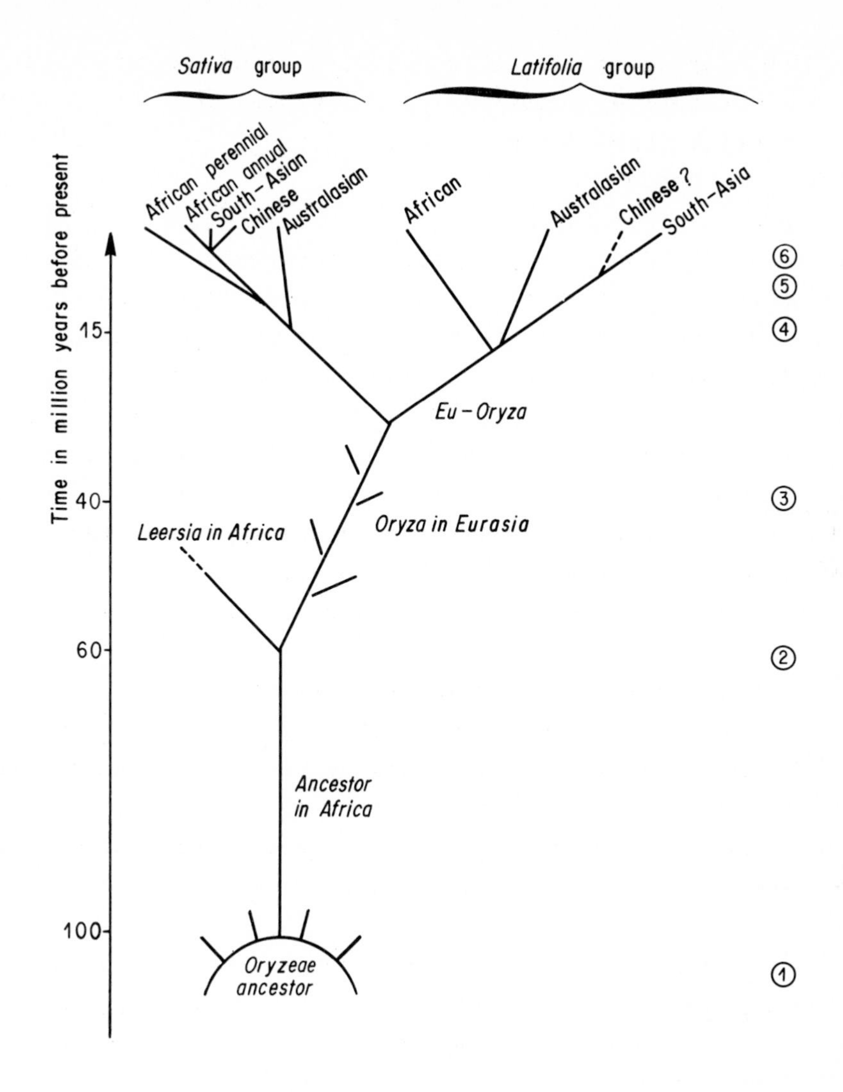

Fig. 5 - A proposed phylcgenetic tree of the Oryza genus related to some important geological events in the Old World (Second 1984c): 1)- Progressive drift of Gondwana fragments with associated Oryzeae ancestors. 2)- Temporary connection between Africa and Eurasia during Paleocene. Migration of the Leersia - Oryza ancestor into Eurasia. 3)- Land connection between India and Eurasia is established. 4)- Collision between Australasia and South-East Asia : formation of the Malesian archipelago. 5)- Progressive appearance of a climatic and sea barrier between Africa and Eurasia. 6)- Progressive emergence of the Himalayas as a barrier to plant migration.

plate tectonics (Wilson et al. 1977). A DNA or immunological approach should allow a comparison with representatives of the Oryzeae in Australia, Africa and South America.

In the present state of data, the picture which emerges for the Oryzeae is one of allopatric speciation. Introgression and allotetraploidization between species or semi-species would occur when genetic isolation has not been fully accomplished in spite of an independant evolution during geological ages and intermixing of diverging populations would happen through tectonic or climatic events and the intervention of animals or man. Obviously, this last factor has been of overwhelming importance for the microevolution and the recent dispersion of rice but also for many other wild or domesticated organisms.

In this scheme, the isozymes which are recognized neutral -or nearly neutral- to direct selection appear to be associated with adaptative traits because independent evolution occurs in different environments. It is thus not surprising that in phylogenetically heterogeneous species or populations, with introgressive hybridization following a long time of divergence, isozymes appear to behave as if selected by environmental factors.

As judged from paleobotanical data, the evolution of plant species has been a slow process at the organismic level since, say, the Miocene epoch, compared to what is observed in many Mammals. There are examples in the literature of plants whose ability to hybridize after a very long time of isolation was renewed when their distribution overlapped or their habitat was disturbed (Anderson and Stebbins, 1954).

Many studies which have showed clear environmental associations of isozyme diversity and environment type were conducted in Triticeae, Hordeae and Aveneae originated in the Near East or recently introduced in California from the Near East. In that connection, it should be kept in mind that the establishment of a Mediterranean climate in the Near East may be as recent as 11000 years ago (Wright, 1976). Mediterranean grasses previously isolated in North Africa, Southern Europe and Central Asia could have been brought together at that time, creating genetically heterogeneous populations.

It could be rightly argued that many cycles of hybridization-isolation may have occured so that the differentiation we observe now has no direct relation with any time of divergence. In that respect, the *Oryzeae*, with evident strong geographic barriers to migration which might have been broken down for some of the species only since the cultural evolution of Man, may be a good material to investigate divergence and convergence in the differentiation of plant population as well as of multigene families within individuals.

REFERENCES

ANDERSON E, STEBBINS GL (1954) Hybridization as an evolutionary stimulus. Evolution 8: 378-388.

CLAYTON WD (1975) Chorology of the genera of *Gramineae*. Kew bulletin 30, 1: 111-132.

HORI H, OSAWA S (1979) Evolutionary change in 5S RNA secondary structure and a phylogenetic tree of 54 5S RNA species. Proc Natl Acad Sci USA 76: 381-5.

KIMURA M (1983) The neutral theory of molecular evolution. Cambridge University Press, p 367.

MORISHIMA H (1969) Phenetic similarity and phylogenetic relationships among strains of *Oryza perennis*, estimated by methods of numerical taxonomy. Evolution 23: 429-443.

OKA HI (1974) Experimental studies on the origin of cultivated rice. Genetics 78: 475-486.

OKA HI (1982) Phylogenetic differentiation of cultivated rice, XXIII. Potentiality of wild progenitors to evolve the *indica* and *japonica* types of rice cultivars. Euphytica 31: 41-50.

PRAGER EM, FOWLER DP, WILSON AC (1976) Rates of evolution in conifers (*Pinaceae*). Evolution 30: 637-649.

SECOND G (1982) Origin of the genic diversity of cultivated rice (Oryza spp.) Study of the polymorphism scored at 40 isozyme loci. Jpn J Genet 57: 25-27.

SECOND G (1984a) (to be published) Evolutionary relationships in the Sativa group of Oryza based on isozyme data. Genet. Sel. Evol. 17-1.

SECOND G (1984b) (to be published) A new insight into the genome differentiation in Oryza L. through isozymic studies. In Sharma AK, Sharma Archana (eds) Advances in Chromosomes and Cell Genetics. New Delhi Oxford and IBH.

SECOND G (1984c) Relations évolutives chez le genre Oryza et processus de domestication des riz. Thèse Doctorat Etat. Univ. Paris XI Orsay.

STEBBINS GL Jr (1956) Cytogenetics and evolution of the grass family. Amer. Jour Bot 43: 890-895.

TATEOKA T (1957) Miscellaneous papers on the phylogeny of Poaceae (10). Proposition of a new phylogenetic system of Poaceae. J. Japan Bot 32,9: 275-287.

TATEOKA T (1963) Taxonomic studies of Oryza. III. Key to the species and their enumeration. Bot Mag Tokyo 76: 165-173.

WILSON AC, CARLSON SS, WHITE TJ (1977) - Biochemical evolution. Ann Rev Biochem 46: 573-639.

WRIGHT HE (1976) The environmental setting for plant domestication in the Near East. Science 194: 385-389.

Breeding system and genetic structure of a Chenopodium album population according to crop and herbicide rotation.

Jacques Gasquez

I.N.R.A., Laboratoire de Malherbologie
B.V. 1540, 21034 DIJON Cédex (France)

ABSTRACT

After the determination of the inheritance of polymorphic Esterases, L.A.P. and acid phosphatases isozymes, in Chenopodium album they have been used to estimate the rate of outcrossing which may vary from 50 % to 0 % according to distance. Afterwards we have studied the structure and the evolution of a population under crop rotation, one part treated with herbicides (D), the other without treatments (T). This population is composed of tetraploïd and hexaploïd plants which are virtually completely isolated. The tetraploïd population has a very low polymorphism and is continuously found even in D part inspite of its herbicide susceptibility (certainly due to a short life cycle). The hexaploïd population is polymorphic especially in the T part ; but there are no real stable differences between plants from the T part and plants from D part. From year to year some alleles differe in frequency. Furthermore there is a lack of heterozygous plants in seedling samples after the second year.

One may think that the seed bank is only reconstituted the year in which there is sugar-beet and that the seed bank is becoming impoverished during the three years in which cereals are grown especially in D part (by the lack of mature plants) and certainly also in T part (maybe due to a limiting rate for allogamy and a harvest before Chenopodium fructification). Thus inspite of such conditions, why are there 11 polymorphic loci out of 13 even in D part ? What chance have these plants of exchanging genes from one end of the field to the other ?

INTRODUCTION

Since the first systematic use of isozymes in order to determine the genetic structure of natural populations, many species have been studied (cf. review of Nevo 1978). However there are few studies on the genetic structure of annual weeds, specially in fields. Some studies have been made but in marginal agricultural areas (Avena fatua, Allard et al. 1968 ; Poa

NATO ASI Series, Vol. G5
Genetic Differentiation and Dispersal in Plants
Edited by P. Jacquard et al.

annua, Law et al. 1977 and Senecio vulgaris, Abbot 1976). In fact weeds in fields support very different conditions and drastic ecological factors from year to year, so if they remain in the station, they should evolve in a few generations because herbicides applied every year are different.

So we have tried to study how a weed population may evolve under herbicide treatments and various farming methods. As weed species produce numerous seeds which may be more or less dormant what is really the population of an annual weed species ? Does a year germination set represent a good sample of the seed population ? On the one hand the seedlings of a particular cohort may have been produced in one peculiar year ; so that from year to year the population could be composed of different annual sub populations. On the other hand, differential dormancy could delay germination thus every year the plants established could be a sample of the seed bank, which is generally 10 or 20 fold greater than an annual emergence set (Barralis 1972).

In a previous study, populations of Chenopodium album appeared to be more polymorphic, with at least 28 phenotypes corresponding to 13 enzymatic loci, in garden populations than in herbicide treated fields where populations were composed of only 4 phenotypes (Al Mouemar et al. 1983) or in the extreme, triazine resistant populations which have only one phenotype (Gasquez et al. 1981). Numerous chromosome counts have shown that all meiosis were absolutely regular with always 27 and 18 bivalents within hexaploid and tetraploid plants P.M.C. respectively (see the review of Uotila 1978). This constancy led various authors to think this species should be an amphidiploid and a very old polyploid with obscure ancestry (Williams 1963)

As the counts recorded in the population we have studied were always consistent with the literature (Al Mouemar 1983) we assumed that these plants should behave like a diploid. As for isozymes we found some plants having, for the three studied enzymes, all the recorded bands which appeared to be polymorphic in the progenies. Within large progenies of these plants the frequency of the numerous phenotypes was only consistent with the hypothesis of every band independent of each other ; so that we assumed these plants should segregate as diploid and every band should correspond to a locus with a dominant allele and a recessive nul allele (Al Mouemar 1983).

The aim of this investigation is to compare, by allozyme studies, the structure and the polymorphism of two subpopulations of one field growing the same way, one in a herbicide treated part, the other in a control part without any herbicide treatment. In addition the rate of allogamy under various conditions was determined in order to provide information on the possible origin of the progenies studied in this experiment.

MATERIAL AND METHODS

The plants studied were grown in the greenhouse (16 H light at 23 C and 8 H dark at 17 C) or were collected as seedlings in the field which is very close to the laboratory. Isozymes have been isolated by disc electrophoresis on acrylamid gels (Gasquez et al. 1977). The staining procedure of the three isozyme systems was carried out as previously described (Al Mouemar 1983). As each studied locus is supposed to be composed of two alleles, a dominant allele and a recessive null allele, at least 11 descendants of each plant were analysed in order to check whether the locus was heterozygous or not.

By the use of plants homozygous for various alleles we measured the outcrossing rate by the frequency of hybrid plants in the progenies i.e. showing an isozyme pattern combining patterns of both parents. In a first experiment with three replications, one plant was used as a female at the center of circles (0.25, 0.5, 2 and 5 m radius) with respectively 7, 16, 17, 32 pollinator plants around. In a second experiment 25 plants of two homozygotes loci were grown in staggered rows evenly spaced out at 25 and 50 cm.

The main experimental has been running since 1964 with a herbicide discard part (D) and a never treated part as test (T). At the beginning of the experiment there was no perennial weed and no annual grass. The field was sown with sugar beet in 1980, winter wheat in 1981 and 1982 and spring barley in 1983. The cereals are treated only with post emergence herbicides. *Chenopodium album* seeds or seedlings were collected during three years ('81, '82 and '83).

In february 1981, we took 200 soil samples (cores of 8 cm in diameter and 20 cm depth) in both D and T parts and we analysed the isozymes of at least 134 plants in each part. In 1982 and 1983 we collected at least 4 plants per square in 24 squares each one of 1 m^2 regularly spaced out in T and as many in D part where soil was avoided from herbicide by a piece of plastic during the treatment as there are no *Chenopodium album* which can support the treatment. This led us to analyse up to 652 plants over the three years and about 7172 descendants for testing whether or not the plants were heterozygous. It is noteworthy that there are no seedling samples in 1981 as all plants were accidentaly destroyed in the field soon after emergence.

RESULTS

From figure 1 it can be seen that the rate of outcrossing decreases with increasing the distance between the plants. After logarithmic transformation a straight line relationship ($y = -43.7 \log x + 17.9$) can be established which shows that plants may have 50 % and 0 % of allogamy from 0.18 m and 2.6 m respectively. So, when two plants are very close, there is a fifty-fifty chance of having a hybrid seed. This data was obtained with either many plants of each genotype or with an excess of pollinators suggesting that pollen competition might not have operated, at least between these genotypes.

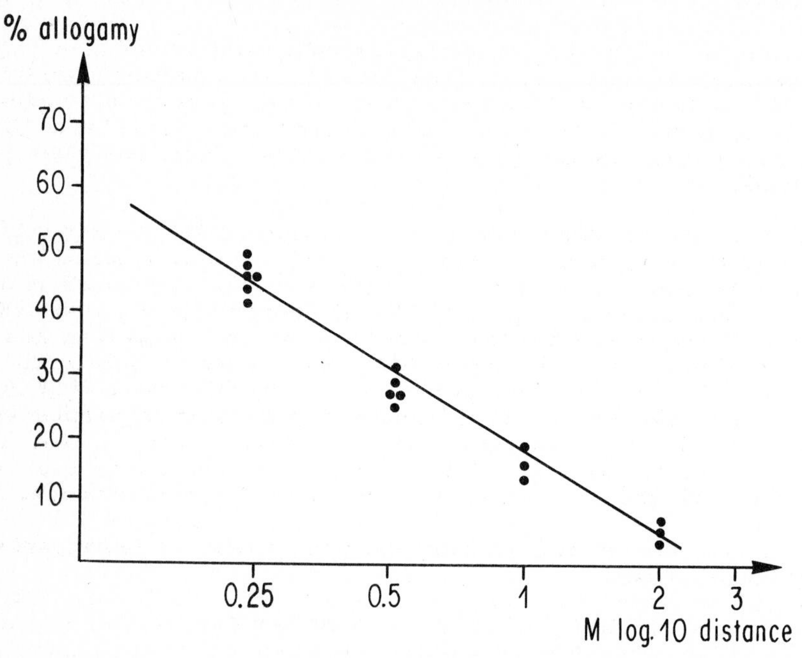

Fig. 1. Rate of allogamy in relation to the distance between two plants in pure stand.

Table 1. Frequency of polymorphic bands within tetraploīd and hexaploīd plants during the three years.

Loci	Esterases E_1	E_2	E_3	E_4	Acid phosphatases Ph_1	Ph_3	Ph_4	LAP LAP_1	LAP_2	LAP_3	LAP_4
6 x											
T 81	0	64	59	90	10	88	13	100	62	53	88
T 82	4	58	62	92	6	67	36	100	51	57	94
T 83	0	42	49	93	23	93	19	100	46	64	100
D 81	6	54	52	96	20	91	9	100	40	64	95
D 82	0	67	35	100	9	84	15	100	47	53	100
D 83	10	40	40	83	35	100	0	100	25	71	93
4 x											
T 81	0	7	51	41	1	1	98	99	100		
T 82	2	27	93	70	0	6	91	98	100		
T 83	0	1	99	28	0	1	99	99	100		
D 81	0	1	99	91	1	4	96	100	100		
D 82	0	9	95	88	0	2	98	100	100		
D 83	0	0	97	65	0	0	100	100	100		

T : non treated part - D : treated part - 81, 82, 83 year of sampling

This experiment was carried out in a pure stand which may be very similar for <u>Chenopodium album</u> to a sugar beet field. However, two plants 0.5 m apart in a maize field have 2 % outcrossing rate whereas in pure stand they may reach 31 %. In a cereal field at such a distance the plants did not exchange genes, because they are smaller.

The seed samples in '81 corresponded to 900 seed/m^2 in T part and 200 seeds/m^2 in D part. In '82 and '83 we observed similar densities in each part : 45 seedlings/m^2 in T part and 4 seedlings/m^2 in D part. This difference between the densities within the two parts is certainly due to the herbicide treatments in D part.

Furthermore we have shown there are two levels of ploidy (4 x + some B and 6 x) within various <u>Chenopodium album</u> population. Fortunately tetraploid plants display only two LAP loci : LAP 1 and LAP 2 (Al Mouemar et al. 1979). Moreover these units are almost entirely genetically isolated because of a time lag between their flowering dates and a very low level of hybridization (Al Mouemar 1983). There are more tetraploïds in T part than in D part. The ratio of tetraploïd to hexaploïd plants is 3.2, 3.1 and 3.2 in '81, '82 and '83 respectively in the T part, whereas it is lower in D part and it decreases regularly from 1.62 to 1.29 and 1.03 in '81, '82 and '83 respectively.

Table 2. Frequency of the tetraploïd phenotypes in both parts during the three years.

	T 81	T 82	T 83	D 81	D 82	D 83
Est						
2-3	73,4	48,9	27,0	89,7	79,5	62,5
2	12,4	21,3	72,0	8,8	11,4	34,4
1-3	4,7	4,2		1,5	2,3	
1-2	9,5	6,4				
3		2,1	0,5		2,3	3,1
0-1-2		2,1				
1-2-3		14,9	0,5		4,5	
LAP						
1-2	99,5	97,8	99,5	100,0	100,0	100,0
2	0,5	2,2	0,5			
Phos						
2-3	0,5	8,5	0,5	4,1	2,3	
2-4	98,5	91,5	99,5	94,5	97,7	100,0
2-3-4	0,5					
1-2-4	0,5					

In fact genetic structures in both parts are very similar. Thanks to the three enzymatic systems we may observe 11 and 13 loci within tetraploïd and hexaploïd plants respectively.

As regards to the polymorphic bands, either within tetraploïd or within hexaploïd plants, the frequencies of almost bands are similar in T and D samples (Table 1). We can only find some differences between T and D samples and some variations from year to year in each part of the field, but no really significant discrepancy (Table 1).

Despite the great number of polymorphic loci, there are only 13 phenotypes within tetraploïds (7/15 Esterases, 3/3 LAP, 4/7 Acid Phosphatases). Moreover there is no significant difference between T and D part (Table 2). Hexaploïds are a bit more polymorphic with 24 phenotypes (12/15 Esterases, 6/15 LAP and

Table 3. Frequency of the hexaploïd phenotypes in both parts during the three years.

	T 81	T 82	T 83	D 81	D 82	D 83
Est						
2-3	25,7	32,0	32,8	39,1	29,4	26,6
2	10,2	6,0	4,9	4,3		6,6
1-3	41,0	36,0	31,1	45,6	61,7	30,0
1-2-3	23,1	20,0	9,8	4,3	5,9	3,3
0-1-2-3				4,3		
0-3				2,2		
0-2-3		2,0				
3		2,0	19,7		2,9	23,3
0-1-2		2,0				
1-2			1,6			
0-1						6,6
0-2						3,3
LAP						
1-2-4	44,1	42,8	34,4	35,5	47,1	19,3
1-3-4	32,3	44,9	50,8	55,5	52,9	61,3
1-2-3-4	8,8	6,1	9,8	4,4		3,2
1-3	2,9	4,1		4,4		3,2
1-2-3	8,8	2,0				3,2
1-4	2,2		4,9			9,7
Phos						
2-3	77,9	59,2	70,5	73,3	78,1	64,5
2-4	10,3	32,6	6,5	6,6	12,5	
1-2-4	1,5			2,2	3,1	
1-2-3	8,8	4,1	16,4	17,8	6,2	35,5
2-3-4	1,5	2,0				
1-2-3-4		2,0	6,5			

Table 4. Genotypic frequencies of tetraploïds for the polymorphic loci

	Esterases												Acid phosphatases									LAP					
	11	10	00	22	20	00	33	30	00	44	40	00	11	10	00	33	30	00	44	40	00	11	10	00	22	20	00
T 81	0	0	100	2	39	59	88	4	8	44	36	20	2	0	98	3	1	96	96	0	4	99	0	1	100	0	0
T 82	0	2	98	5	0	95	93	2	5	73	27	0	0	0	100	7	0	93	93	0	7	100	0	0	98	2	0
T 83	0	0	100	1	0	99	99	0	1	28	0	72	0	0	100	1	0	99	99	0	1	99	0	1	100	0	0
D 81	1	0	99	27	23	50	62	22	16	43	16	41	1	0	99	1	0	99	99	0	1	100	0	0	100	0	0
D 82	0	0	100	65	0	35	94	3	3	77	23	0	0	0	100	0	0	100	100	0	0	100	0	0	100	0	0
D 83	0	0	100	0	0	100	99	0	1	65	0	35	0	0	100	0	0	100	100	0	0	100	0	0	100	0	0

Table 5. Genotypic frequencies of hexaploïds for the polymorphic loci

	Esterases												Acid phosphatases									LAP								
	11	10	00	22	20	00	33	30	00	44	40	00	11	10	00	33	30	00	44	40	00	22	20	00	33	30	00	44	40	00
T 81	0	0	100	53	30	17	30	35	35	78	17	5	15	0	85	88	0	12	33	0	67	50	5	45	38	15	47	83	0	17
T 82	2	0	98	45	5	50	55	0	45	95	5	0	5	0	95	67	0	33	35	0	65	55	0	45	53	2	45	90	2	8
T 83	0	0	100	42	0	58	50	0	50	93	0	7	23	0	77	93	0	7	8	0	92	44	0	56	60	0	40	100	0	0
D 81	11	0	89	57	15	28	39	4	57	63	13	24	28	0	72	93	0	7	65	0	35	33	2	65	67	0	33	93	0	7
D 82	3	0	97	67	0	33	39	3	58	97	0	3	12	0	88	85	0	15	15	0	85	42	0	58	58	0	42	100	0	0
D 83	9	0	91	38	0	62	39	0	61	80	0	20	35	0	65	100	0	0	0	0	100	25	0	75	71	0	29	93	0	7

6/7 Acid Phosphatases). There are some variations which cannot be correlated neither with possible differences between T and D parts nor with a variation from year to year in the same part (Table 3). It is noteworthy that any phenotype is specific for T or D part except some rare phenotypes always represented by one plant, which appear from time to time. Furthermore there is no increase of the frequency of any allele, in D part under the effects of a possible selection.

Whatever the sample, 9 and 10 loci are polymorphic respectively within tetraploïd and hexaploïd plants. But there is no difference between T and D part (Tables 4 and 5). It is noteworthy that, from year to year, the same loci are polymorphic and each allele has similar frequencies either in T part or in D part. However there are heterozygotes only for 6 loci within tetraploïds and 5 within hexaploïds (Tables 4 and 5). But their frequency decreases quickly during the time, so that '83 samples have only homozygous plants. Furthermore the heterozygote frequency for two esterases loci corresponds to the Hardy-Weinberg equilibrium, but only within the '81 sample from T part pointing to a high rate of outcrossing the previous year.

DISCUSSION - CONCLUSION

While it is generally said that *Chenopodium album*, as many annual weeds, is predominantly self-pollinated we have observed that this species may have a high rate of outcrossing, at least between closely adjacent plants. In fact *Chenopodium album* seems to be strictly wind pollinated with reduced dispersal possibilities of pollen. Nevertheless, instead of this low pollen dispersal, the rate of allogamy varies with the environment where this species grows. Because of plant density in the field experiment very few seeds would be produced by allogamy in cereals in T part. In D part, because of the herbicide treatment, we never observed mature *Chenopodium album*. On the other hand, in sugar beet culture, *Chenopodium album* can produce seeds by inbreeding as well as by outcrossing in both parts of the field.

One of the few effects of the herbicide treatment seems to be the decrease of the weed density in D part (4 to 5 fold the T part). Likewise variation of the ratio of tetraploïds to hexaploïds may be related to the treatment. The lower density of tetraploïds in D part and their decrease every year is certainly due to a greater susceptibility to herbicides used in cereals. Though tetraploïd plants are smaller than hexaploïd ones their higher frequency in T part could be due to their biological characteristics. Actually these plants have a delayed germination, so that they may escape from some herbicides; they are small and have a short life cycle so that they may produce, at least, some seeds before harvest time.

Despite a lower polymorphism within tetraploïd plants there

are surprisingly some heterozygous plants within the two types of plants, whereas there is no allogamy three years out of four.

Furthermore the herbicide treatment seems to be very important as there are overall fewer plants in D part and no seed produced in cereals. But there is no significant difference between the genetic structure in each part. The same loci are polymorphic, the same alleles and the same phenotypes have the higher frequency, within tetraploïd as well as hexaploïd plants.

This situation could be explained by the fact that many seeds can be produced specially by allogamy in both T and D parts one year out of four when sugar beet is grown. During the three years when there are cereals, there is only a reduction of the seed bank without seed production at least in the D part. So this population would largely rebuild its seed bank and its genetic variability one year out of four, the other years are of little importance for the evolution of the population. This population should evolve four times less than an annual species in pure stand, thanks to a large seed bank and an important dormancy which allows it to pass several years without seed set.

Surprisingly, within this population, heterozygous plants decrease very quickly so that they disappeared after three years. On the one hand they could be counter selected and they would disappear directly in the soil or as young seedlings. This would be very surprising for heterozygous plants. On the other hand most seeds should germinate during the first year following their production because they should have a higher rate of germination and a lower dormancy.

In my opinion such a situation is certainly very common within annual weeds and the rapid selection for resistance such as for triazine chloroplastic resistance (Gasquez et al. 1981) is certainly very rare. Generally, herbicides should strongly reduce the density of plants, especially when the weed is very susceptible. In a crop rotation, when herbicides vary from year to year, it cannot lead to simple selection. In fact within *Chenopodium album* the population seems to be really selected one year out of four since in cereals the weed is likely entirely destroyed. Such a population should rather fluctuate periodically because of alternations between explosive developments and recessions under the effects of cultivation or herbicide treatment.

REFERENCES

ABBOT R.J., 1976. - Variation within common groundsel *Senecio vulgaris* L..I. Genetic response to spatial variation of the environment. II. Local differences within different populations of Puffin Island. *New Phytol.*, *76*, 53-72.

ALLARD R.W., JAIN S.K., WORKMAN P.L., 1968. - The genetics of inbreeding populations. Adv. Genet., 14, 55-131.
AL MOUEMAR A., 1983. - Etude de l'évolution de populations de Chenopodium album L. en fonction de facteurs phytotechniques : structure génétique, résistance aux herbicides. Thèse doct. Etat Besançon.
AL MOUEMAR A., GASQUEZ J., 1979. - Variations caryologiques et enzymatiques chez Chenopodium album L.. C.R. Acad. Sc. Paris, T 228 (D), 677-680.
AL MOUEMAR A., GASQUEZ J., 1983. - Environmental conditions and isozyme polymorphism in Chenopodium album L.. Weed Res., 23, 141-149.
BARRALIS G., 1972. - Evolution comparative de la flore adventice avec ou sans désherbage chimique. Weed Res., 12, 115-127.
GASQUEZ J., 1984. - Approche génétique des mauvaises herbes : Variabilité infraspécifique. Evolution. Résistance. Schweiz-Landwirt. Forschung., 23 (1-2), 77-88.
GASQUEZ J., COMPOINT J.P., 1977. - Mise en évidence de la variabilité génétique infrapopulation par l'utilisation d'isoenzymes foliaires chez Echinochloa crus galli (L.) P.B.. Ann. Amélior. Plantes, 27 (2), 267-278.
GASQUEZ J., COMPOINT J.P., 1981. - Isoenzymatic variations in populations of Chenopodium album L. resistant and susceptible to triazines. Agroecosystem, 7, 1-10.
LAW R., BRADSHAW A.D., PUTWAIN P.D., 1977. - Life history variation in Poa annua. Evolution, 31, 233-246.
NEVO E., 1978. - Genetic variation in natural populations : Patterns and theory. Theoret. Popul. Biology, 13 (1), 121-177.
UOTILA P., 1978. - Variation distribution and taxonomy of Chenopodium suecicum and C. album in N. Europe. Acta Bot. Fennica, 108, 1-35.
WILLIAMS J.T., 1963. - Biological flora of the British isles : Chenopodium album L.. J. Ecol., 51 (3), 711-725.

Does cytoplasmic variation in *Plantago lanceolata* contribute to ecological differentiation?

Jos M.M. van Damme and Jaap Graveland
Institute of Ecological Research
Dept. of Dune Research "Weevers'Duin"
Duinzoom 20a, 3233 EG Oostvoorne
The Netherlands

ABSTRACT

Differences between populations in the cytoplasmic genome have been demonstrated in several plant species, indicating that, beside the nuclear genes, cytoplasmic genes may also contribute to ecological differentiation. In the gynodioecious species *Plantago lanceolata* two cytoplasmic types, R and P, exist, which can be distinguished by the anther morphology of male sterile plants. A wide intra-populational polymorphism for cytoplasmic type exists in this species, but cytoplasm R seems to be frequently absent in wet hayfields, the vegetations of which are dominated by the alliances *Calthion palustris* and *Junco-Molinion*.

Several experiments were carried out to investigate whether the ecological ranges of the cytoplasms R and P differ. In a field experiment R and P seedlings were transplanted to a site where R was absent. In addition, in a garden experiment R and P plants were grown at various levels of ground-water and competition.

No clear differences in growth and biomass production were found between the cytoplasmic types at the end of one season. The consequences of the results for the dynamics of the cytoplasmic polymorphism are discussed in relation to the observed spatial distribution of R and P within and between populations.

NATO ASI Series, Vol. G5
Genetic Differentiation and Dispersal in Plants
Edited by P. Jacquard et al.

INTRODUCTION

In the majority of papers dealing with genetic variation reference is made to nuclear genes. This is not surprising, as over 99.9 per cent of the genetic information in the cell is located in the chromosomes. A tiny fraction of the genome, however, lies in cytoplasmic particles, mostly mitochondria and chloroplasts. Cytoplasmic genes, though low in number, are important in determining the plant phenotype and fitness. Since in most angiosperm species cytoplasmic genes are transmitted to the next generation as an entity via the ovule, we will refer to a cytoplasmic gene or a set of genes as cytoplasmic type or, simply, as cytoplasm. Variation in cytoplasmic type has been known for some time (for a review see Grun, 1976). Michaelis and Bakker (1948), for example, crossed plants of _Epilobium hirsutum_ from different geographical regions. The crosses were followed by repeated backcrossing of offspring on the father plant. Transferring the nucleus of the father in this way into the alien, maternal cytoplasm often resulted in lethality, chlorosis or male sterility. Michaelis' experiments made it clear that genetic variation exists between cytoplasms of widely distant populations. Recently Kheyr-Pour (1981) showed in _Origanum vulgare_ that such differences also existed between cytoplasms from populations that were only ten kilometers apart. It is clear that genetic differences between plant populations include differences in cytoplasmic genes. Cytoplasmic variation may therefore be important in ecological differentiation.

In _Origanum vulgare_ the lack of cooperation between cytoplasmic and nuclear genes is expressed as male sterility. The literature shows that this is a very widespread phenomenon. Nuclear-cytoplasmic inheritance of male sterility is commonly found after hybridisation of species or races of crop plants (Gottschalk and Kaul, 1974), and probably also applies to gynodioecious species, although in the latter case evidence is often circumstantial (Ross, 1978; Charlesworth, 1981). Apparently, from a genetic point of view, it is a general characteristic of angiosperms that genetic information from both the nucleus and the cytoplasm is necessary for a normal stamen development. It is of interest that in several species a correlation has been found between male sterile morphology and cytoplasmic type (Chaplin, 1964; Morelock, 1974; Van Damme and van Delden, 1982). In these cases male sterility may serve as a morphological marker for cytoplasmic type, which makes it useful in the study of cytoplasmic variation. In this paper we have used male sterility as a tool to study ecological differentiation between cytoplasmic types in *Plantago lanceolata*.

SPATIAL DISTRIBUTION OF CYTOPLASMIC TYPES

Gynodioecy in *Plantago lanceolata* has long been known to be complex (Correns, 1906). Three male sterility types can be distinguished which differ in morphology, mode of inheritance, distribution and in the way they are maintained in natural populations (Van Damme, 1983a). Two of these types, MS1 and MS2, are of interest here because each type is expressed only in the presence of a particular cytoplasmic type. MS1 plants always have cytoplasmic type R and MS2 plants cytoplasmic type P. The former male sterility type is characterised by small brown anthers without any pollen, whereas the latter lacks anthers entirely and has petaloid stamens (Van Damme and van Delden, 1982). Male sterility in each case may be partial in some plants, but, as the morphological differences remain visible in these plants, partial and complete male steriles will, throughout this paper, simply be referred to male steriles. In the inheritance of each MS type a number of nuclear genes is involved (Van Damme, 1983b). This implies that a plant with a particular cytoplasmic type is not necessarily male sterile. It may also be a hermaphrodite depending on the nuclear genotype.

Because MS1 and MS2 can easily be distinguished by their flowers, they serve as morphological markers for the cytoplasmic type. In hermaphrodites, however, which have either cytoplasm R or P (there is no evidence for a third 'non-sterility' cytoplasm), no distinction between R and P can be made on the basis of anther morphology. Nevertheless, polymorphism for cytoplasmic type within populations can be assessed qualitatively whenever both MS types are found to be present together. In this way a previous study has shown that most populations are polymorphic for cytoplasmic type (Van Damme and van Delden, 1982). However, in five out of 27 populations only MS2 and no MS1 plants were found. It is noteworthy that these five populations all came from one particular habitat type. This habitat type differs strongly in vegetation and in several soil characteristics from all the other types used in the study. It concerns very wet hayfields on peat soils and with high ground water levels. The vegetations are dominated by species from the alliances Calthion palustris and Junco-Molinion. More recently counts have been made in a further six wet hayfield populations. A total of thirteen populations have now been visited, seven of which have been found to lack MS1.

For one of these populations, the Merrevliet, a silted up river bed at Voorne in the Netherlands, circumstantial evidence suggests that it is fixed for cytoplasmic type P and nearly so for the nuclear genes determining MS1 (Van Damme and van Delden, 1982). If a MS1 plant immigrates into this type of population, it will be pollinated by hermaphrodites that have mainly MS1 genes. (The only reason that these hermaphrodites are not MS1 themselves is the presence of cytoplasm P in these plants.) Inheritance of male sterility in this situation is nearly cytoplasmic and most of the progeny of the immigrant will be MS1. Due to the extreme vulnerability of the vegetation in these wet hayfields the sex type counts were made in relatively small areas varying in size from 200 to 500 m^2. Recently the populations, examined by Van Damme and Van Delden, were revisited and searched for MS1 plants in much larger areas. The findings generally agreed with earlier counts, but in the Merrevliet population two small patches, of approximately ten square meters, with high MS1 frequencies (up to 50 per cent) were found along the border of the population although the overall frequency of MS1 in the Merrevliet was close to zero. Such a distribution pattern is exactly what is to be expected if MS1 plants migrate into a population that is fixed for cytoplasm P and nearly so for the nuclear MS1 genes, as was suggested to be the case in the Merrevliet by the previous study.

Thus more than half the wet hayfield populations examined lack MS1 (and probably also cytoplasm R), whereas MS1 was not lacking in the populations of other habitat types. The probability the observed distribution occur purely by chance is very low (P = 0.0004). The cause of this distribution might be that plants with cytoplasm R are on average less fit in the wet hayfield habitat than those with P. Therefore several experiments were carried out to test whether any ecological differentiation exists between cytoplasmic types R and P.

FITNESS DIFFERENCES BETWEEN CYTOPLASMIC TYPES

In a comparison between cytoplasmic types R and P for fitness characteristics ideally the nuclear background of both types should be identical, in particular with respect to the male sterility genes. In spite of an elaborate study of the genetics of male sterility (Van Damme, 1983b), not enough is known to fulfill this condition. This is due to the extreme complexity of nuclear inheritance of male sterility in *Plantago lanceolata*. The nearest to this condition that can be achieved at present is to equalize the general nuclear background by using R and P plants from the same population. This can be achieved simply by collecting seeds from MS1 and MS2 plants in populations where both types occur in sufficiently high frequencies. A consequence of this procedure is that any existing interaction between nuclear and cytoplasmic genes with respect to the fitness traits under study may obscure the results. In order to estimate the effect of using a particular nuclear background R and P seeds were collected from plants in each of two different populations: (1) the Westduinen (Wd), an old dune grassland with a patchy distribution of MS1 similar to that in some of the wet hayfields (Van Damme, in prep.), and (2) the campus of the Biological Centre of the University of Groningen (Gr). (The abbreviations used are identical to the ones in Van Damme and van Delden, 1982).

Several experiments have been carried out, two of which are reported here. The purpose of these experiments was to find out whether any fitness differences between R and P exist and

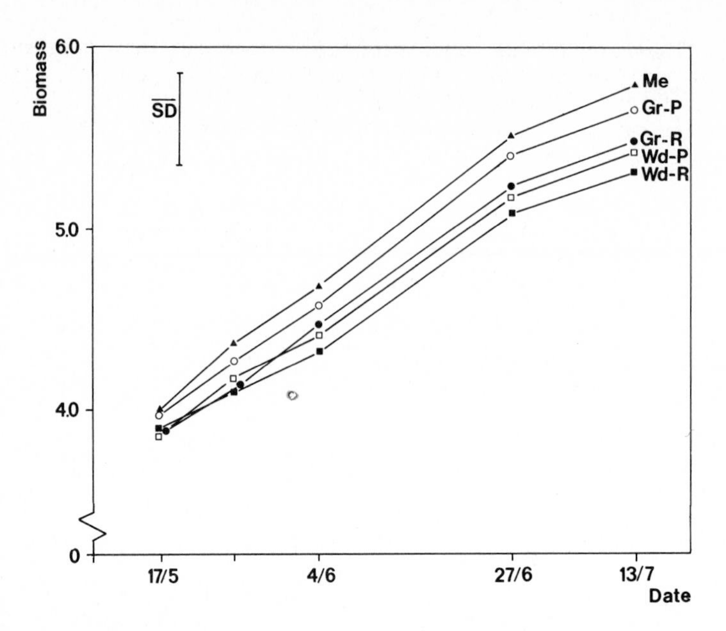

Figure 1. The development of biomass of transplanted seedlings in the Me hayfield. Seedlings raised from seeds of the Gr or the Wd population, have either cytoplasmic type R or P. Me seedlings probably have cytoplasm P. Biomass is expressed as ln (sum of length of leaves in mm.).

to prepare further experiments to analyse the causes should any differences be found. Only the main results will be given. Full details of the experiments will be published elsewhere.

In a field experiment precultured seedlings were transplanted into the Merrevliet (Me) population at the end of March 1983. Five seed sources were used: R and P seeds from both the Wd and the Gr populations and seeds that had been collected from the Merrevliet in the previous year. The last group probably consisted entirely of P seeds and served as a control. Two plots were used. In each plot 40 seedlings per source were planted (spacing ten cm) in groups of five using a block design. In total 400 seedlings were planted. The plots were visited regularly until mid July, when the field was mown. At each visit the relative biomass per plant was estimated non-destructively by measuring the length of all green leaves. In this way the relative growth rate per individual could be calculated. In addition, on the date of mowing the plants were

Table 1. The t-values of comparisons of class means after analysis of variance on Relative Growth Rate and dry weight of the transplant experiment in Fig. 1. Average sample sizes per class means are also given. *** denotes P < .001, ** P < .01 and NS not significant.

	Comparison		
	Me-(Wd + Gr)	Wd-Gr	R-P
RGR	1.28 NS	2.57 **	0.49 NS
Dry weight	9.20 ***	4.91 ***	1.32 NS
Sample size	173	138	138

clipped and the individual dry weights determined.

Since the preliminary analysis showed no differences between plots, the results are presented together (Fig. 1). There were significant differences between populations in relative growth rate as well as in dry weight after clipping (Table 1). Seeds from the 'home' population Me grew better than those from the 'alien' populations and the Gr population did better than Wd. In contrast to other reports (Antonovics and Primack, 1982; Van Damme and van Delden, 1984) this indicates that in Plantago lanceolata selective differences can occur in the juvenile stage. The nuclear genome of plants from Me seems to be better adapted to the wet hayfield environment than that from the other populations. The most important comparison, however, is between cytoplasmic types R and P. Although figure 1 shows consistent, small differences in biomass in favour of P, they are not significant (Table 1). Similarly there were no differences in mortality between R and P plants which averaged 14 per cent during the period of observation, nor were there any differences between populations.

In order to be able to vary environmental variables another experiment was set up in a garden. In order to keep it as natural as possible twelve sods were dug out from wet hayfields in the nature reserve Vlaardingse Vlietlanden. The sods, cubic in form with sides of 28 cm, were put in buckets which were

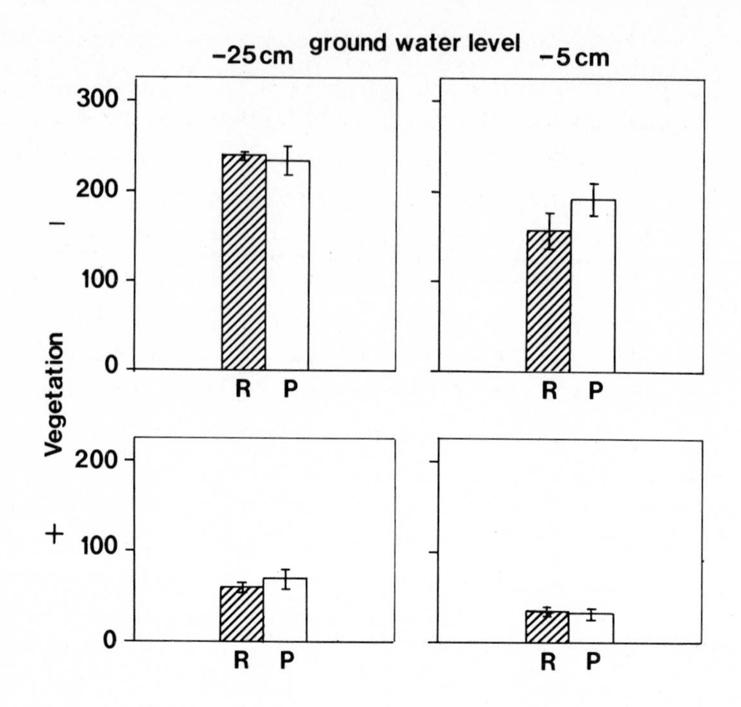

Figure 2. The biomass (in mg dry weight) and standard error of R and P plants two months after transplantation in sods, depending on ground water level and on the presence (+) or absence (-) of the vegetation. Biomass was estimated using multiple regression on leaf measurements.

dug into the soil of the garden. In each sod twelve R and twelve P seedlings of similar size, coming from the Gr population, were transplanted in an alternating grid. With the aid of a siphon system a ground water level of -5 cm was maintained in half the buckets and of -25 cm in the others. Both levels fall within the range normally found in wet hayfields. In addition, the above ground vegetation of six of the sods was removed. So there were four treatments, each replicated trice with a total of 288 plants. The experiment started in June and in August the biomass per plant was estimated non-destructively, as in the previous experiment. Except for two plants shortly after the start of the experiment, there was no mortality.

The results in Fig. 2 show that again no differences were found between cytoplasmic types ($P > .05$). The treatment effects were highly significant ($P < .001$): a higher biomass was obtained when the ground water level was lower and, particularly when the surrounding vegetation had been removed. The treatment effects are probably the result of respectively a higher availability of nutrients and reduced competition. Any interactions between cytoplasmic type and either of the treatments would be of interest because they indicate the factors involved in any existing ecological differentiation between cytoplasmic types R and P. However, they were not significant, which is not surprising in view of the absence of a difference between R and P.

DISCUSSION

In neither of the two experiments were there any differences found between cytoplasmic types in the growth of juveniles. This seems contrary to our initial expectation which was based on the distribution of populations lacking MS1 plants in different habitat types. One should however be aware of the limitations of the data. Small differences might exist that could not be detected in our experimental set up. Small differences are of interest because in a population with a polymorphism for cytoplasmic type such differences may be sufficient to allow the fitter type to become fixed, provided selection exceeds any effect of drift (Lewis 1941; Lloyd, 1974). Alternatively, differences between cytoplasmic types might be expressed in later stages of the life cycle (cf. Antonovics and Primack, 1982; Van Damme and van Delden, 1984). For example, the difference in male fertility between hermaphrodites with cytoplasm R and P, that has been observed in reciprocal crosses, may be cytoplasmically determined (Van Damme, 1984).

If, in accordance with the results presented here, there are no fitness differences between cytoplasmic types R and P, then our initial expectation needs adjusting. To this end we should reconsider the distribution of R and P in the field. In Fig. 3 the distribution data, referred to earlier in this paper,

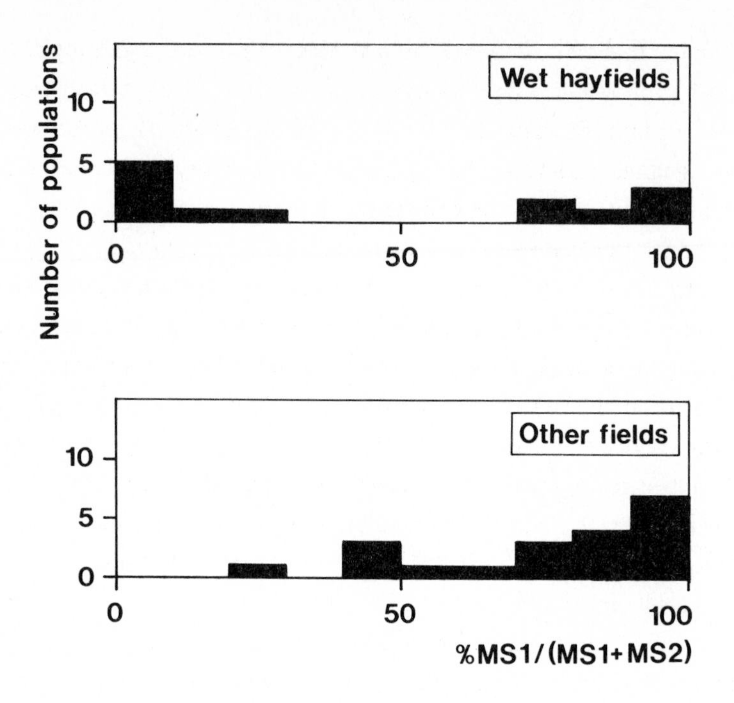

Figure 3. The distribution of the percentage MS1 plants of all male steriles (MS1 + MS2) over populations in wet hayfields and in 'other' habitats (see text).

are represented as the percentage MS1 plants of all male steriles in the population. Since MS1 and MS2 carry cytoplasmic type R and P respectively, this ratio estimates the percentage of the plants in the population with cytoplasm R. The bias of the estimate depends on how well male sterile frequencies represent cytoplasmic freqencies. Since hermaphrodites are always in the majority in populations, the bias will not be known until methods have been developed to determine routinely the cytoplasmic type in these plants (like in *Zea mays*, see Kemble, 1980). Let us assume that, at least in the wet hayfields, the bias is small. Then in Fig. 3 approximately half of the populations of the wet hayfield group is (nearly) fixed for cytoplasm P and the other half for R, whereas the situation in the other fields is totally different, virtually all populations being on one side of the figure. The picture that emerges is that of joint polymorphism, with variation in nuclear genes as well as cytoplasmic type, in all populations except in the wet hayfields where there is nuclear inheritance of either MS1 or MS2. We suggest that the following might be the case. In most situations joint polymorphism is realised but for some reason this is not possible in the wet hayfield habitat. It is not clear what stages wet hayfield populations go through, but possibly an intermediate stage is involved with polymorphism for cytoplasmic type only. Cytoplasmic polymorphisms are known to have neutral equilibria, which are easily disturbed by drift as well as by small fitness differences between the cytoplasmic types (Lloyd, 1974). In the long run therefore, one or the other type will become fixed. The fact that half the wet hayfield populations are (nearly) fixed for cytoplasmic type R and the other half for P, is then consistent with the absence of any detectable fitness difference between the two types. Charlesworth (1981) found in a simulation model that in self-incompatible species like _Plantago lanceolata_ a fitness difference between cytoplasmic types is _necessary_ to maintain joint polymorphism. This would imply that the absence of a fitness difference is itself the cause of the failure to maintain joint polymorphism in wet hayfields. Clearly, more knowledge of actual cytoplasmic frequencies and their changes in time is required in order to understand the situation fully.

This explanation requires fitnesses to be environmentally dependent in the sense that a fitness difference between R and P can not be expected in wet hayfields, whereas it can be expected in other habitat types. Recent theoretical studies (Charlesworth, 1981; Delannay et al. 1981; Ross and Gregorius, 1984) have shown that the maintenance of joint polymorphism is a delicate matter, depending on specific fitness effects of sex genotypes. A failure to maintain joint polymorphism in particular situations may not only be due to environmental dependent fitness effects of cytoplasmic types, but also of nuclear male sterility genes or of their interaction with cytoplasmic type. As explained earlier nuclear genes were not considered in this study because of the extreme complexity of nuclear inheritance of male sterility. We are convinced that the prospects for this type of study would improve greatly if fitness effects could be studied per nuclear-cytoplasmic genotype instead of per sex phenotype. Then the limitations in interpretation, inherent in the experimental set up used in this study, could be removed. Therefore high priority should be given to the study of the genetics underlying this sex polymorphism.

REFERENCES

Antonovics J. & Primack R.B., 1982. - Experimental ecological genetics in Plantago VI The demography of seedling transplants of P.lanceolata. J.Ecol., 70, 55-75.

Chaplin J.F., 1964. - Use of male sterile tobacco in the production of hybrid seed. Tobacco Sci, 8, 105-109.

Charlesworth D., 1981. - A further study of the problem of the maintenance of females in gynodioecious species. Heredity, 46, 27-39.

Correns C., 1906. - Die Vererbung der Geschlechtsformen bei den gynodiöcischen Pflanzen. Ber. Deutsch Bot. Ges., 24, 459-474.

Delannay X., Gouyon P.H. & Valdeyron G., 1981. - Mathematical study of the evolution of gynodioecy with cytoplasmic inheritance under the effect of a nuclear restorer gene. Genetics, 99, 169-181.

Gottschalk W. & Kaul M.L.H., 1974. - The genetic control of microspore-genesis in higher plants. Nucleus (Calcutta), 17, 133-166.

Grun P., 1976. - Cytoplasmic genetics and evolution. Columbia U.P., New York.

Kemble R.J., 1980. - A rapid, single leaf assay for detecting the presence of "S"-male-sterile cytoplasm in maize. Theor. Appl. Genet., 57, 97-100.

Kheyr-Pour A., 1981. - Wide nucleo-cytoplasmic polymorphism for male sterility in Origanum vulgare L. J.Hered., 72, 45-52.

Lewis D., 1941. - Male sterility in natural populations of hermaphrodite plants. New Phytol., 40, 56-63.

Lloyd D.G., 1974. - Theoretical sex ratios of dioecious and gynodioecious angiosperms. Heredity, 32, 11-34.

Michaelis P. & Bakker D., 1948. - Über reziproke verschiedene Sippenbastarde bei Epilobium hirsutum VIII Vergleichende Untersuchungen über den Plasmon mehrer Epilobium hirsutum - Sippen, die bei reziproker Kreuzung Unterschiede der Pollenfertilität zeigen. Z. indrukt. Abstamm. u. Vererblehre, 82, 384-414.

Morelock T.E., 1974. - Influence of cytoplasmic source on expression of male sterility in carrot, D.carota. Ph. D. Thesis, Univ. of Wisconsin.

Ross M.D., 1978. - The evolution of gynodioecy and subdioecy. Evolution, 32, 174-188.

Ross M.D. & Gregorius H.R., 1984. - Selection with gene-cytoplasm interactions II Maintenance of gynodioecy. Manuscript.

Van Damme J.M.M., 1983a. - On gynodioecy in Plantago lanceolata. Thesis, Univ. of Groningen.

Van Damme J.M.M., 1983b. - Gynodioecy in Plantago lanceolata L. II Inheritance of three male sterility types. Heredity,

50, 253-273.
Van Damme J.M.M., 1984 - Gynodioecy in *Plantago lanceolata* L. III Sexual reproduction and the maintenance of male steriles. *Heredity*, 52, 77-93.
Van Damme J.M.M. & Van Delden W., 1982. - Gynodioecy in *Plantago lanceolata* L. I Polymorphism for plasmon type. *Heredity*, 49, 303-318.
Van Damme J.M.M. & Van Delden W., 1984. - Gynodioecy in *Plantago lanceolata* L. IV Fitness components of sex types in different life cycle stages. *Evolution* (in press).

Grassland Research Group Publication No. 82

Genetic Differentiation: Variation in Phenotype and Fitness

Adaptation, Differentiation and Reproductive Systems in *Lolium perenne*

M.D. Hayward

Welsh Plant Breeding Station,
University College of Wales
Aberystwyth,
Wales, United Kingdom.

ABSTRACT

The *Lolium* genus comprises a number of species which include inbreeding annuals and perennial outbreeders. Ecotypic diversity within the perennial species may be related to edaphic and biotic factors. The genetic basis of population differentiation has been ascertained for a number of populations of *Lolium perenne*. Isozyme analyses have shown that variability is highest in populations from the centre of origin around the Mediterranean. Allele frequency at some loci is correlated with soil nutrient status. The biometrical assessment of quantitative variation has shown that population differentiation was controlled by predominantly additive gene action with little evidence of dominance effects. In addition an important maternal/reciprocal plasmon effect was operative particularly for adult plant characters. Somatic selection experiments revealed the presence of a further dimension of variability associated with vegetative reproduction by tillering. It is suggested that population differentiation in the persistent perennial ryegrasses is based on control by both nuclear and plasmon determinants. This dual control is predisposed by the sexual and asexual modes of reproduction to be found in the persistent perennial *Loliums*.

INTRODUCTION

The *Lolium* genus is a group of diploid (2n=14) grass species whose centre of origin is around the northern and eastern Mediterranean but with a distribution which is now world wide as a consequence of

NATO ASI Series, Vol. G5
Genetic Differentiation and Dispersal in Plants
Edited by P. Jacquard et al.

some of them being of major importance in temperate grassland agriculture. Within the genus a range of life forms may be encountered from the short-lived annual, generally inbreeding members such as *Lolium temulentum* to the longer lived outbreeders like *L.rigidum* and *L.multiflorum* or the extreme perennials, such as *L.perenne*. These latter forms are well adapted to outbreeding being anemophilous and possessing a two locus gametophytically determined incompatibility system (Cornish *et al*., 1979; Fearon *et al*., 1983). Between the two extremes of annual inbreeders and perennial outbreeders a virtual continuum of morphological types can be found such that no clear species barriers exist; hybrids between the various types being readily achieved (Jenkin, 1959; Terrell, 1966).

Variation and adaptation within the group is closely associated with their evolution in conjunction with the grazing. The long-lived forms are a major constituent of permanent grassland in the oceanic regions of Europe, whilst the annual types are to be found in the Mediterranean climatic zone of Europe and the near East, often in disturbed habitats, or as weeds of cereal cropping systems. As a result of their importance to agriculture the longer lived *L.multiflorum* and *L.perenne* have been the subject of numerous detailed investigations of their genetic adaptation, organisation and reproductive biology in order to more easily exploit their attributes by practical plant breeding (see e.g. Breese, 1983).

Ecotypic Differentiation in *Lolium perenne*

Many studies have shown that considerable ecotypic differentiation exists within the perennial section of the *Lolium* genus and it was this variation that predisposed the success of the early ryegrass breeders (Beddows, 1953). This differentiation can often be related to ecological factors known to be operative on the natural pastures where *Lolium perenne* abounds. This is clearly shown by e.g. the survey of a series of natural populations collected along the Monmouthshire Moors area of the Severn estuary (United Kingdom) (Breese and Charles, 1962). Both morphological and physiological characteristics of the populations when assessed in a common environment could be related to edaphic, and biotic/management factors operating on the patures. For example, the timing of inflorescence emergence was correlated with grazing intensity; lax grazing by cattle was associated with early heading whereas intensive sheep grazing lead to late flowering. Similarly the productivity of spaced plants was related to the percentage ryegrass within the pasture. We have investigated the genetic basis of this and other collections of ryegrass showing ecotypic differentiation by two procedures; at the biochemical level by the use of isozymes and at the quantitative genetic level by the application of biometrical techniques.

Isozyme Variation Within and Between Populations of *Lolium perenne*

An electrophoretic survey of 46 natural populations of *L.perenne* from various parts of the United Kingdom and continental Europe, particularly Italy, using four enzyme systems, phosphogluco-isomerase, PGI, glutamate oxaloacetic transaminase, GOT, acid phosphatase, AcP, and sodium oxide dismutase, SOD, has been carried out. These enzymes provide five polymorphic and one monomorphic loci within perennial ryegrass. The survey has shown considerable allelic diversity both within and between populations (Table 1).

Table 1. Genetic variation in United Kingdom and Italian populations of *Lolium perenne**

	Average heterozygosity	Average numbers alleles/locus	Average genetic distance	Number of populations
UK	0.3723	3.09	0.0455	37
range	0.219-0.490	2.6-3.75		
Italy	0.4245	3.266	0.1102	9
range	0.339-0.524	2.6-4.0		

Average genetic distance between United Kingdom and Italy = 0.1041

*Information based on five loci, with parameters calculated according to the methods of Nei (1978)

The average levels of heterozygosity vary between populations revealing differing potential for change. In general it can be seen that variability is greater in the Italian populations than in those on the extremities of the distribution.

Whilst large differences in levels of heterozygosity between populations at individual loci were found, no clearly defined relationships could be established between geographical factors and levels of variability. Within the United Kingdom however, the frequency of certain alleles at some loci may be related to specific ecological factors. Thus the frequency of the 'b' allele at the PGI/2 and AcP/1 loci are correlated with soil calcium and pH status (Balls, unpubl.), whilst no relation could be established for the other loci. Unlike in _Dactylis_ where a relationship between allelic status at the GOT/1 and temperature at site of origin could be explained by differing thermal properties of the allozymes (Lumaret, 1984) no simple explanation can account for the present set of observations. They may of course reflect selection operating on adjacent linked loci and not on the isozyme locus itself.

The outbreeding nature of all the populations surveyed was shown by the occurrence of Hardy-Weinberg equilibria for all loci examined, except in a few instances at the AcP locus. These minor discrepancies may be accounted for by the presence of null alleles which have been detected in other investigations of this locus.

Quantitative Variation Between Populations

The genetic organisation of natural populations of _Lolium perenne_ for quantitative characters has been determined in a number of studies at the Welsh Plant Breeding Station and elsewhere. They have all utilized the techniques of biometrical genetics for assessing the occurrence of genetic variability and the nature of the gene action responsible. Details of the various methods we have utilized may be found in Breese and Charles (1962); Hayward and Breese (1966) and Hayward (1973). In general, they have involved an initial clonal comparison of populations using vegetative propagules of usually sixty individuals collected from their natural habitat. The collection procedures aimed at sampling individual plants and this was later confirmed to be the case by a study of their isozyme genotype. The clonal analysis allowed the partitioning of variation into both between and within population components together with a residual representing environmental variance. The collections were assessed for a number of morphological and physiological characters, generally over a two year period. In the majority of instances the analyses revealed differences between populations to be greater than within thus establishing foundations for further detailed genetic analyses.

For the determination of population differentiation the assessments have all followed a comparable scheme in that a sub-sample of populations has been taken from a region such as the Monmouthshire Moors and an appropriate biometrical technique like the diallel cross applied (see Mather and Jinks, 1982 for further details and other methods). The assessment has involved the measurement of seedling characters including rate of germination, leaf and tiller production, and leaf area before transplanting into the field, where the material was established as a spaced plant randomized block experiment and maintained for two years. Over this period, a range of characters were again measured and included timing and number of inflorescences produced, and four biomass determinations; two during the establishment year and a hay and aftermath cut during the first harvest year.

Three major surveys have been carried out and all have revealed a similar pattern of genetic control of population differentiation (Hayward and Breese, 1966; Hayward, 1967; Hayward and Breese, 1968; Thomas, 1967). The main features to emerge from these studies were that for the majority of characters during the juvenile phase of the plants' life, variation was predominantly additive in nature with only seedling characters such as rate of germination showing any evidence of dominance or epistatic effects. Some maternal effects were detected which could be attributed to seed weight differences. For adult plant characteristics, i.e. after the material had passed through one winter in the field a simple picture of additive genetic control was shown to be operative accounting for up to 45% of the total phenotypic variation for characters such as total annual yield. In addition, however, the analyses revealed an important reciprocal component of variation. This took the form of an 'average maternal' component during early adult life followed later by a 'specific reciprocal' effect. The former reflect the average effect of using a population as a female parent in a set of crosses compared with its use as a male whilst the latter can be accounted for by specific nuclear/cytoplasmic interactions of individual crosses. The relative magnitude of these various components for some of the yield characters measured in the assessment of populations from the Monmouthshire Moors and North Wales are shown in Figure 1.

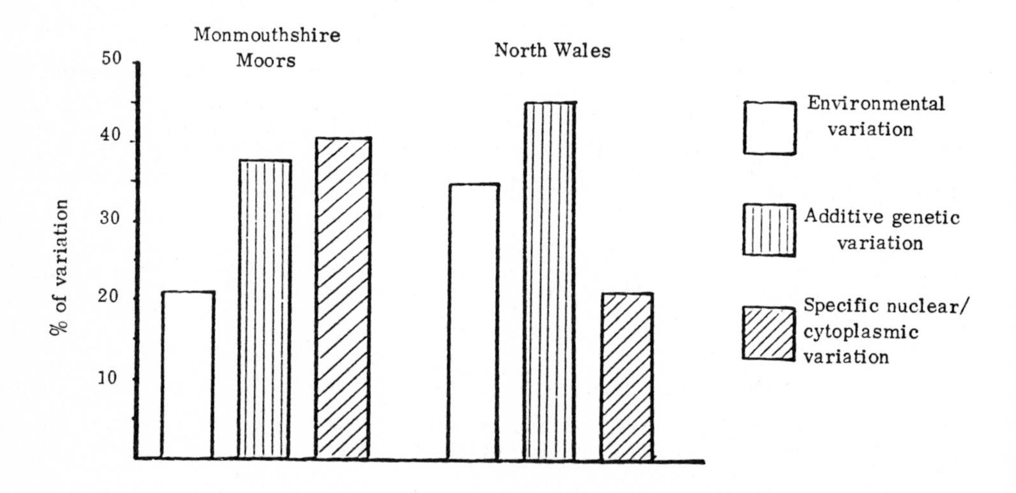

Fig. 1. Components of variation : production of green matter

It would thus appear that population differentiation in these long-lived perennial ryegrass pastures has not been accompanied by the evolution of dominance and epistatic gene combinations at the nuclear genetic level as a result of strong direction selection. In contrast it has led to the evolution of coadapted combinations between the nuclear and extranuclear genome.

Quantitative Variation Within Populations

The outbreeding nature of Lolium perenne would suggest the possible presence of genetic variation within as well as between populations. This has been confirmed by a considerable number of experiments covering a wide range of materials and characters and of course has been extensively exploited in practical breeding programmes.

The heterozygous nature of individuals and the heterogeneous structure of populations of L.perenne was clearly revealed by the classic selection experiments of Cooper (1959) for timing of inflorescence emergence where the response achieved extended well beyond the range exhibited by the original populations. More detailed studies of the genetic architecture within populations have shown that in long-lived pastures variability follows a comparable pattern to that between populations, i.e. simple additive variation together with an extranuclear component (Hayward and Nsowah, 1969; Balls, pers comm.). In contrast however, short-lived populations, particularly those characterized by early heading and high seed production show evidence of dominance and an absence of reciprocal effects (Breese, 1960).

Here again, like the result of the between population analyses, within population variability in the more persistent perennial ryegrasses would appear to be under the dual control of nuclear and extranuclear components. This mode of control may be accounted for by the reproductive strategy of perennial ryegrass (see also Breese, 1965).

Survival Strategy and Population Differentiation

The ability of perennial ryegrass to survive under intensive grazing is dependent on its capacity for continual vegetative reproduction by tillering. Inflorescence production, flowering, seed set and establishment from seed occurs at a low frequency under these conditions. In view of these restrictions on the sexual processes it was considered that variation may occur at the tiller level which could be of adaptive significance.

The potential for adaptive change during vegetative reproduction was examined by means of a somatic selection experiment for rate of tiller production (see Breese *et al.*, 1965). The most significant feature to emerge from this selection programme was that it was possible to select for high and low rates of tiller production from within a single clone of *Lolium perenne*. A response could only be achieved however, when that clone was relatively young, i.e. derived from a seedling, as no response was realized by selection within clones of indeterminate age collected from natural populations some ten years previously. A further feature to emerge was that the magnitude of the response was genotypic dependent; a greater response being elicited from the more persistent perennial genotypes than from relatively short-lived material (Breese *et al.*, loc cit). Selection has continued for fifteen years with an average of three tillering cycles per year. In addition to the main lines, reversed selection or relaxed selection has been applied to some lines. The main results to emerge from this continuation are that whilst initially, i.e. two years after establishment a response to back selection could be achieved, when applied later in the course of the experiment, at four years after initiation, no consistent response resulted (Shimamoto and Hayward, 1975). Similarly, under relaxed selection the differences between the lines have been maintained (Thomas, pers.comm.). These results would suggest that with increasing age, variability at the somatic level is restricted and the determinants responsible for the initial changes have become 'fixed', thus no further response can be obtained. A comparable situation may well apply in other grass species for the persistence of clonal variation has been established in *Pannicum* (Pernes *et al.*, 1970), and *Phleum* (Palenzona *et al.*, 1973).

Genetic Architecture of Lolium perenne

These several different investigations of the genetic organization of natural populations of perennial ryegrass have revealed the presence of a dual control of variation. The main control is the responsibility of nuclear genes, but in addition there is an extranuclear or 'plasmon' component. This latter component includes both the extranuclear effects revealed by the genetic analyses and the somatic variability selectable at the vegetative level. Thus although natural selection must have operated at the nuclear level it has not been sufficient to lead to the coadaptation of the nuclear genes necessary for the maintenance of differences over sexual generations (Breese and Mather, 1960). In contrast however, we have good evidence of coadaptation between the nuclear and plasmon genotype, differentiation of the latter being of a form which persists through sexual reproduction. This coadaptation would appear to have evolved as a consequence of survival being ensured by the continual production of vegetative tillers, each of which may be relatively short-lived, whilst the clone itself may be extremely persistent. Population differentiation in Lolium is thus dependent on both sexual and asexual modes of reproduction, the latter being of increasing importance in the more perennial types.

ACKNOWLEDGEMENTS

The results presented here are a synopsis of a continuing programme of research at the Welsh Plant Breeding Station into the genetic organization of Lolium under the stimulating guidance of Dr E.L. Breese. His cooperation and encouragement, together with that of other colleagues and research students is most gratefully acknowledged.

REFERENCES

Beddows AR (1953) The ryegrasses in British agriculture: a survey. Bull Ser H. 17, Welsh Pl Breed Stn

Breese EL (1960) The genetic assessment of breeding material. Proc Eight Int Grassld Cong 1960, 45-49

Breese EL (1965) Reproduction in ryegrass. BSBI Conf Report 9:51-60

Breese EL (1983) Exploitation of the genetic resource through breeding: Lolium species. In: J.G. McIvor & R.A. Bray, eds., Genetic resources of forage plants. CSIRO, Melbourne, pp 275-288

Breese EL, Charles AH (1962) Population studies in ryegrass. Rep Welsh Pl Breed Stn for 1961, pp 30-34

Breese EL, Mather K (1960) The organisation of polygenic activity within a chromosome in Drosophila. II. Viability. Heredity 14:375-401

Breese EL, Hayward MD, Thomas AC (1965) Somatic selection in perennial ryegrass. Heredity 20:367-379

Cooper JP (1959) Selection and population structure in Lolium. III. Selection for date of ear emergence. Heredity 13:461-479

Cornish MA, Hayward MD, Lawrence MJ (1979) Self incompatibility in ryegrass. 1. Genetic control in diploid Lolium perenne L. Heredity 43: 95-106

Fearon CH, Hayward MD, Lawrence MJ (1983) Self incompatibility in ryegrass. V. Genetic control, linkage and seed set in diploid Lolium multiflorum Lam. Heredity 50:35-46

Hayward MD (1967) The genetic organization of natural populations of Lolium perenne. II. Inflorescence production. Heredity 22:105-116

Hayward MD (1973) Practical approach to the biometrical evaluation of breeding material. In: M.A. do Valle Ribero & P. O'Donnell, eds., Evaluation of Breeding Material in Herbage Crops. Eucarpia Fodder Crops Section, Dublin 1972

Hayward MD, Breese EL (1966) The genetic organisation of natural populations of Lolium perenne. 1. Seed and seedling characters. Heredity 21:287-304

Hayward MD, Breese EL (1968) The genetic organisation of natural populations of Lolium perenne. III. Productivity. Heredity 23:357-368

Hayward MD, Nsowah GF (1969) The genetic organisation of natural populations of Lolium perenne. IV. Variation within populations. Heredity 24:521-528

Jenkin TJ (1959) The ryegrasses. In: H. Kappert, W. Rudorf & P. Parey, eds., Handbuck der Pflanzenzuchtg. Band IV, Futterpflanzen. pp 435-452

Lumaret R (1984) The role of polyploidy in the adaptive significance of polymorphism at the GOT 1 locus in the *Dactylis glomerata* complex. Heredity 52:153-170

Mather K, Jinks JL (1982) Biometrical Genetics. Third Edition. Chapman & Hall, London

Nei M (1978) Estimation of average heterozygosity and genetic distance from a small number of individuals. Genetics 89:583-590

Palenzona DL, Cavicchi S, Macardi L (1973) Irreversible phenotypic changes in *Phleum* associated with cloning in different environments. Proc International Meeting on Quantitative Inheritance, Polymorphism Selection and Environment, Bologna, Italy 1972, pp 133-142

Pernes J, Combes D, Rene-Chaume R (1970) Differentiation of natural populations of *Panicum maximum* Jacq. on the Ivory Coast by acquisition of modification some of which are transmissible by apomictic seed and others by vegetative propagation. CR Acad Sci, Paris 270:1992-1995

Shimamoto Y, Hayward MD (1975) Somatic variation in *Lolium perenne*. Heredity 34:225-230

Terrell EE (1966) Taxonomic implications of genetics in ryegrass (*Lolium*). Bot Rev 32:138-164

Thomas RL (1967) Interpopulation variation in perennial ryegrass. 1. Population Means. Heredity 22:481-498

PREFERENTIAL ASSOCIATIONS AMONG CHARACTERS IN CROSSES BETWEEN PEARL MILLET (PENNISETUM Typhoides) AND ITS WILD RELATIVES.

H. JOLY and A. SARR

G.P.D.P. - C.N.R.S.
F91190 GIF-sur-YVETTE

ABSTRACT

The study of the domestication of pearl millet using controlled crosses between spontaneous and cultivated forms has shown that :

1. The differences between cultivated and spontaneous forms involve few characters, mostly concerning the structure of the spike. An association among characters exists, called the "domestication syndrome".

2. It is very easy to obtain cultivated phenotypes having incorporated genes from spontaneous forms ; this is notably true as early as the first backcross.

In this paper, an analysis of F_2 and backross progenies issued from crosses between 5 cultivated and 3 spontaneous forms is presented, for 27 characters. The results confirm, on a wider basis, the existence of preferential associations among characters, particularly for the domestication syndrome.

Considering that pearl millet is an outcrosser and that cultivated and spontaneous forms coexist in normal field conditions in Africa resulting in gene flow between them, how might these preferential associations function in the field ? Results of experiments on reproductive biology, involving cultivated forms only, indicated some possible explanatory pathways. They also furnish a general methodology of approach for this type of problem.

NATO ASI Series, Vol. G5
Genetic Differentiation and Dispersal in Plants
Edited by P. Jacquard et al.

INTRODUCTION

This paper is a report on studies performed on plant domestication, or more specifically the investigation of relations between cultivated and related wild forms when they are present in the same ecological zone. The mechanisms were sought which regulate gene flow between forms, permitting each of them to conserve their phenotypic homogeneity and adapt to different selective pressures. Gene flow can be limited by many factors ; different flowering seasons, breeding systems, etc. We have tried to find out some of the relationships existing between reproductive biology and population structure.

Pearl millet, P. typhoides (Burm.) Stapf & Hubb., is a cereal species grown in semi-arid regions. It is a good material for population genetic research in relation to gene pools and population dynamics : in sub-Sahelian Africa, the principal zone of cultivation, the cereal is grown in the presence of related wild forms. Pearl millet is protogynous (anthesis occurs 3 to 5 days after the emergence of stigmas), which establishes a tendancy towards outcrossing (Rao et al. 1949, Burton 1974). In natural conditions, no apparent reproductive barrier exists to prevent genetic exchanges between the two forms. This was confirmed by the analysis of progeny obtained from "N'Douls" or intermediate forms (P. americanum s.sp. stenostachyum according to nomenclature of Brunken et al. 1977) collected in West Africa (Rey-Herme 1982). Several authors have found it easy to cross cultivated and wild forms (Bilquez and Lecomte 1969, Pernès et al. 1980, Belliard et al. 1980, Niangado 1981, Beninga 1981). The two forms differ by certain characters of plant architecture and spike and seed structure which constitute what has been called the "domestication syndrome" (Harlan et al. 1973). Studying progeny from crosses between a cultivar and wild P. mollissimum, Rey-Herme (1982) showed that the genetic control of spikelet morphology was relatively simple in pearl millet, and that some of the corresponding loci were geneticaly linked. It is not difficult to obtain cultivated-type phenotypes from a wild x cultivated cross ; they can be found in particular as early as the first backcross or in F_2 progeny. This indicates the presence of preferentially associated characters. Are these associations typical of confrontations between wild and cultivated genomes ? What mechanisms can explain them ? Two parallel approches were used to answer these questions. On the one hand, progeny of crosses between wild and cultivated forms of diverse origins were studied to see to what extent they showed the same preferential associations among characters. Secondly, aspects of reproductive biology were investigated to attempt to find mechanisms responsable for maintaining the associations.

METHODS AND MATERIALS

The genotypes studied are indicated on Table 1.

Table 1 - Pearl millet genotypes studied.

genotype	origin	form	degree of fixation	approach
Chine	China	cultivated	pure line	reprod. biology
Ligui	Chad	"	"	"
Massue	Mauritania/ oasis	"	"	" wild/cult.crosses
Tiotandé	Senegal	"	"	reprod. biology wild/cult.crosses
$23d_2B$	Ghana	"	pure line selected in Tifton USA	reprod. biology wild/cult.crosses
Drôo	Tunisia	"	trad. cultivar	wild/cult.crosses
Zongo	Niger	"	"	"
P.mollissimum	Mali	wild	5 selfs	"
P.violaceum C	Cameroon	"	1 full sib	"
P.violaceum N	Niger	"	1 self	"

Table 2 - Crosses between wild and cultivated forms of pearl millet.

wild \ cultivated	Massue	Tiotandé	Drôo	$23d_2B$	Zongo
P. mollissimum	x	x	x	x	
P. violaceum C				x	
P. violaceum N			x	x	x

Table 3 - Characters observed in progenies of wild/cultivated crosses.

	character	sign		scoring for qualitative characters
1	Internode colour	COE	pigmentation	0-1
2	Internode pilosity	PIE	and	0-1
3	Leaf pilosity (abaxial)	PIF		0-1
4	Leaf pilosity (adaxial)	PSF	pilosity	0-1
5	Time to flowering	ISE		
6	Aspect of tillering	AST		0-1-2
7	Height at flowering	HAE		
8	Height at maturity	HAM		
9	Number of basal tillers at maturity	NTB		
10	Number of fertile spikes	NEU	plant	
11	Rank of flagleaf	RGD	architecture	
12	Length of flagleaf sheath	LGD		
13	Length of flagleaf blade	LOD		
14	Width of flagleaf blade	LAD		
15	Width of spike peduncle	LAP		
16	Width of spike	LAC		
17	Length of spike	LOC		
18	Length of involucral pedicel	LPI		
19	Degree of shedding	NOC	domestication	0 to 4
20	Functional abscission layer	CAF		0-1
21	Enclosure of seeds in lemma and palea	ENG		0 to 4
22	Seed starchiness	VIT	syndrome	0 to 4
23	Spike aspect	ASP	for	0-1-2
24	Length of bristles	LOS	spikelet	
25	Length of longest bristle	LS+	and seed	
26	Presence of one very long bristle	PS+	morphology	0-1
27	Bristle morphology	MOS		0-1-2

Table 4 - Characters measured for reproductive biology studies

Growth characters :

V30	Vigour score 30 days after planting
HP30	Plant height 30 days after planting
NT30	Number of tillers at heading stage
DT	Tiller initiation date
NTE	Number of tillers at heading stage
HTE	Plant height at heading stage

Flowering characters :

Ep	Main tiller heading date
LOD	Flagleaf length
LAD	Flagleaf width
DFF	Starting date of female flowering
FFF	Date of the end of female flowering
DFM	Anther emergence date
FFM	Date of the end of male flowering
SFM	Staminate flowering direction
Exer	Exertion type
FE	Spike shape
EpG	Protogyny duration

Maturity characters :

HAM	Main tiller height
LAC	Spike width
LOC	Spike length
NE/P	Number of spikes per plant
PCP	Main spike weight
PG/C	Main spike seed weight
GAF	Seed production under selfing
PMG	1.000 seeds weight
PAANT or AANT	Presence or absence of anthocyanin on the stalk
Pil	Hairiness of nodes

Enzyme markers :

EST	Esterase
PER	Peroxidase

1. Confrontations between Wild and Cultivated Forms.

Eight reciprocal crosses between the two forms were carried out in G.P.D.P. greenhouses during the year 1981-1982. Backcrosses (BC) were performed using cultivated parent pollen. Segregations in F_2 and backcross progenies were observed during the 1982-1983 dry season at the C.N.R.A.* Station, Bambey, Senegal. Ninety plants per F_2 were observed during growth, at flowering and at maturity. Table 2 indicates the crosses observed ; the characters measured are shown on Table 3.

2. Reproductive Biology.

The different aspects of reproductive biology investigated were studied in cultivated forms only.

a. Protogyny : was characterized by an index $IPG = \frac{\Delta(f♀ - f♂)}{df♀}$

i.e. the difference in days between the beginning of female flowering and the beginning of male flowering, divided by the duration of the female flowering. The index of protogeny permits evaluation of the possibility of self-pollination ; when $IPG > 1$ all the stigmas of the inflorescence will have emerged before anthesis, when $IPG \leqslant 1$ self-pollen as well as out-pollen can be present on stigmas at the same time.

b. Pistil Receptivity : Each genotype was taken as female and pollinated with saturating quantities of pollen of its own or other genotypes. At regular intervals, spikelets were removed and fixed as described by Martin (1958). Pollen tubes were observed under ultraviolet light (400 nm.). The following parameters were observed :

- time elapsed before germination
- speed of pollen tube growth in the pistil
- efficiency of penetration into the ovaries

$$E.P.O. = \frac{\text{number of tubes in the ovary x 100}}{\text{number of tubes at the base of the style}},$$

measured 3 hours after pollination.

---

* Centre National de la Recherche Agronomique (Senegal)

c. Pollen Competition : Equal quantities of pollen of two genotypes were thoroughly mixed, and each female was pollinated by pollen taken from the five genotypes, two by two. The relative proportions of the hybrids resulting from these pollen mixtures were determined by electrophoresis of esterases and peroxidases extracted from young seedlings. The genetic control of these enzymes was given by Sandmeier et al. (1980).

d. Reproductive Biology and Progeny Structure : taking account of the results of the pistil receptivity and pollen competition experiments, the following material was created to study the implications of the autofertility of Ligui upon its progeny : an F_1 (Ligui x Massue), the F_2 obtained by selfing the F_1, and the two reciprocal BC (F_1 x Ligui and Ligui x F_1). The characters measured are indicated on Table 4.

e. Data Analysis : The statistical methods used included multivariate analysis, principal component analysis, stepwise discriminant analysis and χ^2 tests.

RESULTS

1. Existence of preferential associations.

The correlation matrices calculated from the wild x cultivated F_2 segregations showed that groups of strongly correlated characters were present (see Fig. 1). The first group was composed of characters of plant architecture : the aspect of the tillering, basal tillers, fertile tillers, the number of leaves on the principal tiller and the height at maturity. Characters describing the form of the spike and the foliage predominated in the second group : spike and flagleaf length and width and the diameter of the spike peduncle. Spikelet structure was found in a third group : shedding, abscission layer, length of the involucral pedicel, and the enclosure of seeds in lemma and palea. These groups were the same for all the wild and cultivated genotypes crossed, although the correlations were more intense in crosses with *P. mollissimum* than with *P. violaceum*. It is therefore evident that preferential associations did exist. The first two components of the principal component analysis on all the F_2's accounted for 30.7% of the total variability. The first

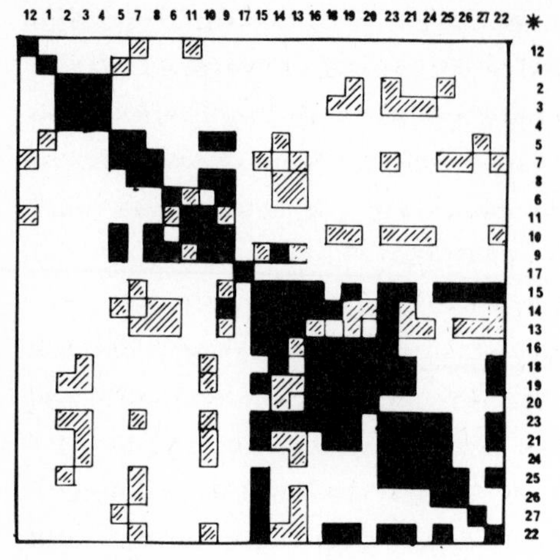

Massue x P.mollissimum

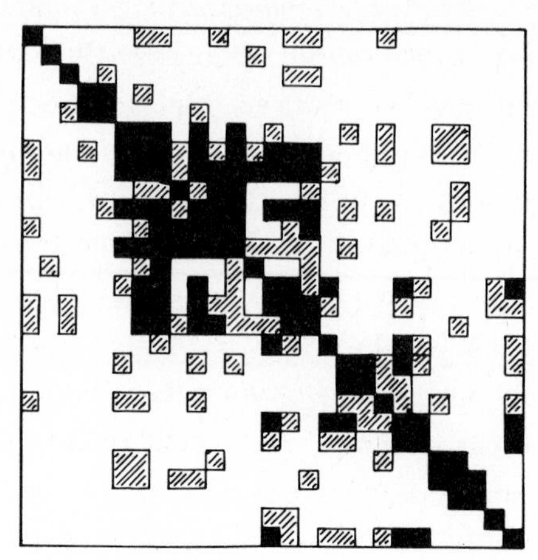

Drôo x P.violaceum

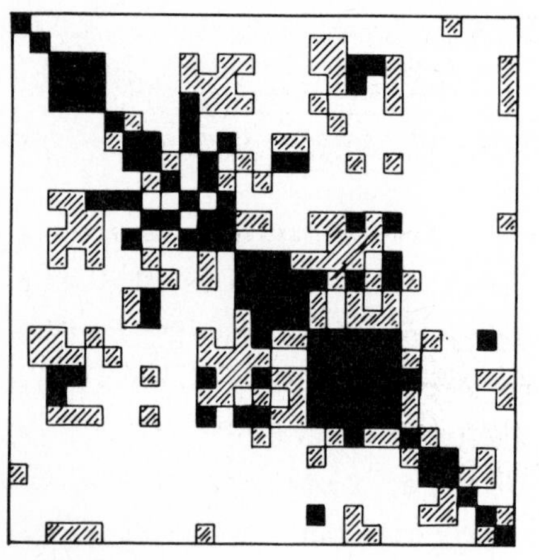

Tiotandé x P.mollissimum

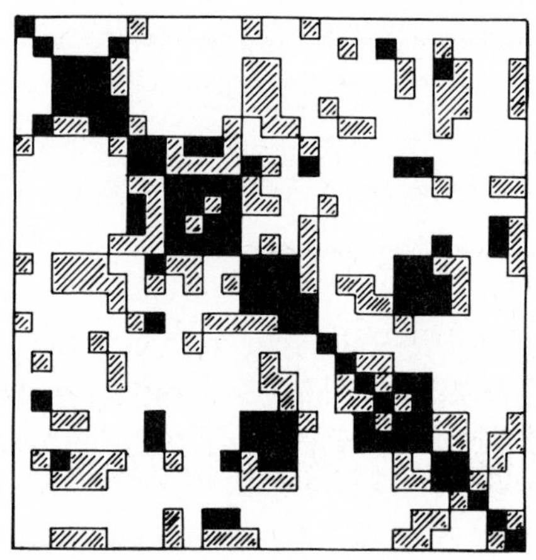

Zongo x P.violaceum

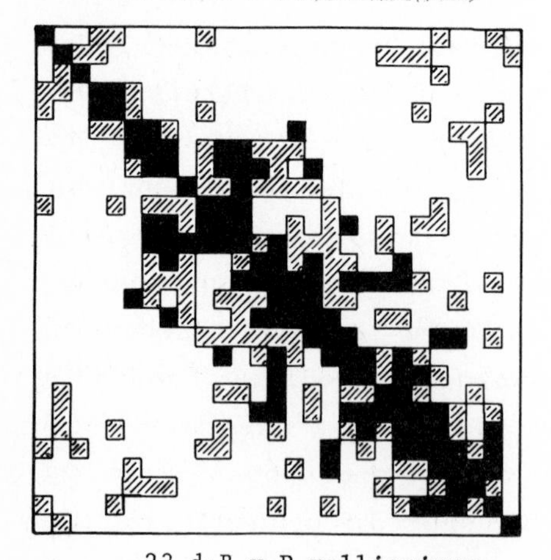

23 d_2B x P.mollissimum

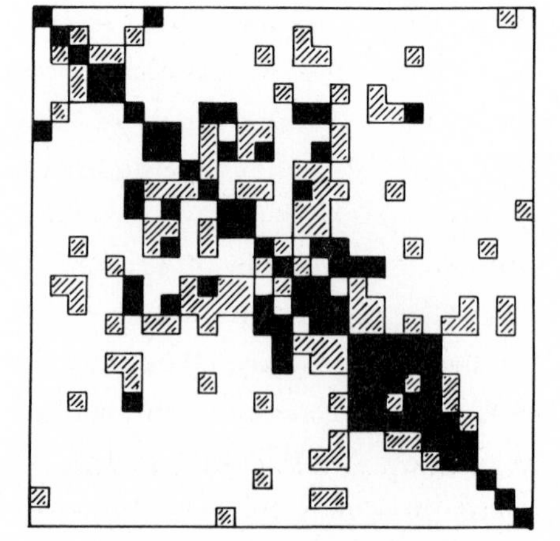

23 d_2B x P.violaceum

* See Table 3 for characters corresponding to numbers

Intervals of correlation : ■, $r \geq 0.40$; ▨, $0.25 \leq r < 0.40$; □, $r < 0.25$.

Fig. 1 - Correlated characters in six wild-cultivated crosses.

component was defined by spike characteristics : LAP, ASP, NOC, LAC, LAD, LOD and LPI, by order of decreasing importance. The second component was defined by plant architecture : NTB, AST, RGD, VIT and NEU. The structure of the variability in this plane confirmed the existence of preferential associations. Recombinant forms, however, were also found.

The genetic control of the spikelet characters was simple ; most of the corresponding loci were contained within 35 units of recombination at most (Joly, 1984). This confirmed on a wider basis results obtained previously (Beninga 1981, Rey-Herme 1982). There were important distortions in BC, compared to those expected according to the F_2's, in favour of cultivated forms. Such phenomena were also observed by Rey-Herme (1982). Distortions in favour of wild alleles have been reported for enzyme loci (Marchais and Tostain, 1984).

2. Reproductive biology.

a. Protogyny : was extremely variable in different genotypes and environmental conditions. For instance, 1PG was 0.5 for Tiotandé and 5.0 for Chine, which indicated, notably in the former case, a strong probability of coincidence of male and female flowering on the same spike thereby producing a situation favourable for self-pollination. In other experiments (Sarr and Sandmeier in preparation), estimations of the rate of self- and cross-pollination in our experimental conditions confirmed the importance of self-pollination in the reproductive strategy of pearl millet. Outcrossing estimates (obtained from an alternate row planting procedure with esterase as marker locus) were quite low, considering that the species is reputed to be allogamous. Chine, for exemple, was scored only 40% cross-pollinated. Pistils were found to be receptive immediately upon emergence and the receptivity lasted largely until anthesis, reinforcing the possibilities for self-pollination.

b. Pistil Receptivity and Pollen Competition : the biological signification of the E.P.O. parameter, studied by Sarr et al. (1983), is linked to the correlation between the success of fertilization and the number of pollen tubes (Ter-Avanesian 1978). A significative difference existed between genotypes according to whether they were used as male or female. This aptitude for association between pollen and pistil was called "differential receptivity". In most cases, self-pollen was as effective as out-pollen, and sometimes even more so (see Table 5). This was especially true for the Ligui genotype, which presented a phenomenon called "mass effect" : it was necessary to saturate the female flowers with pollen in order to achieve good fertilization. Since in natural conditions this situation can only occur after male flowering on the same spike, this mechanism favours self-pollination. The same genotype presented pistil receptivity favouring self-pollen, characterized by faster pollen tube growth for self-pollen than out-pollen.

Table 5 - Average E.P.O. 3 hours after pollination. (a)

	$23d_2B$	Ligui	Chine	Massue	Tiotandé
$23d_2B$	44	55	33	40	70
Ligui	40	98	57	48	96
Tiotandé	33	98	77	31	85

(a) taken and simplified from Sarr et al. (1983).

Table 6 - Examples of pollen competition taking $23d_2B$ and Ligui as female.

pollen mixture	χ^2	signif.	dominant genotype
$23d_2B$ x (Ligui + Chine)	0.42	NS	neither
x (Ligui + Tiotandé)	14.04	***	Tiotandé
x (Chine + Massue)	10.84	***	Massue
x (Tio. + Chine)	21.32	***	Chine
x (Tio. + Massue)	70.04	***	Tiotandé
Ligui x (Tio. + Massue)	20.18	***	Tiotandé
x (Tio. + Chine)	-	-	Tiotandé(75% of hybrids)
x (Chine + Ligui)	84.72	***	Ligui
x (Ligui + $23d_2B$)	26.58	***	Ligui
x (Chine + Massue)	-	-	Massue

NS non-significant

*** significant at 1‰ level

- too much selfing to calculate χ^2

Pollen competitions (Table 6) expressed the relative efficiencies of the different pollen sources. There was no clear hierarchical tendancy, but rather a phenomenon interpretable in terms analogous to combining ability with specific interactions of the type (♂x♂ , ♀x♂). Two exceptions were noted, one of them concerning Ligui. In all combinations of the type Ligui x (Ligui + out-pollen), Ligui was "preponent". This tendancy was confirmed in experiments when out-pollen was applied a certain length of time (a half hour, 3 hours, 24 hours) before self-pollen.

<u>c. Implications of the Characteristics of Reproductive Biology on the Structure of Progenies</u> : This aspect as well was illustrated by the particular case of Ligui. Segregations distorded in Ligui's favour appeared systematically (Valverde 1982). Table 7 shows several instances for qualitative characters.

<u>Table 7</u> - Segregations of qualitative characters in a Ligui x Massue F_2.

character	segregation tested	dof	χ^2	
esterase (seeds)	1-2-1	2	0.8	NS
esterase (leaves)	1-2-1	2	1.09	NS
peroxidase	9-7	1	0.32	NS
colour base stem	3-1	1	10	*
anthocyanines	3-1	1	5.9	*
node pilosity	9-6-1	2	3.80	
spike emergence	9-7	1	4.51	*

NS non significant
* significant at 5% level

As concerns qualitative and quantitative characters in general, two BC were studied by principal component analysis. The signification and inertia of the components and the important parameters are shown on Table 8 . In Fig. 2, the

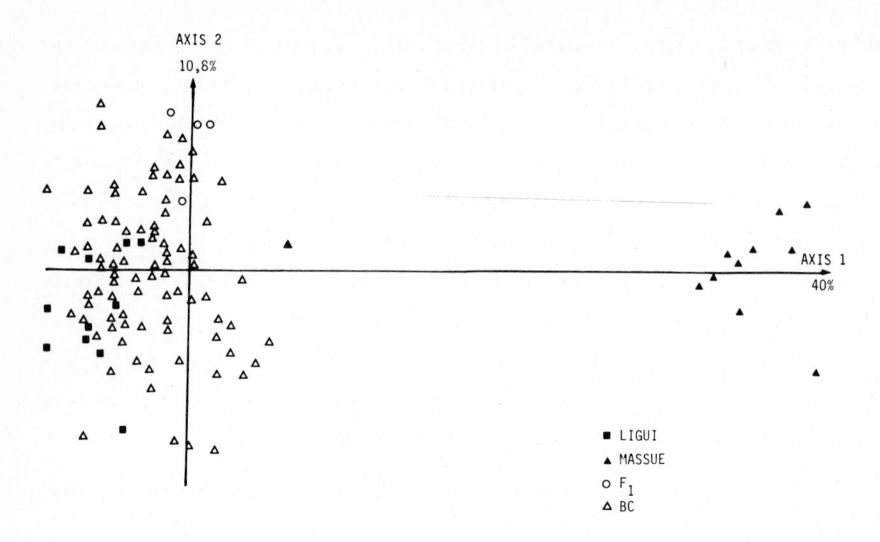

Fig. 2 - Principal component analysis on reciprocal backcrosses involving Ligui (recurrent parent) and Massue.

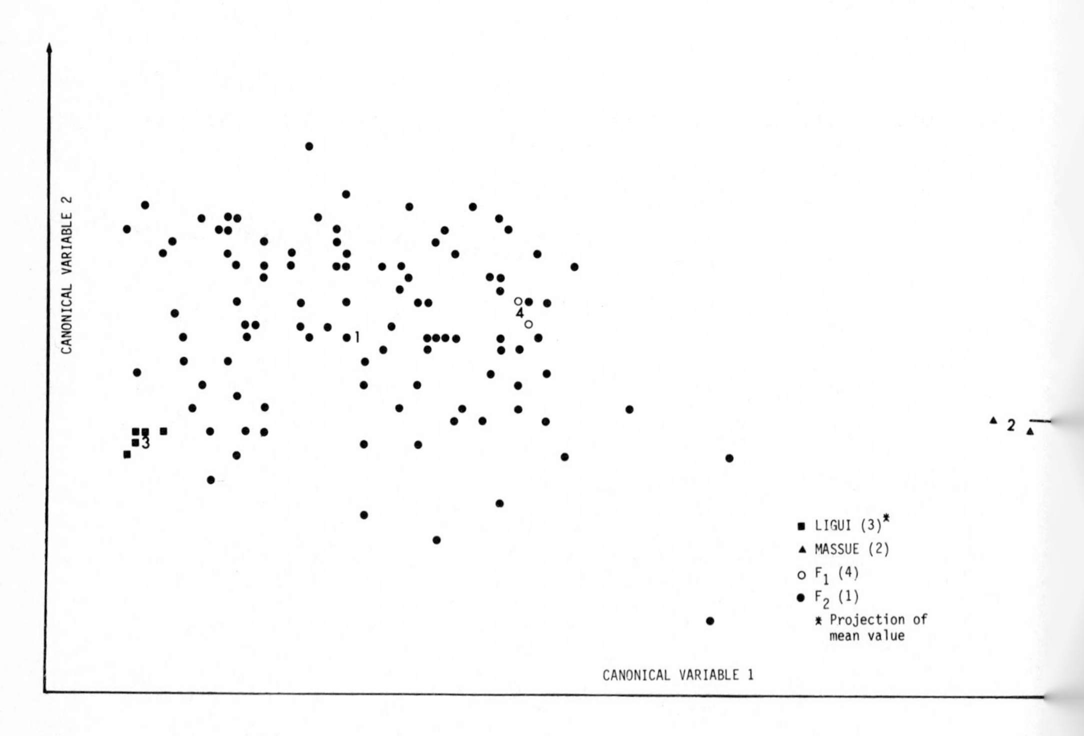

Fig. 3 - Discriminant analysis involving Ligui, Massue, F_1 and F_2.

projection on the plane defined by the first two components showed the BC progenies (F1 x Ligui and Ligui x F1) grouped around Ligui in total opposition to Massue. It would seem that preferential choice of gametic combinations lead to these distortions. Stepwise discriminant analysis on F_2 and BC progeny confirmed the tendancy towards preferential association of Ligui-type characters. Only 3 F_2 plants and no BC were classed with Massue (see Table 9 and Fig. 3).

Crosses between cultivated and wild genotypes have given similar results. The preferential associations among characters found in F_2's of crosses between wild P. mollissimum and cultivated Massue, $23d_2b$ and Ligui were much more intense for Ligui (Beninga 1982).

Table 8 - Principal component analysis on backcrosses involving Ligui and Massue.

component	% inertia	description	important parameters
1	23.4	vigour and floral biology	+ V30, H30 - ETP, DFF, DFM, FFF, FFM
2	14.2	productivity	+ PGP, PG/C, HTE, LAC, HAM
3	10.1	tillering	+ NTE - DT

Table 9 - Discriminant analysis on F_2 and BC progenies.

	F_2						BC				
characters	DFM	LAD	PCP	LAD	HAM	HP30	LAC	LOC	FFM	NE/P	LAD
% correct class.	71.4	100	100	100	100	100	100	100	100	100	100
plants classed: Massue	68	2	1	1	3	0	0	0	0	0	0
plants classed: Ligui	13	51	35	33	34	35	38	32	32	35	36
plants classed: F_1	35	63	80	82	79	78	60	66	66	63	62

The results of pistil receptivity and pollen competition experiments lead to the hypothesis of preferential gametic associations, originating from specific pollen-pistil recognition phenomena. Analogous cases have been described by Plonka (1971), Mulcahy and Mulcahy (1983), Ottaviano et al. (1983) and Emerson (1933).

DISCUSSION - CONCLUSION

The results presented on the reproductive biology of pearl millet clearly demonstrate the existence of preferential associations among characters in confrontations between cultivated varieties. The underlying causes of this situation can be identified as arising from diverse gametophyte x sporophyte interactions : pollen competition, differential receptivity, and the particularities of pearl millet floral biology such as the influence of the protogyny index. In addition, the impact of the environmental factors should not be ignored. Robert (1983), using the same material as the authors, showed that the temperature regime affects segregations in F_2's.

The correlation matrices and the principal component analysis described for the wild x domesticated crosses shows that preferential associations among characters were present here as well, and especially so in the case of characters related to the domestication syndrome.

The existence of these phenomena evokes the question of what processus are involved in recombination. The phenotypic differences between wild and cultivated forms are the result of diverse selective pressures (artificial and natural) exerted throughout the history of the domestication of this crop. It is possible to imagine that the stability that has been observed in domesticated genotypes, in the presence of wild forms, was made possible by the elaboration of mechanisms preserving gene blocks or coadapted chromosomes from random rearrangements.

Recombination can be controlled in many ways. Often this is accomplished through fertility barriers, which preserve genotypic associations by preventing exchanges between genomes. However, to our knowledge, crosses between cultivated and wild forms of pearl millet have never revealed any controls of this type. Our results rather suggested that the preferential associations we observed were not necessarly governed by specific accidents engendred by the confrontation of genomes. The same kinds of distortions, for both enzyme and phenotypic characters, occurred in both cultivated x cultivated and wild x cultivated crosses. The link established between the floral biology and progeny structure, as well as the results obtained on the mechanisms that influence the outcrossing rate (pollen competition, pistil receptivity) all indicate the importance of reproductive biology in controlling recombination (Darlington 1939, Stebblins 1950, Grant 1958). Pollen competition, and its unterlying genetic mechanisms (Mulcahy and Mulcahy 1983, Ottaviano et al. 1983 and Pfahler 1967), show that the gametophyte can be a target for selection at a lower cost to individual plants and to the population. The phenomenon of genetic overlapping described by Mulcahy (1983) and corroborated by the results obtained by Zamir (1981) on Lycopersicon amply illustrate this idea.

Explaining the preferential associations among characters by the mechanisms related to the reproductive strategy of cultivated forms will have important consequences on the understanding of the species complex and the use of wild germplasm in pearl millet breeding. In the future, the reproductive strategy of wild forms will have to be evaluated in the light of the results obtained on cultivated forms. This will complement investigations in progress on the cohesion of the groups of genes involved in the domestication process and in gene flow in the pearl millet complex.

ACKNOWLEDGEMENT

The authors thank Brad Fraleigh for the English translation.

REFERENCES

BELLIARD J., NGUYEN VAN E., PERNES J., 1980. - Analyse des relations génétiques entre formes spontanées et cultivées chez le mil à chandelle. I. Etude des parents et des hybrides de première génération (F1) entre un écotype de Pennisetum mollissimum Hochst (forme spontanée) et différentes formes cultivées de Pennisetum americanum (L), Leeke. Ann. Amélior. Plantes, 30 (3), 229-251.

BENINGA M.B., 1981. - Structure génétique du complexe des mils penicillaires : analyse des descendances issues d'hybrides entre formes cultivées et formes spontanées. Thèse 3ème cycle. Amélioration des Plantes. Univ. Paris-Sud. Orsay.

BILQUEZ A.F., LECOMTE J., 1969. - Relations entre mils sauvages et mils cultivés : Etude de l'hybride Pennisetum typhoides (Stapf et Hubb) x Pennisetum violaceum L (Rich). Agronomie tropicale XXIV (3) 249-257.

BRUNKEN J., DE WET J.M.J., HARLAN J.R., 1977. - The morphology and domestication of pearl millet. Economic Botany 31, 163-174.

BURTON G.W., 1974. - Factors affecting pollen movement and natural crossing in pearl millet. Crop Science 14, 802-805.

DARLINGTON C.D., 1939. - The evolution of genetic systems. Cambridge. Cambridge University Press.

DEMARLY Y., 1979. - The concept of linkat. Proc. Conf. Broadening genetic base of crops. Wageningen, 1978.

EMERSON R.A., 1933. - Relation of the differential fertilization genes, Ga, ga to certain other genes of the Su Tu linkage group of maize. Genetics, 19, 137-156.

GRANT V., 1958. - The regulation of recombination in plants. Cold Spring Harbour Symp. Quant. Biol. 23, 337-363.

HARLAN J.P., DE WET J.M.J., GLENPRICE E., 1973. - Comparative evolution of cereals. Evolution 27, 311-325.

JOLY H., 1984. - Hérédité du syndrome de domestication chez le mil Pennisetum typhoides : Etude comparée de descendances (F2 et RC) issues de croisements entre plusieurs géniteurs cultivés et spontanés. Thèse 3e cycle. Univ. Paris-Sud. Orsay.

MARCHAIS L., TOSTAIN S., 1984. - Genetic divergence between wild and cultivated pearl millets (Pennisetum typhoides). II. Characters of domestication (in press, Z. Pflanzenzuchtung).

MARTIN F.W., 1958. - Staining pollen tubes by means of fluorescence. Stain Technol. 34, 125-128.

MULCAHY D.L. and MULCAHY G., 1983. - Pollen selection : an overview. In D.L. Mulcahy and E. Ottaviano (eds) Pollen : Biology and implication for plant breeding. p. XV-XVII.

NIANGADO O., 1981. - Utilisation des rétrocroisements chez le mil Pennisetum americanum (L.) Leeke 1/ pour changer le régime de floraison, 2/ pour exploiter la variabilité génétique des formes spontanées. Thèse 3e cycle. Amélioration des Plantes. Univ. Paris-Sud. Orsay.

OTTAVIANO E., SARI GORLA M. and ARENARI I., 1983. - Male gamete competitive ability in maize selection and implications with regard to the breeding system. In D.L. Mulcahy and E. Ottaviano (eds) Pollen : Biology and implication for plant breeding. Elsevier Biomedical New York.

PERNES J., NGUYEN VAN E., BENINGA M.B. and BELLIARD J., 1980. - Analyse des relations génétiques entre formes spontanées et cultivées chez le Mil à chandelle (Pennisetum americanum (L) Leeke, P. mollissimum Hochst). II. Etude de 3 familles F2 issues d'hybrides entre une plante d'un écotype de Pennisetum mollissimum Hochst et 3 lignées de Mil cultivé, Pennisetum americanum (L.) Leeke. Ann. Amélior. Plantes 30 (3), 253-269.

PFAHLER P.L., 1967. - Fertilization ability of maize pollen grains. II. Pollen genotype; female sporophyte and pollen storage interaction. Genetics 57, 513-521.

PLONKA F., 1971. - Compétition pollinique chez le lin cultivé. Ann. Amélior. Plantes 21 (2), 179-220.

RAO P.K., KUNCHIKORON A. and KUSHNAMURTHY I.V.G., 1949. - Natural crossing in cumber Pennisetum typhoides. Madras Agr. J. India 36, 526-529.

REY-HERME C., 1982. - Les relations génétiques entre les formes spontanées et cultivées chez le mil (Pennisetum sp.). Thèse 3e cycle. Amélioration des Plantes. Univ. Paris-Sud. Orsay.

ROBERT T., 1983. - Etude des sélections gamétophytiques chez le mil. D.E.A. Amélioration des Plantes. Univ. Paris-Sud. Orsay.

SANDMEIER M., BENINGA M.B. and PERNES J., 1981. - Analyse des relations entre formes spontanées et cultivées chez le mil à chandelle. III. Etude de l'hérédité des estérases et des peroxydases anodiques. Agronomie 1 (6), 487-494.

SARR A., FRALEIGH B. and SANDMEIER M., 1983. - Aspects of reproductive biology in pearl millet. In D. L. Mulcahy and E. Ottaviano (eds) Pollen : Biology and implication for plant breeding. Elsevier Biomedical. New York p. 381-388.

SARR A. and SANDMEIER M., (in preparation) - Estimation du taux d'allogamie chez le mil.

STEBBINS G.L., 1950. - Variation and evolution in plants. New York. Columbia University.

TER-AVANESIAN D.V., 1978. - Significance of pollen amount for fertilization. Bull. Torrey Bot. Club 105 (1) 2-8.

VALVERDE LANDROVE V., 1982. - Contribution à l'étude de la biologie de la reproduction chez le mil Pennisetum typhoides : Recombinaison et distorsions de ségrégation. D.E.A. Amélioration des Plantes. Univ. Paris-Sud. Orsay.

ZAMIR D., 1981. - Low temperature effect on selective fertilization by pollen mixtures of wild and cultivated tomato species. Theor. Appl. Genet. 59, 235-238.

Genetic Diversity of Foxtail Millet (*Setaria italica*)

E. Nguyen Van and J. Pernès

G.P.D.P. - C.N.R.S.
F91190 GIF-sur-YVETTE

ABSTRACT

The huge diversity of a world collection of foxtail millets can schematically be described through the intersection of subsets : either forage (moharia subspecies) or cereals (maxima subspecies), and tropical or temperate origins. The biochemical diversity (10 isozyme loci) was very much smaller, but, however, clearly structured by regions. The comparison with populations from wild species (*S. viridis*) and either within or between species hybridizations enabled testing various hypotheses explaining that diversity ; the discussion will mostly be focused on the "multiple or single" domestication problem.

INTRODUCTION

The domestication of cereals affords many good examples of diversification in plants ; the initial variability of wild populations (especially in the case of multiple domestications) was subjected to diverse adaptative trends imposed by traditional farming methods and chanelled by various techniques and ecological conditions. Actually, domestication was a long term experiment in plant genetics. This was the case for foxtail millet, *Setaria italica*, the cereal plant studied here.

Several studies have clearly established that the closest ancestor species of this self-pollinated diploid (2n = 18) cereal was *Setaria viridis*. The crop plant is strongly

NATO ASI Series, Vol. G5
Genetic Differentiation and Dispersal in Plants
Edited by P. Jacquard et al.

differenciated into two subspecies : moharia and maxima (de Wet et al. 1979). Archeological records established the presence of Setaria seeds in China 5000 years B.C. and in Swiss lakes 2000 years B.C.. There is no complete reproductive barrier between wild Setaria viridis and cultivated Setaria italica. Gritzenko (1960) tentatively suggested the existence of some differenciation within the maxima subspecies (Manchurian, Korean and Mongolian types). Chikara and P.K. Gupta (1979) described some chromosomal differenciation within Setaria italica when they studied their karyotypes. However, their paper did not give any morphological information, thus it was impossible to link these variations either to particular foxtail millet forms or to partial hybrid sterilities which sometimes occur. Miyagi, (p.c.), Kokubu et al. (1977) and Kawase and Sakamoto (1982, 1984) analyzed several collections of foxtail millet and were mostly interested in the Asian north-south cline from Korea-North Japan to Taïwan, either for flowering earliness (clearly depending on the latitude origin of strains) or for the frequencies of polyphenoloxydase and esterase alleles.

By human migration, the dissemination of Setaria seeds all over the world probably occurred very soon after the initial domestications. Over a period beginning several thousand years ago, and continuing until modern times, seeds could have been exchanged southwards and eastwards from North China to Korea and the Japanese islands, to the west towards Europe via Caucasia, or from India northeastward to Taïwan and Japan and to the northwest towards Europe. Succeeding waves of migration lead to introgréssions with local cultivated or wild Setarias. Setaria viridis has been observed in every continent and its spontaneous dissemination millions of years ago anticipated domestication. Dissemination of the crop plant over short distances occurred through seed commerce in marketplaces. Although to our knowledge pollen flow has never been measured, it is certainly less than several hundred metres (wind pollination is responsible for a small allogamy rate generally less than 1%). The principal factors to be evaluated are the simultaneous consequences of active long distance dissemination (migration) and short distance dissemination by hybridization and introgression.

The present paper is part of a series of analyses aimed at clarifying the evolutionary history of foxtail domestication, by the means of genetic analyses. In the first step described here we tried to obtain a reliable classification of a worldwide foxtail millet strain collection, using morphological and developmental characters and marker genes determined by isozyme electrophoresis. Other papers (Brabant et al. (1980) ; Darmency et al. (1984), Darmency and Pernès, (1984), and unpublished data) analyze interspecific and intraspecific hybridizations within the Setaria complex, the inheritance of various markers and cytogenetical variations. The present paper ought to be useful as a reference point to situate which strains various authors were dealing with.

MATERIAL AND METHODS

121 different strains were described simultaneously. They came from our basic collection of 731 lines obtained as follows :

1. a world collection obtained from Miyagi (468 strains) with description of the origin (country, often towns) and sometimes vernacular names. Asiatic, African and European strains were represented.
2. varieties collected in Japan by Sakamoto (15 strains).
3. varieties collected by one of the authors (J.Pernès) in Kagoshima Island (Japan) (21 strains).
4. strains collected in Taïwan by G. Second (12 strains).
5. various strains from China (Shenxi, Kirin, Liaoning, Heilongjang, Shandong, Shanxi, Hopei) either given by Chinese breeders in different provinces or directly collected by J. Pernès (1978, 1979, 1983) (64 strains).
6. cultivated varieties collected regularly in France by various workers in the G.P.D.P. laboratory (Pernès, Poirier-Hamon, Barreneche, de Cherisey, Brabant) (128 strains).
7. various strains sent to us from several European botanical gardens (England, Denmark, Belgium,...) (17 strains).
8. varieties from Botel Tobago Island, given by J. Arnaud (6

strains).

All these strains are stored in cold rooms at the G.P.D.P. laboratory. The samples described as strains and the varieties directly collected in the fields were most of the time pure lines. All strains analysed here appeared to be homozygous pure lines for all the characters studied.

Among these 731 different strains in our collection, 121 were chosen in order to avoid duplications (preliminary field trials allowed to eliminate from analysis all strains which were not significatively different for quantitative measurements and had the same electrophoretic pattern). We also tried to balance the geographical origins of our samples despite the fact that the great majority of strains came from Japan and China, and many from France (where the sampling was not at all proportional to the economic significance of foxtail millet !). Moreover few strains from India or Africa could be observed in our field conditions in France (Pau and Gif/Yvette).

Morphological Descriptors

More than 40 different characters were observed in a half-hectare 3 randomized blocks experiment, 20 plants per plot, measured at Pau (1982). Other trials at Gif/Yvette (Essonne) and Marquèze (Landes) enabled us to confirm some of the observations by reference to common check lines. Among these characters, only the following will be used here for numerical taxonomy : time of heading (days) (ETP), number of tillers (2 months after sowing) (NT2M), number of well developped spikes (NEP), height of the main tiller, measured from ground level to the last developped leaf blade base (2 months after sowing) (H2M), length and width of spikes (LOE and LAE), weight of 1000 seeds (PMG), width of the stalk just below the main spike (LAP), length and width of the flag leaf blade (2 months after sowing) (LO2M and LA2M). We limited ourselves to only these ten descriptors because the series of measurements was completed with all strains under study and because some redundancies of descriptions (as shown in preliminary works) allowed us to concentrate on easy, simple and generally used observations.

Electrophoretic Data

De Cherisey *et al.* (in preparation) analyzed several isozyme series genetically and biochemically and were able to establish the links between banding patterns and allelic states in specific loci. In this study, the following loci with their alleles were used : EST1 (esterase 1, alleles a,b,c), EST2 (esterase 2, alleles a,b,c), EST3 (esterase 3, alleles a,b), PHAC (acid phosphatase, alleles a,b), GOT1 (glutamate-oxalo-acetate-transaminase 1, alleles a,b), GOT2 (glutamate-oxalo-acetate-transaminase 2, alleles a,b), MDH1 (malate-dehydrogenase 1, alleles a,b,c), MDH2 (malate-dehydrogenase 2, alleles a,b) PGD1 (6-phospho-gluco-dehydrogenase 1, alleles a,b), PGD2 (6-phospho-gluco-dehydrogenase 2, alleles a,b,c). Some other marker genes (polyphenol oxydase, alcohol dehydrogenase, glutinous), described by Jusuf *et al.* (in preparation) were not analyzed in the present work but can easily be added. We used only the markers which were polymorphic, because our purpose was more to establish a classification than to measure some definite population parameters such as the absolute polymorphism rate or the genetic distance (in time units).

Each homozygous strain was described by the series of its alleles at the ten loci studied. After regrouping strains either on a geographical basis or a morphological basis, other descriptions begin again using frequencies of alleles at each locus within groups.

Multivariate Methods

Classical univariate statistical analyses (analysis of variance, comparisons of means) were only used to search out duplications and environmental effects (within plot variations, between blocks effects) and to determine how safely the means could be used to describe and compare the different strains.

The *morphological characters* were studied in a principal component analysis, followed by a dendrogram description. We then looked for substructures to start a taxonomy by Diday's "Nuées Dynamiques" algorithm (1980). This algorithm starts grouping from a limited number (n) of randomly chosen strains and through successive iterations it defines a partition in an _a priori_ given number of clusters, so that each strain is closer to the center of the group to which it is attributed than to any other center. These centers are progressively established during the iterations and a measurement of the quality of partition (ratio between/within group distances) is given. The distances calculated are Euclidian distances obtained from the coordinates of the strains as given by the principal component analysis (they are independent variables). Several (here 3 or 5) random runs of n strains lead to possibly different groups. These different groups were compared for common aspects, i.e. sets of strains that always grouped together. Several strains remained isolated at this step. The sets of strains, isolated strains excluded, are called "formes fortes" and were used as starting points to continue the job by applying another classical method : discriminating factorial analysis. The "formes fortes" having been determined _a priori_, we tried to recognize them by discriminating analysis. By changing some of the strain attributions to groups, we tried to obtain one hundred percent good classification by trial and error. When this result was acquired we used the same analysis to classify the strains not yet integrated within the basic "formes fortes". Either there were other unclassified strains observed close to them and a new group was created, or there were not and they remained isolated and we did not go any further. The final calculation of Mahalanobis distances and the projection of the clusters on to the first planes of the canonical variable space closed the taxonomy.

With the _qualitative observations of electrophoretic allelomorphs_, we proceded in a similar way. However there were some differences. First, many of the strains shared exactly the same zymogram for all 10 loci. Thus we chose to study only the different zymogram patterns (26) themselves, without any weight, thus emphasizing zymogram pattern differences. These 26 zymograms were described either through a principal component analysis (PCA) or a factorial analysis of correspondances (FAC) (Benzecri, 1971). Compared to Euclidian distances calculated as if coordinates allowed to alleles at each locus were independent, the Mahalanobis distance (principal component) reduces the overall distance between zymograms because of the existence of some correlations, and Benzecri's X^2 distance (factorial correspondances) increases distances because of overweighting rare alleles or exceptional associations of alleles. Using factors as variables (either from FAC or PCA), the "Nuées Dynamiques" procedure gave us the opportunity to study the clustering of the zymogram patterns.

Simultaneous analyses of morphological and isozyme data were made either graphically or by studying genetic distances (Nei and Cavalli-Sforza) between groups and the isozyme polymorphism within groups, the groups having been obtained by morphological classificiation.

RESULTS

Morphological Data

The eigenvalues of the first 3 factors and the coefficients of the variables (Table I) clearly exhibited the general trends of the differenciation of our millet collection. Axis 1 described tillering and plants with short spikes and narrow stalks in negative values, axis 2 put high, wide-leaved and early plants in positive values, and axis 3 put heavier seeds and early plants in positive values. It can clearly be seen on Fig. 1a (plan 1,2) and Fig. 1b (plan 2,3) that the moharia lay on left of Fig. 1a and the tropical and warm temperate origins were in the negative quadrant for both axes in Fig. 1b. Points were scattered all over both figures without strong evidence of clustering. Some clear regionalization was seen for Taïwanese, Japanese and Korean, and Indian and African strains. Chinese strains were mostly scattered along axis 2 ; the scattering of European strains mostly along axis 1 was concordant with the existence of both moharia and maxima millets in that region ; some homogeneity for flowering earliness gave them their positive values in axes 2 and 3. Despite this extensive scattering, the dendrogram and "Nuées Dynamiques" for two *a priori* groups suggested a strong differenciation into two clusters, mainly moharia opposite to maxima. Asking for 4 *a priori* groups, "Nuées Dynamiques" after 3 different random trials suggested the 7 "formes fortes" described on Fig. 2 (PCA

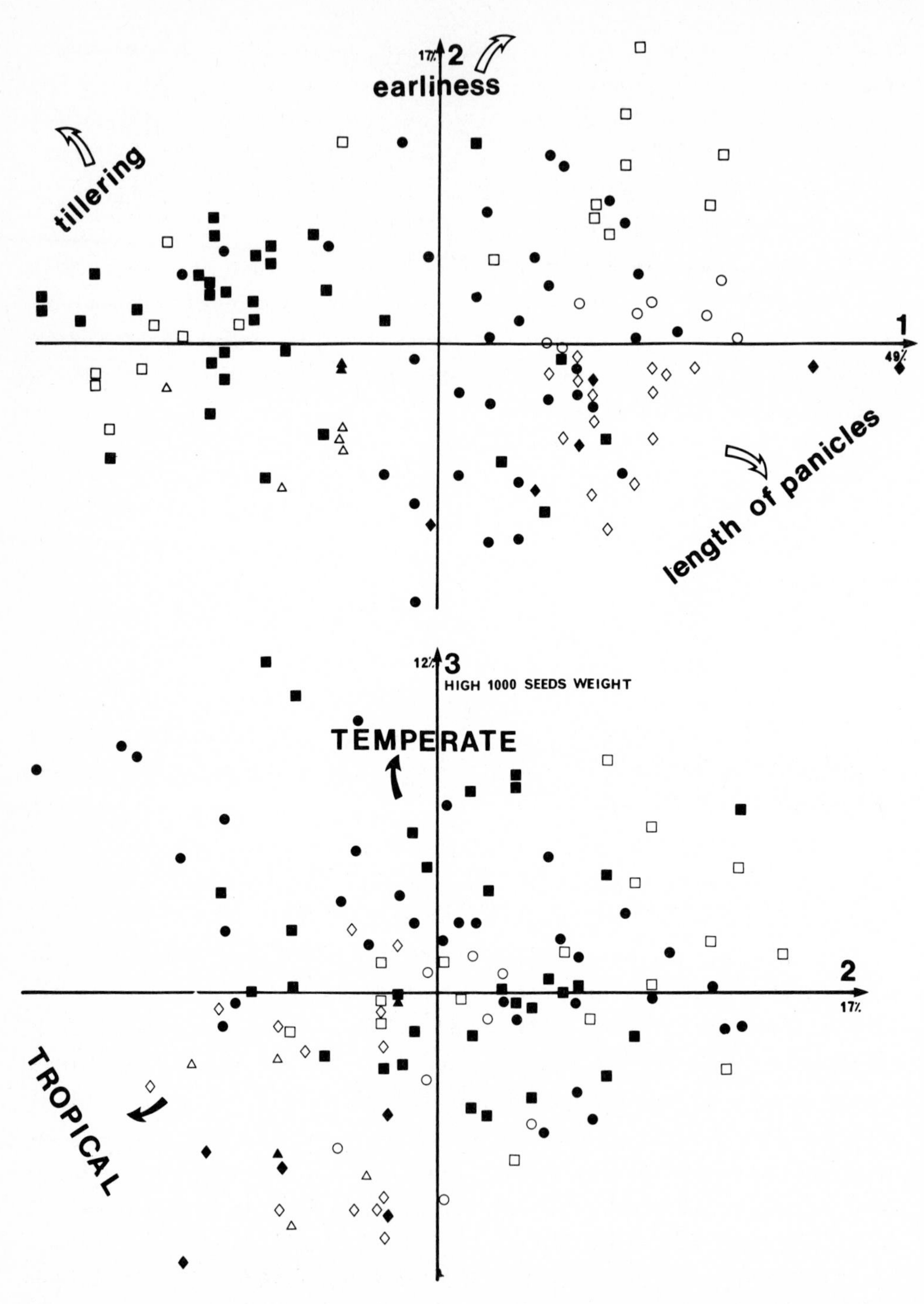

Fig. 1 - Principal component analysis of 121 strains (10 morphological characters). Strains represented by following symbols according to their geographical origin.

□ France, ○ Korea, ◇ Japan, ◆ Taïwan, ● China, ■ Europe (except France), ▲ Africa, △ India.

plan 1,2).

Table 1 - Eigenvalues, coordinates of the most important characters defining the axes, and biological interpretation of the first 3 axes given by principal component analysis of the 121 strains (10 morphological characters).

Eigenvalues in %	Axis 1 49.7		Axis 2 16.6		Axis 3 12.5	
		coord.		coord.		coord.
	NT2M	-792	H2M	+887	PMG	+828
coordonate	NEP	-862	LA2M	+619		
of characters	LOE	+811	ETP	-519	ETP	-401
	LAE	+909				
	LAP	+934				
biological significance of axis	tillering and short spikes asssociated		earliness		size of seeds	
inter-pretation	moharia(-)		early temperate origins(+)		early, large seeded temperate maxima(+)	
	maxima (+)		late tropical and warm temperate(+)			

NT2M : number of tillers, NEP : number of well developped spikes, LOE and LAE : length and width of spikes, LAP : width of the stalk just below the main spike, H2M : height of main tiller, LO2M and LA2M : length and width of the flag leaf blade, ETP : time of heading (days), PMG : weight of 1000 seeds.

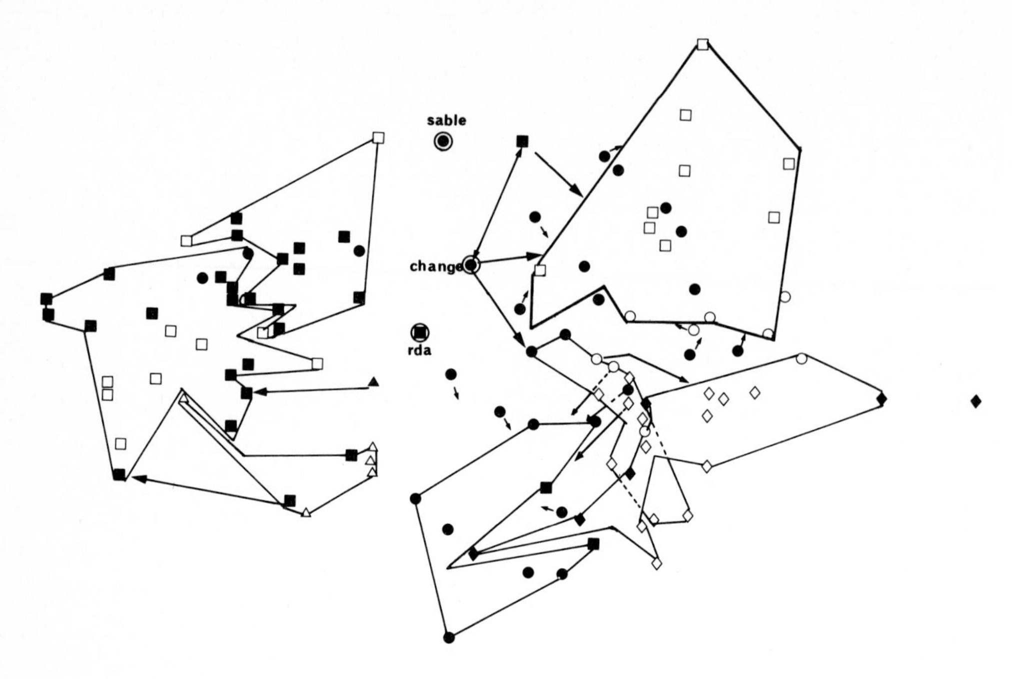

Fig. 2 - Principal component analysis (same plan 1-2 as in Fig. 1a) with the 7 suggested Diday's "FORMES FORTES" described.

Starting from these 7 clusters the factorial discriminant analysis lead to 100% good classification of 7 fundamental types plus 10 isolated strains. The 7 types can be described as follows, after examination of the mean values of characters in Table II, and the positioning in the canonical 1-2 plane on Fig. 3.

I. "macro" types : tillering, short erect spikes, early, small yielding ability (named macro because several strains among this group were described as 'macrostachya' or 'macrochaete' varieties by some botanical gardens).

II. "moha" types : more tillers than group 1, short erect spikes, early, better yielding ability.

III."tropical moha" types mostly from Africa or India.

IV. early cereal types mostly from France or China, with long spikes.

V. high yielding cereal types, with large seeds, moderate

earliness, and with largest spikes ; mostly from China and some from Eastern or Central Europe.

VI. warm temperate cereal strains, mostly from Japan and Korea, badly adapted to French conditions, with shorter spikes.

VII. warm temperate or tropical cereal strains coming from Taïwan or the south of Japan.

Table II - Mean values of six (of 10) characters for the 7 basic groups obtained after 100% good classification in discriminating analysis.

	I Macro Moha1	II Moha Moha2	III InAf	IV $FrCh_1$	V CH_2	VI JA-KO	VII TA
NT2M	5.1	8.2	9.5	1.2	1.4	1.5	2.1
NEP	4.9	8.4	7.1	1.1	1.5	1.4	2.1
ETP(d)	51.6	56.6	123.8	70.2	79.0	105.3	124.4
LOE(cm)	11.3	12.5	14.8	25.7	22.9	20.9	24.2
LAE(mm)	14.1	12.1	13.2	28.8	25.2	30.1	31.
PMG(g)	2.4	2.4	2.5	2.6	2.7	2.3	1.7

InAf : India and Africa, $FrCh_1$: France and China 1, CH_2 : China 2, JA-KO : Japan and Korea, TA : Taïwan, MOHA : Moharia, MACRO : Macrochaete.

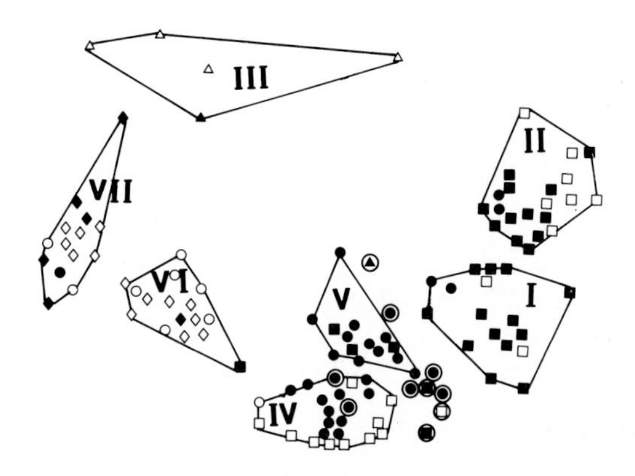

Fig. 3 - First plane canonical variables 1-2, describing the 7 basic groups after 100% good classification in discriminating analysis, and the unclassified isolated strains. Symbols for geographical origin are as decribed in Fig. 1.

The unclassified strains were particularly interesting and had systematically been used as parents in breeding programs, before this classification was achieved.

Electrophoretic Data

As opposed to morphological analysis, where there were no duplications, the zymogram patterns exhibited an astonishing uniformity, as described in Table III.

Table III - Number of strains having a zymogram pattern differing by, 0, 1, 2, 3 or 4 electromorphs from the most frequent pattern. Only 121 strains of this series were completely described for morphological characters and zymograms.

NUMBER OF DIFFERENCES	0	1	2	3	4
NUMBER OF STRAINS	86	106	24	5	1

However, this reduced variability was not random. The 26 different zymogram patterns were analysed through PCA and FCA analysis.

a. first, the central zymogram pattern in PCA analysis (Fig. not shown) was also the most frequent although there was no weighting by pattern frequencies.

b. second, the FAC exhibited some clustering leading to four different groups of patterns, more or less associated with geographic origin (European, north and central Asia, tropical).(Remember that FAC puts some weight in favor of rare alleles or rare associations of alleles) (Fig. not shown).

Comparison between Morphological and Electrophoretic Data and Geographic Origins.

The main zymogram pattern could be found within all the morphological groups. However, as clearly demonstrated in Fig. 4, the electrophoretic compositions of the groups were different. The dendrogram obtained from genetic distance (Fig. 5) after separating group n°IV into two subgroups (China and

France) gave a similar organization of the clustering. The within polymorphism of the groups was not equivalent ; the moharia groups seemed to be more variable. The temperate cereal group IV was mainly a mixture of two different patterns ; one from the central China pattern, the other characterizing the French cereal strains. A similar mixture also seemed to be suggested within the tropical warm temperate group (VII), but with closer geographic proximities (Fig. 4).

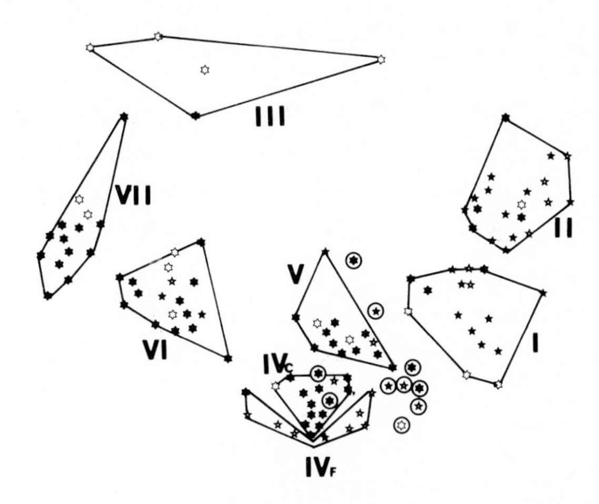

Fig. 4 - First plane canonical variable 1-2 (same as Fig. 3) with symbols describing the 4 clusters of zymogram patterns obtained after using Diday's algorithm to FAC of the 26 zymogram patterns.

✸ ★ ✮ ✡ Codes for the 4 groups of zymogram patterns determined by multivariate analysis.

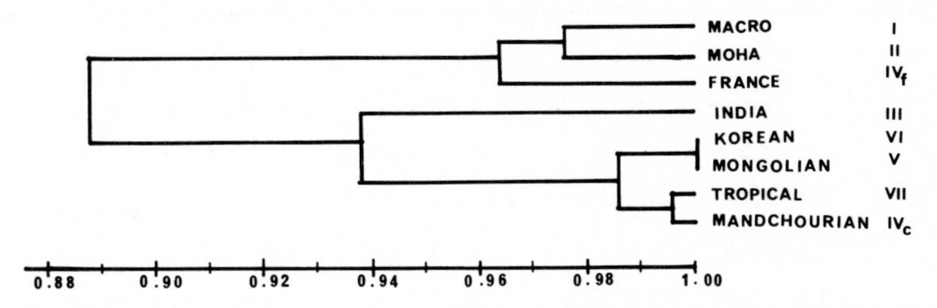

Fig. 5 - Dendrogram describing the 8 final groups by means of Nei's identity. Other genetic distances lead to similar dendrograms (10 electrophoretic loci). The eight groups were I, II, III, V, VI, VII obtained from morphological data and IV_f, IV_c from partitioning of group IV.

DISCUSSION

Comments on the Clustering

Moharia and maxima were clearly confirmed to be fundamentally distinct. However there was more genic (isozymic) community between tropical moharia and maxima on the one hand and between European maxima and moharia on the other hand than within either the overall group of moharia or the overall group of maxima.

The temperate cereal group had to be separated into two groups, French-European (IV_f)and Chinese (IV_c) (this last one could be the Mongolian subgroup described by Gritzenko). The tropical maxima group (VII) and the warm temperate group (VI) exhibited some reciprocal introgressions, in good agreement with Miyagi's (p.c.) and Sakamoto's hypotheses that Kagoshima Island was the site of introgression between tropical (East Asian) millets and continental millets from China-Korea.

The high yielding maxima group could be identified as the Manchurian-Mongolian group of Gritzenko. We find here some particular strains which could recombine genes from western and eastern parts of the temperate maxima variability (Caucasian, center of differenciation, de Wet).

Finally, among the temperate moharia, some authors have suggested establishing a macrostachya (macrochaete) subspecies group. This is in concordance with the qualitative aspect of the plants (not described here) and particular zymogram patterns. Thus we suggest this final classification :

I.	MOHARIA-MACROCHAETE	TEMPERATE MOHARIA
II.	MOHARIA	
III.	TROPICAL MOHARIA (INDIA)	
IV_f.	FRENCH	TEMPERATE MAXIMA
IVc.	CHINESE (MANCHURIAN)	
V.	CHINESE (MONGOLIAN)	
VI.	KOREAN	
VII.	WARM TEMPERATE-TROPICAL MAXIMA	

Some Perspectives on Millet Domestication

Numerous introgressions, either recent or ancient, certainly occurred several times among foxtail millets. The processes of migration, hybridization, differenciation and adaptation have quite often been at work. Therefore it is difficult to suggest a general history on such a limited basis. Genetic analyses, papers in preparation, and the present data suggested that there could be more genetic differenciation in the west-east/tropical partition than in the moharia/maxima partition. For instance, within the homogeneous looking Chinese Manchurian-French maxima group several phenomena of sterility were observed in F_1 hybrids and through following selfed generations.

Another point of interest is the gap between the 7 morphological groups at the center of the canonical variables graph (Fig. 3 and 4) ; why was no strain found here : because of genetical limits to recombination or some bias in our sampling which missed a major world center of diversity (Caucasian ?) ? Is it possible that the agronomically interesting strains came from exceptional recombination which meant that they were very difficult to classify into the main groups ?

More information about genetic differenciation came from preliminary results giving evidence that, on the average, wild S. viridis from China or France resembled less each other than S. italica from its respective geographic origin.

Despite the central position of the main zymogram pattern (in the PCA analysis) which suggested some radiating differenciation from North China, we felt that most of the results and descriptions are in agreement with a multiple domestication hypothesis, directed towards the same morphological type or different types (moharia/maxima) sometimes even at the same localities. Migration and introgression certainly followed but some barrier to recombination ought to have limited gene flow and maintained the peripheric position of groups described by the canonical variables (Fig. 3). Finally, inheritance of characters such as the tillering habit (recessivity towards high tillering) and earliness suggested that moharia (temperate or tropical) was not an intermediate domestication towards the maxima type but was another particularly improved ideotype.

ACKNOWLEDGMENTS

We are particularly thank Drs Miyagi, Sakamoto and G. Second for giving us samples of foxtail millet and B. Fraleigh for correcting the English.

REFERENCES

BENZECRI J.P., 1973. L'analyse des Données. II, "L'analyse des Correspondances". Dunod (ed.); Paris.

BRABANT P., BELLIARD J., METAILIE G., NGUYEN VAN E., POIRIER S., POIRIER B. et PERNES J., 1981. Données préliminaires pour la réintroduction et la culture du Millet Setaria en France. Journ. d'Agric. Trad. et de Bot. Appl. XXVIII 309-328.

CHIKARA J. and GUPTA P.K., 1979. Karyological studies in the genus Setaria. I. Variability within Setaria italica (L.) Beauv. J. Cytol.Genet. **14** : 75-79.

DARMENCY H. and PERNES J., 1984. The use of wild Setaria viridis (L.)Beauv. after hybridization. Weed Res. (in press).

DARMENCY H., JUSUF M., NGUYEN VAN E , POIRIER-HAMON S., BARRENECHE T., et PERNES J., 1984. Relations génétiques dans le complexe Setaria viridis 7e colloque Colum-EWRS., octobre 1984, Paris (in press).

DIDAY E. et al, 1980. Optimisation en classification automatique - 2 tomes. INRIA.

GRITZENKO R.J., 1960. Chumiza (Italian millet) taxonomy (Setaria italica) (L.) subsp. maxima Alef). Bull. Apll. Bot. Genet., Pl.Breed. **32** : 145-182.

KAWASE M. and SAKAMOTO S., 1982. Geographical distribution and genetic analysis of phenol color reaction in Foxtail Millet, Setaria italica (L.) P. Beauv. Theor. Appl. Genet. **63** : 117-119.

KAWASE M. and SAKAMOTO S., 1984. Variation, geographical distribution and genetical analysis of esterase isozymes in foxtail millet, Setaria italica (L.) P. Beauv. Theor. Appl. Genet. **67** : 529-533.

KOKUBU T., ISHIMINE Y. and MIYAGI Y., 1977. Variations of growth-period of Italian Millet strains, Setaria italica Beauv. and their responses to day-length and temperature. II. Changes of growth-period of strains gathered from different districts, both native and foreign, due to the different seeding dates. Reprinted from the Memoirs of the Faculty of Agriculture, Kagoshima University, Vol.XIII (Whole Number 22).

de WET J.M.J., DESTRY-STID L.L.and CUBERO J.I., 1979. Origins and evolution of foxtail millets (Setaria italica). Journ. d'Agric. Trad. et de Bot Appl. XXVI (1) 53-64.

Morphological variation, breeding system and demography at populational and subpopulational levels in *Armeria maritima* (Mill.) Willd.

C. Lefèbvre
Laboratoire d'Ecologie Végétale et de Génétique
Université Libre de Bruxelles
Chaussée de Wavre 1850, 1160 Brussels - Belgium.

ABSTRACT

On a large geographical scale *Armeria maritima* follows in North Western Europe a classical variational trend : two morphologically and geographycally well distinct groups of populations (subsp. *maritima*, coastal and subsp. *elongata*, continental) separated by a hybrid swarm (subsp. *intermedia*). In the area of subsp. *intermedia* (The Netherlands to East-Denmark), *A. maritima* shows for polymorphic characters an interpopulational variation sometimes without any environmental determinism : different characters frequencies are found in very similar habitat and inversely. The processes of variations are thus rather erratic. Although these populations are strictly outbred gene flow is so strongly restricted that different sub-populations can coexist in close vicinity through generations in the same population. These subpopulations are panmitics units for the polymorphism of a monogenic character (hairiness of flower stalk).

In a zinc-lead mine tolerant population the demographical features are rather static and conservative when compared with other perennials.

In *Armeria maritima* the populations and even adjacent subpopulations appears as reproductively isolated units where drift is a major component in the differenciation process and in which demographical dynamics is slow.

GENERAL TRENDS OF MORPHOLOGICAL VARIATION

The specific complex *Armeria maritima* (Mill.) Willd. is in North, West and Central Europe highly variable for morphological characters assumed to have a taxonomic value. In general terms the trends of variation are classical : two morphologically and eco-geographically well differenciated groups of populations (the atlantic subsp. *maritima* and the north-eastern continental subsp. *elongata*) separated by a hybrid swarm made of polymorphic populations combining in many ways the features of the putative parental groups.

Populations of *A.m.* subsp. *maritima* show an atlantic coastal distribution from

NATO ASI Series, Vol. G5
Genetic Differentiation and Dispersal in Plants
Edited by P. Jacquard et al.

South of France to North of Norway and have as main morphological features : leaves narrow ($<$1mm), 1-nerved, glabrous or hairy; flowerstalk hairy, outer involucral bracts short, blunt.

Populations of *A.m.* subsp. *elongata* are found on acidic sandy soils in continental Europe and around the Baltic Sea; they are as follows : leaves broad (1,5-2mm), 3-nerved, hairy, flower stalk glabrous, outer involucral bracts elongated with a sharp end.

The hybrid swarm populations extend from Noord-Holland (Texel Island) to the western part of the Baltic Sea and are formaly referred as subsp. *intermedia*.

Additionaly, populations found on old zinc, lead or copper mines scattered from East-Belgium to South Poland must be included in this hybrid swarm (Lefèbvre 1974 and Tab. 1). Those ecogeographical data are pictorialized in figure 1.
More precisely a large number of individuals scored for their "maritima" (M), "intermedia" (I), and "elongata" (E) morphotype all along the european coastal range shows there is, on a large geographical scale, a gradual shift from West to East of groups of populations successively dominated by *maritima*, *intermedia* and *elongata* morphotypes (Tab. 1).

Tab. 1 : Frequency (%) of maritima (M), intermedia (I) and elongata (E) phenotypes in the main coastal areas in Europe.
n = nb of individuals scored

	M	I	E	n
Britany-Sealand (G.B. included)	82	18	0	745
Nothern Holland - West Denmark	32	59	9	405
West Baltic	15	72	15	386
East Baltic	0	18	82	207
Metallicolous populations	11	82	7	240

This general trend is in fact an oversimplification of what is occuring at the populationnal level. From table 2 it can be seen that populations do not follow a transitional pattern : for instance, populations at the opposite ends of the West-East transect pictorialized at figure 2 are very similar (Naarden-Amager) in M,I,E distribution. On the contrary populations dissimilar in M,I,E distribution can exist in close vicinity as in Texel (The Netherlands) and Rømø (Denmark islands) where the differences between almost adjacent populations can be related to ecological situations, populations form seaside beeing more "maritima" than the inland ones (Tab. 2).
This simple statement encompass complex populationnal evolutive processes which are still obscure. From table 3 it can be seen that the balance between "elongata" and "maritima" characters in populations is reached through different contribution of each peculiar character : in Naarden an Amager populations, the same balance (45% of "maritima" characters) is attained with different frequencies of individual characters.

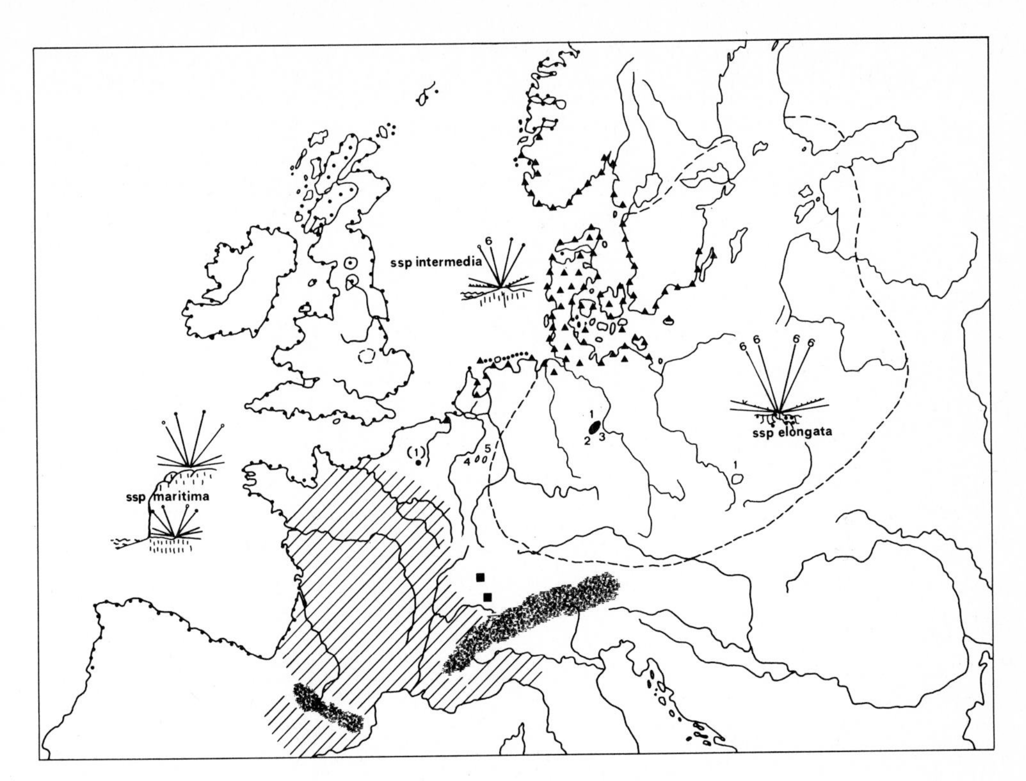

Fig.1. Distribution of Armeria maritima ssp. maritima (...), ssp. elongata (---), ssp. intermedia (▲▲▲), ssp. alpina (▒▒); Armeria alliacea (////) ; 1, 2, 3, 4, 5 : metallophytes.

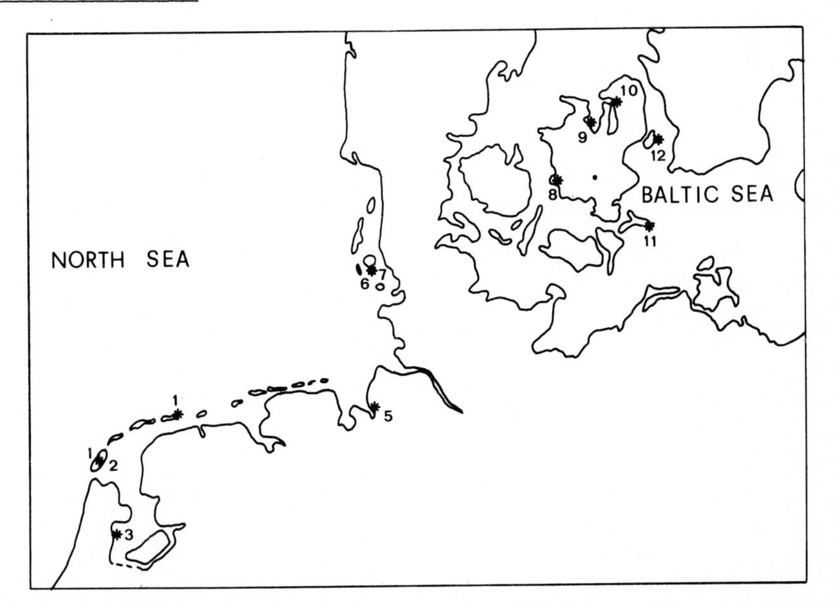

Fig. 2. Geographical position of A. maritima ssp. intermedia populations scored for morphological features (see table 2).

Tab. 2 : Frequency (%) of maritima (M), intermedia (I) and elongata (E) phenotypes in populations in the range of distribution of the hybrid swarm between A.m. subsp. maritima and subsp. elongata.
% mar. = frequency of maritima characters viz elongata characters.
RGP = Relative geographical position; 0 = extreme West; 1 = extreme East.

	M	I	E	%mar	RGP
1. Texel (coast)	55	45	0	68	0,00
2. Texel (inland)	32	64	4	52	0,00
3. Naarden	0	88	12	44	0,04
4. Ameland	58	42	0	39	0,13
5. Christianskoog	88	12	0	64	0,46
6. Rømø (coastal)	47	53	0	55	0,48
7. Rømø (inland)	0	96	4	36	0,48
8. Nyborg	0	95	5	39	0,82
9. Boegnes	60	40	0	55	0,88
10. Jaegerpris	17	83	0	55	0,92
11. Ulvshale	0	47	53	15	0,98
12. Amager	0	97	3	46	1,00

Tab. 3 : Frequency (in %) of "maritima" characters in populations of A. maritima subsp. intermedia.
1 = flower stalk hairy; 2 = leaves glabrous; 3 = calyx hairy all around; 4 = involucral bracts short.

	1	2	3	4	%mar	n
Texel dunes	55	54	64	100	68	30
Texel inland	53	0	76	80	52	30
Rømø dunes	54	46	33	90	55	30
Rømø inland	0	18	33	93	36	30
Naarden (NL)	7	62	14	93	44	29
Amager (DK)	4	32	53	94	46	30

The fact that the same ecological transition on a small geographical scale is reflected in Texel and Rømø by a different equilibrium between the main "elongata" and maritima features is difficult to explain in adaptive terms. Moreover when looking at the flower stalk hairiness (a monogenic character, individual with glabrous flower stalk beeing recessive, Lefèbvre 1977) distribution among populations it appears that this character varies in some cases in parallel with ecological factors and in some other cases appears unrelated to apparent ecogeographical factors. From table 4 it appears that in Texel the same frequencies of hairy flowers stalk viz glabrous are found in different ecological situations and inversely in apparent very similar environmental conditions along the northern coast of Ijsselmeer different frequencies are found between populations (Oude Mirdum and Laaksum).

Tab. 4 : Frequency of hairy and glabrous flower stalks in Armeria maritima populations

	Hairy	Glabrous	% hairy
Texel Island			
Inland grasslands	76	69	52,4
Beach sands	67	77	46,5
Salted meadows	86	89	49,2
Salt marshes	42	33	56,0
X_3^2 = 2,15 N.S.			
North Coast Ijselmeer			
Laaksum	7	47	12,9
Oude Mirdum	20	37	35,1
X_1^2 = 7,37 $p < 0{,}01$			

These observations suggest the neutrality of the character. But in Rømø island the variation in frequency is in the direction of the general ecogeographical trends of variations observed for the morphology of the two subspecies.

This study on the hybrid swarm populations of Armeria maritima reveals that this species is made of a high number of various populations differing in the frequency of obvious morphological characters. The population differenciation can occur on small distance almost adjacently. The morphological polymorphism appears sometimes over-printed on ecological situation and sometimes it appears as neutral, this is probably the result of an unseizable equilibrium between drift and direct or undirect selection.

STUDIES ON A POPULATION FROM AN OLD ZINC-LEAD MINE IN BELGIUM

In Belgium, inland Armeria maritima is only found on very few (4) old zinc-lead mine wastes among which Plombières is the largest. This population is thus spatially well delimited and is isolated from its nearest neighbour by 4 km. The populations size is about 1 million individuals.

The breeding system of Armeria maritima which is quite variable has been deeply investigated by Baker (1966). In European populations the sexual reproduction is regulated by a dimorphic selfincompatibility system governed by a supergene containing a subunit (C) for stigma morphology and an other subunit (A) for pollen morphology. Natural populations contain 50% of AC.ac (A type : smooth stigma and reticulate pollen) and 50% of ac.ac. (B type : papillous stigma and punctate pollen) so that strict allogamy maintains the 50/50 ratio. Artic populations are monomorphic and self-fertile, all the individuals beeing AcAc. Intermediary situations are found in dimorphic European populations in nothern maritima populations (Baker 1966) and in heavy metal mine populations (Lefèbvre 1970) where some individuals shows a relatively high level of selffertility (till 30%) when bagged in experimental plots (Tab. 5).

Tab. 5 : Potential self-fertility in metallicolous and non-metallicous Armeria maritima. % of potential self-fertility : nb seeds/nb flowers in bagged flower heads (100 individuals tested).

% self-fertility	0-5	5-10	10-15	15-20	20-25	25-30	30
Metallicous plants (5 populations)	50	16	8	5	5	6	10
Non-metallicous plants (5 populations)	73	19	8	-	-	-	-

X^2 test between metallicous (M) and non-metallicous (nM) plants

% self-fertility	<10%	>10%
M	76	24
nM	92	8

$X_1^2 = 9{,}52 \quad p < 0{,}001$

Tab. 6 : A and B type distribution in Plombières population of Armeria maritima. Overall samples through the mine site.

	A type	B type	$X^2_{50:50}$	
Seedling from 93 individuals	1329	1281	0,88	N.S.
Adults	198	202	0,04	N.S.

This experimental self-fertility must result, if occuring in natural population, in a shift of the 50 A types : 50 B types ratio to another proportion where the recessive B type must predominate as it segregates additionaly from the dominant A type. Samples of progeny and adults collected all over the mine (Tab. 6) do not show any statistical departure from 50 : 50 ration and none Ac.Ac type was detected; from that it can be concluded that the breeding system in Plombières is strictly allogamous. The significance of such a potential self-fertility which has been found also in tolerant Anthoxanthum odoratum and Agrostis tenuis (Antonovics 1968) where it has been assumed to act as an isolating mechanism against gene flow from non-tolerant adjacent populations remains obscure in metal tolerant Armeria maritima and is presumably a relictual characteristics from the first rare colonizers of the zinc-lead mine allowing them to produce a reliable progeny.

On the mine site of Plombières Armeria maritima occurs discontinously as groups of individuals, which will be named here subpopulations, separated by small environmental barriers (bogs, steep slopes, river). Sampling at the subpopulationnal level revealed (Lefèbvre 1976) differences in the distribution of the gene for hairiness of the flower stalk (Tab. 7).

Tab. 7 : A : Frequency of glabrous individuals in 5 subpopulations of Armeria maritima (X^2_4 = 55,0^{XXX}) from Plombières.
B : X^2 test for conformity with random mating reproduction using progeny test on recessive (glabrous = hh) parents (for explanation see text).

	A Parental subpop.(n=200)	B Progeny of glabrous Exp.	Obs.	N	X^2_1	
1.	0,700 (140)	0,837	0,832	387	0,07	N.S.
2.	0,624 (125)	0,790	0,800	385	0,23	N.S.
3.	0,667 (133)	0,817	0,822	241	0,00	N.S.
4.	0,828 (166)	0,910	0,924	184	0,50	N.S.
5.	0,894 (179)	0,946	0,932	177	0,65	N.S.

Tab. 8 : Frequencies of hairy and glabrous individuals in 3 subpopulations through years of Armeria maritima in Plombières

Subpop.	Year	Nb. hairy	Nb. glabrous	Freq. hairy	X^2 year
1	1966	4	36	0,100	
	1978	11	96	0,103	0,04 N.S.
	1980	10	96	0,093	
2	1974	60	140	0,300	0,01 N.S.
	1980	31	74	0,295	
3	1974	11	194	0,054	
	1976	6	216	0,027	1,99 N.S.
	1980	6	149	0,038	

Simple theoritical calculations from the Hardy-Weinberg law show that the frequency of the recessive type (glabrous = hh) in the progeny of the recessive fraction of a population equals the root of the frequency of the recessive in a random mating populations. As an example let us take subpopulation 1 from table 6.

1 : frequency of glabrous = f (hh) = 0,700 - 2 : in a random mating population = f (hh) = q^2 = 0,700 thus q = 0,835 and p = 0,165 - 3 : the genotypes frequency in the subpopulation are : hh = q^2 = 0,700 , Hh = 2 pq = 0,273 , HH = p^2 = 0,027 - 4 : the progeny of hh genotypes is then :

hh x HH	0,027 Hh	(a)
hh x Hh	0,136 Hh	(b)
	0,136 hh	(c)
hh x hh	0,700 hh	(d)

(c) + (d) = 0,836 hh = $\sqrt{0,700}$ hh = $\sqrt{f}$ (hh) in parents hh

From table 7 it can be seen that the random mating expectations are strictly encoutered in each of the 5 subpopulations.

The subpopulations maintain their own polymorphism through years : from table 8 it can be seen that for periods of 6 years and 14 years the hh frequencies does not change although, as the half life of Armeria on the mine is 7 years at least, half of the individuals has been replaced during the period of sampling.

These results indicate that :

a. The subpopulations behave as panmatic units maintaining their polymorphism through generations;

b. The gene flow is apparently absent between adjacent subpopulations and is thus very limited;

c. The subpopulational polymorphism could have originated independantly in each subpopulation by hazard or could be the result of balancing selection.

In this third context the subpopulational divergences for the hairy gene frequencies has been tested in relation with the most obvious ecological factors occuring on the mine : zinc toxicity, lead toxicity, drought.

Tab. 9 : Relations between the polymorphism in flower stalk hairiness and the main environmental factors at the subpopulational level in the Plombières population of *Armeria maritima*.

A. Zinc tolerance to 10 ppm of Zn in H_2O

Subpop.	1	2	3	4	5	6	7	LSD 5%
It Zn (1)	114	107	95	92	90	86	79	19
Freq.glabrous(%)	97	71	97	81	93	61	67	

r = 0,435 N.S.

$$(1) : It = \frac{\text{Root increment in toxic solution}}{\text{Root increment in water}} \times 100$$

B. Lead tolerance to 1 ppm of Pb in H_2O

a.

Freq. hh in subpop.	It Pb	n	C.L.5%	"t" test subpop.
0,07	76,7	131	11,4	
0,83	70,3	130	17,0	N.S.
0,90	75,4	79	8,2	

b.

Subpop.	It glabrous ± C.L.5% (n)	It hairy ± C.L.5% (n)	"t" test glabrous-hairy
1	82,2 ± 5,7 (33)	83,4 ± 8,8 (25)	N.S.
2	72,4 ± 12,9 (42	76,7 ± 15,7 (29)	N.S.
3	76,1 ± 6,9 (52)	74,3 ± 11,3 (27)	N.S.

n = nb of plants tested C.L. = confidence limits

C. Drought

	soil moisture (2)	nb glabrous	nb hairy
Open turf	14,1%	24	143
Dense turf	20,8%	11	96

X_1^2 glabrous-hairy = 0,98 N.S.

(2) : soil water content after 5 days without rain.

The results are compiled in table 9 from which the following points can be sorted out :

a. No statistical relation between the Zinc tolerance and the frequency of the hairy gene of the 7 subpopulations tested.

b. No statistical differences for lead tolerance in subpopulations differing in their hairy gene frequencies.
   No statistical differences between hairy and glabrous individuals for lead tolerance.

c. No statistical differences between hairy gene frequency in relation with the drought in the subpopulations.

The variation in subpopulational polymorphism is thus independent from the main ecological factors occuring on the mine. This reinforces the arguments already developed about the neutrality of hairiness polymorphism in some populations of Armeria maritima especially in the North of the Netherlands. It seems clear that small groups of individuals like subpopulations growing in very close vicinity can be maintained different through generations in the absence of evident selective forces.

What the demographic structure of the population of Plombières tell us in this context has been discussed extensively in a recent paper (Lefèbvre & Chandler-Mortimer 1984) so that a brief and synthetic account will be exposed.

The population sample studied in the open turf from Plombières appears as being demographically rather static. The survivorship curve indicated a high mortality for seedlings and 1 year-old plants followed enough rapidly, when plants are 2-3 years, old, by a constant and relatively low annual death rate of approximately 10%. The corresponding half-life estimates for adult plants is 6,6 years which result in a 30-35 years period for population complete turnover. The half-time of Armeria maritima is relatively long when comparing with the data of Antonovics (1972) on Anthoxantum odoratum (half-life of 2 years) on a zinc mine and more generaly when comparing with most (80%) of the perennials for which Harper (1977) gave estimates. The age of flowering which is delayed till 3-5 years and the fact that only 3-4% of the potential seeds reach the reproductive stage are two other conservative characteristics.

In Armeria maritima there is a statistically highly significant ($r=0,907^{xxx}$) relationship between plant size (and thus age)and the number of flowerheads sothat the older and individual is the best reproducer it is.

The oldest individuals ($>$ 15 years) despite their low frequency ($<$ 10%) have a high contribution to the future of the population in producing approximately the third of the sexually adult plants.

Moreover from table 10 it can be seen that the coefficient of variation for mortality recorded for a 4-years period decreases dramatically when the age of the cohort increases. This indicates that the oldest individuals are the less sensitive to general climatic hazards sothat their predominance in seed setting may increase population fitness if there is a genetic component for longevity. But this relative insensibility could be due to plant size, especially roots sothat investigations on the genetics of longevity and on relative fitness of young and old plants are needed.

Tab. 10 : Mean, variance and variation coefficient of survival (in %) of 5 cohorts during a 4-years period (1976-79) in an open community subpopulation of Armeria maritima in Plombières.

Cohort age	seedlings	1year	2-3 years	4-5 years	5 years
survival (%)	29,4	64,1	75,9	87,1	88,1
variance	325,8	283,5	180,8	51,8	7,8
variation coefficient	61,4	26,3	17,7	8,3	3,2

CONCLUSIONS

In the putative hybridization area between A. maritima subsp. maritima and A. maritima subsp. elongata, almost each population has its own characteristics related to the frequency of morphological characters used to distinguish the two parental subspecies. These variations could be in relation with environmental conditions but without an adaptative explanation as it is often found in enzyme polymorphic systems. The variations in the polymorphism of hairiness of flower stalk appear in some areas neutral and in others environmentally linked.

The polymorphism found between populations for stalk hairiness occurs in the same population at the subpopulational level without apparent selection and in a completely outbred system of reproduction. This show that subpopulational units can coexist adjacently as panmatic unit between which gene flow is undetectable.

Populations and subpopulations are thus isolated genetic systems their isolation being renforced by conservative and dynamically slow demographic characteristics.

REFERENCES

Antonovics J., 1968. - Evolution in closely adjacent plant populations. V. Evolution of self-fertility. Heredity, 23, 219-238.

Antonovics J., 1972. - Population dynamics of the grass Anthoxanthum odoratum on a zinc mine. J. Ecol., 60, 351-365.

Baker H.G., 1966. - The evolution, functionning and breakdown of heteromorphic incompatibility systems. I. The Plumbaginaceae. Evolution, 20, 349-368.

Harper J.L., 1977. - Population biology of plants. Academic Press. London.

Lefèbvre C., 1970. - Self-fertility in maritime and zinc populations of Armeria maritima (Mill.)Willd. Evolution, 24, 571-577.

Lefèbvre C., 1974. - Population variation and taxonomy in Armeria maritima (Mill.) Willd. with special references to heavy-metal tolerant populations. New Phytol., 73, 209-219.

Lefèbvre C., 1976 - Breeding system and population structure of Armeria maritima (Mill.) Willd. on a zinc-lead mine. New Phytol., 81, 187-192.

Lefèbvre C. & Chandler-Mortimer A., 1984. - Demographic characteristics of the perennial herb Armeria maritima on zinc-lead mine wastes. J. Appl. Ecol., 21, 255-264.

Gynodioecy in Thyme, *Thymus vulgaris* L.: Evidence from Successional Populations

B. Dommée, P. Jacquard
Unité de Biologie des Populations et des Peuplements
Centre Louis Emberger,C.N.R.S., BP 5051
F-34033 Montpellier Cedex, France.

ABSTRACT

The breeding system of Thyme seems to be profoundly influenced by environmental disturbances. In natural populations, the distribution of the sexes (hermaphrodite and female) is characterized by a proportion of female plants varying from only a few per cent up to 95%. A secondary succession has been studied to relate environmental disturbance and female plants. The genetic structure of 7 populations was studied using allozymes. A predominance of females occurs in unstable biotic environments. The populations were further examined to specify the implicated mechanisms : fecundity of the 2 sexual forms, their density, germination of their seeds. Offsprings of the 2 forms were studied for 35 mother plants : size, sex. The gynodioecy of *Thymus vulgaris* with a variable proportion of females depending on the environment constitutes a good example of a relationship between a breeding system and the environment. The situation described incorporates the heterozygous advantage brought by the female form (male sterility), the decrease in inbreeding depression during the succession, and the spreading of cytoplasmic genes by maternal inheritance.

NATO ASI Series, Vol. G5
Genetic Differentiation and Dispersal in Plants
Edited by P. Jacquard et al.

INTRODUCTION

Numerous hypotheses have been put forward to explain the presence of females and hermaphrodites in plant populations (Gregorius et al. 1982).Some of them are:

1-Females might have a greater longevity (Van Damme 1984) ;
2-They might benefit from superdominance (Ross 1978 ; Gregorius et al. 1982);
3-Females might redistribute their resources towards a female function which would increase their ability to colonize ; this would then be maintained due to the action of cytoplasmic genes (Gouyon and Couvet 1984),and the spreading of these genes by maternal inheritance (Couvet 1984);
4-Heterosis might favor the females which are not self-fertilizing (Valdeyron et al. 1973 ; Dommée et al. 1983).

Recent studies have shown that the breeding system of Thyme seems to be profoundly influenced by environmental disturbances (Dommée et al. 1983;Gouyon et al. 1983).Not all authors draw the same conclusions from this but they do agree in saying that after a disturbance,there is a preponderance of individuals carying genetic information which turns them into females.

The aim of this work is to test the last hypothesis by integrating it into an overall representation of the evolution of sex and by showing:
1- by what mechanisms the proportion of females is modified ;
2- in what way the presence of those females allows the species Thymus vulgaris to survive.

THE DISTRIBUTION OF SEXES IN NATURAL POPULATIONS

In natural populations, the proportion of female plants varies from a few per cent up to 95 % of the population (Dommée 1976). Populations in which only the hermaphrodite form is present have never been found. Yet, frequent sampling has been carried out in different populations in the French Mediterranean region, in particular in the departments of Herault and Drôme (Massif of Mt Ventoux). A consistent correlation is that as the texture of the top soil becomes coarser the proportion of female plants diminishes. This relationship has been tested in the very different environments of the plains of the Herault and Mt Ventoux (which suggests that climatic factors do not play a major role in this correlation). The study of the communities which develop in the environments considered here has made it possible to define their relationship to particular "associations". Substrata with a fine structure carry a vegetation influenced by man ; those with coarse texture are associated with a vegetation characterized by older undisturbed stages (Dommée et al. 1978).

What is the significance of the abundance of female plants ? The female state restricts individuals to reproduction by cross-fertilisation only, while hermaphrodites may self-fertilize. Experiments over several years have shown that this self-fertilisation can occur in the fields (up to 80 % for certain plants) (Valdeyron et al. 1977 ; Dommée 1981). Thus male sterility in strongly disturbed populations reduces self-fertilisation to very low levels. From this one can deduce that, where females are abundant, the populations are predominantly outbreeding vs inbreeding where hermaphrodites dominate the populations. It is known that outcrossing leads to a predominantly high individual heterozygosity,while self-fertilisation results in genetic homozygosity. Numerous examples drawn from experiments (especially with <u>Drosophila)</u> prove that an environment with widely varying conditions (heterogeneous) results in greater genetic variance while a stable environment results in an homozygosity of the studied genes. These aspects have led us to investigate the changes associated with male sterility during succession.

ENVIRONMENTAL DISTURBANCE AND FEMALE PLANTS

A secondary succession, that is the recolonisation by vegetation of abandoned agricultural land (old fields), is a sequence from destabilized environments on one hand to environments in relative equilibrium on the other. The pioneer and early stages consist of species the abundance of which varies strongly from year to year. After this stage perennial species become established, giving a more and more persistant profile to the vegetation (Fig. 1). Preliminary

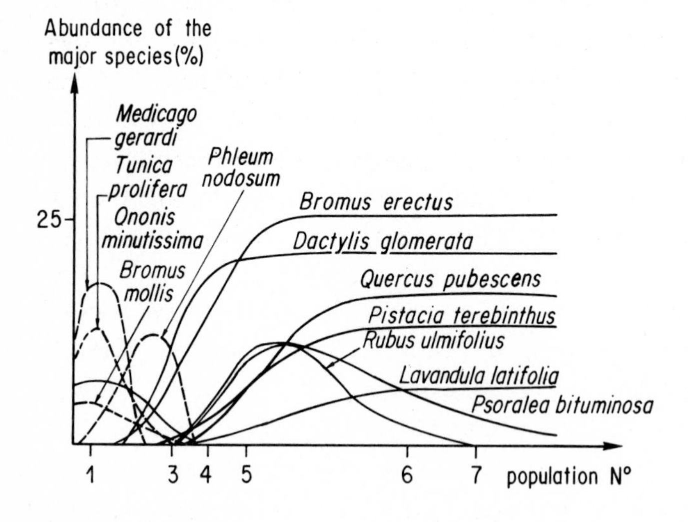

Fig 1. Abundance of principal species in postagricultural succession in the Montpellier region (from Dommée and Guillerm 1980).

studies on the sexual forms of Thyme tended to show that low proportions of females were found in less cultivated zones (rocky substrate). One succession in particular, situated in the north of the Herault (Le Vialaret, Causse de la Selle) was studied. The last stages of the succession consist of white oak forest (Table 1 and Fig 1). In 1978 a sampling of populations and analyses of the flora on the same sites were carried out at 7 situations which had been abandoned for 9 to 60 years.

The greatest distance between 2 habitats was 600 m.They were on similar soils, a red Mediterranean soil of dolomitic origin, with slight variations according to the age of abandonment : increase of organic matter, carbon and nitrogen, of C/N, increase then decrease of K, decrease of P and of pH. The landscape (a high flat plain) was chosen because the large range of stages of the vegetation succession allows to date accurately the period of abandonment (Escarré 1979).In view of estimating the proportions of ♀ and ⚥ and of analyzing the enzymatic polymorphism, a sample of 100 individuals was considered as representative of the whole population. Starting from a central point, all individuals were tagged ; the survey was conducted on concentric areas until the completion of the sample, so the survey can cover an area from 6 m^2 to 7000 m2. The youngest populations contained the highest proportion of female plants (0.93 to 0.79) the oldest showed the lowest proportions : 0.38, 0.45 and 0.48 (Table 1) (Dommée et al. 1983).

ALLOZYMATIC STRUCTURE OF THYME POPULATIONS

The genetic structure of the populations was studied using allozymes, revealed by starch gel electrophoresis. The analysis of the offspring from controlled crossing enabled 4 genes to be identified (3 controlling esterases, one controlling a polyphenol-oxydase). The study of the structure throughout the succession showed that the genotypic composition clearly varied in the direction of a lower

Table 1.Proportion of male-steriles (=females), index of heterozygosity (see text), age and principal characteristics of the sites (from Dommée and Guillerm, 1980).

Site	Proportion of male-steriles	Index of heterozygosity	Age (years)	Short description of the vegetation
1	.93	.98	9	Recently abandoned vineyard ; vines still present ; grasses and forbs (81 %) ; one or several spots of thyme with some spaced individuals.
2	.83	a	10	
3	.87	1.04	20	Predominance of thyme in a large homogeneous spot.
4	.79	.87	26	Old olive-tree orchard with spots of thyme intricated with *Rubus ulmifolius*, *Brachypodium phoenicoides*, *Psoralea bituminosa*.
5	.38	.69	30	Protected plot surrounded by low walls and trees (oaks) ;old olive-tree orchard ;stand of Thyme intricated with *Bromus erectus*, *Festuca duriuscula*.
6	.45	.50	52	Heaths of *Genista scorpius*, *Lavendula latifolia* ; sparse stands of thyme.
7	.48	.70	60	White and green oak forest (*Quercus pubescens*, *Q. ilex*) ; sparse stands of thyme.

a. A more detailed study of the index of heterozygosity is at present taking place at site 2 consisting of subpopulations (see later).

heterozygosity in the oldest plots . Heterozygosity was characterized (Dommée et al. 1983) by the index (Fig. 2) :

$$\frac{\text{observed number of heterozygous plants}}{\text{expected number from random mating}}$$

The results confirm that an abundance of females is associated with maximum heterozygosity, and this occurs in an unstable biotic environment. On the other hand, the abundance of hermaphrodites is associated with an increased homozygosity, in a relatively constant biotic environment.

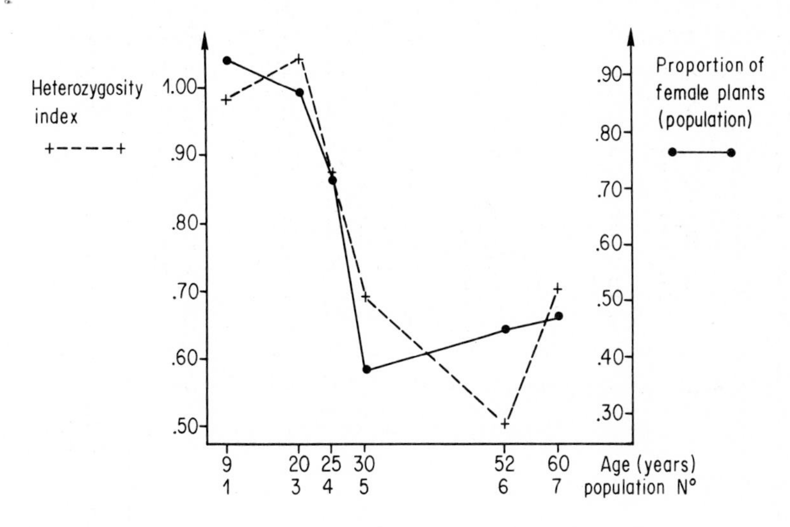

Fig. 2 . Proportion of female plants (male-steriles) (●----●) and index of heterozygosity (+----+) of the studied succession.

In a ten years course (Fig. 1, site 1 to 3) the floristic composition changes completely while in the older sites (5 to 7) constant species are maintained (Dommée et al 1980).

THE IMPLICATED MECHANISMS

The seven Thyme populations in the succession were further

examined to specify the fecundity (seed production) of the two sexual forms, their germination capacity and the size and sex of their offspring.

The Natural Populations

Fecundity of the 2 sexual forms. As an estimate of fecundity (that is, strictly speaking, the seed/individual), we use the seeds/fruit, the fruit of a Labiatae being a "tetrachene", with a maximal number of 4 seeds (Chadefaud and Emberger 1960). In fact, using data from Bonnemaison (1980), we have established that there is a good relation between seeds/fruit and overall reproductive output per individual ($r=0.82$, $P < 0.02$) and that there is no significant differences between the flower numbers (estimated by the number of inflorescences) of females and hermaphrodites (♀ = 20.05 ± 6.29 infl./ind., ⚥ = 22.18 ± 2.85 infl./ind).
The seeds/fruit of the females and the hermaphrodites ($r=-0.745$, $P < 0.05$) are negatively correlated. The extremes meet in site 1 (minimum fecundity for the ♀ and maximum for the ⚥) and site 3 (opposite situation). Thus one finds (Fig.3) that the pioneer stages are marked by an equivalent

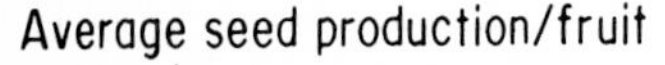

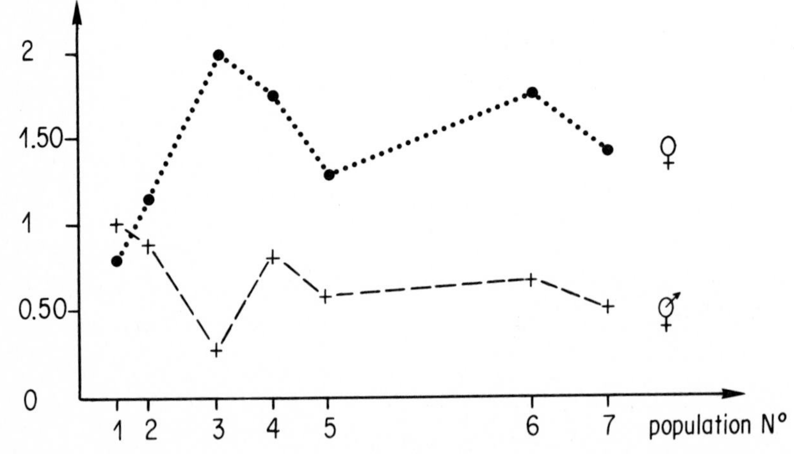

Fig. 3. Number of seeds per fruit of female and hermaphrodite plants according to the sites of the postagricultural succession.

fecundity of ♀ and ⚥ forms ;therefore one cannot invoke a superiority of the ♀ plants to explain their domination of the sex ratio which is clearly to their advantage (proportion of ♀ : 0.93 and 0.83). A very clear superiority of the ♀ (producing notably more seeds than the ⚥) appears in the phase of stabilisation and finally stage 7 show a superior ♀ fecundity compared with the ⚥. The relative fecundities of ♀'s and ⚥ 's change gradually in populations 1-3 but then are stabilized in populations 4-7.

Density of the 2 sexual forms. The density was evaluated from the areas needed to collect a sample of 100 individuals. The density of ⚥ plants varies little in the course of the succession : 1 to 2 per m2 (Fig. 4). The density of ♀ however shows large variations with a high density early in succession and lower density in later stages. In the older sites, where the density of females is comparable to that of the hermaphrodites, the coefficient of fecundity ♀/⚥ is stabilized at values from 2.1 to 2.5.

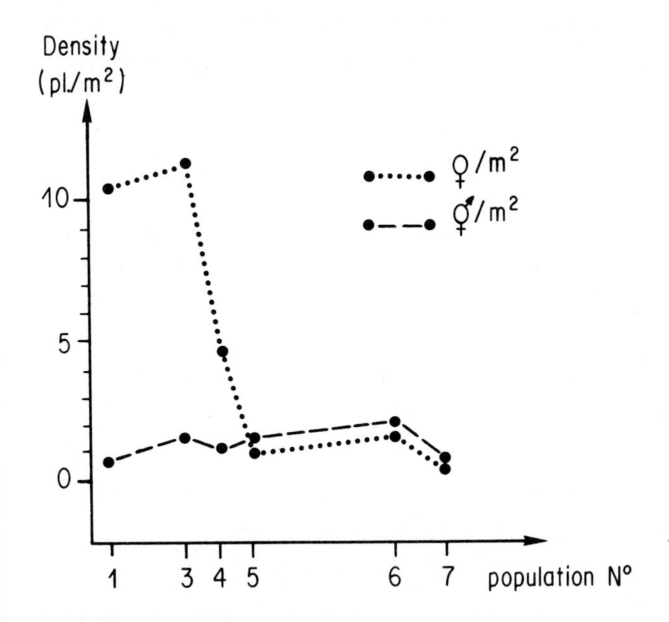

Fig. 4. Comparison of density of ♀ and ⚥ plants in the succession sites (site 2 is a special case, cf Fig. 5).

Site 2, considered until now as one population, is a special case, for it consists of "islets", involving different numbers and sex ratios, separated on average by about 20 meters (Fig. 5). The "islets" with low numbers contain relatively more hermaphrodite plants, the others (with numbers of 6 to 37) contain a majority of female plants. The relationship is consistent (rank correlation : $r = 0.858$, $t=4,426$, $P < 0.01$) (Table 2).

This non random distribution, on the basis of the 2 groups defined in pooling the patches III, VIII, X and II vs the others ($\chi 2=14.62$ $P < .0001$) or the same patches + I vs the others ($\chi 2=5.97$, $P < .02$), seems to indicate that the hermaphrodites are good "founders" but not good "spreaders", if we assume founding effects can be carried out by single individuals ; no single individuals of females are found. Hermaphrodites do not well after founding !

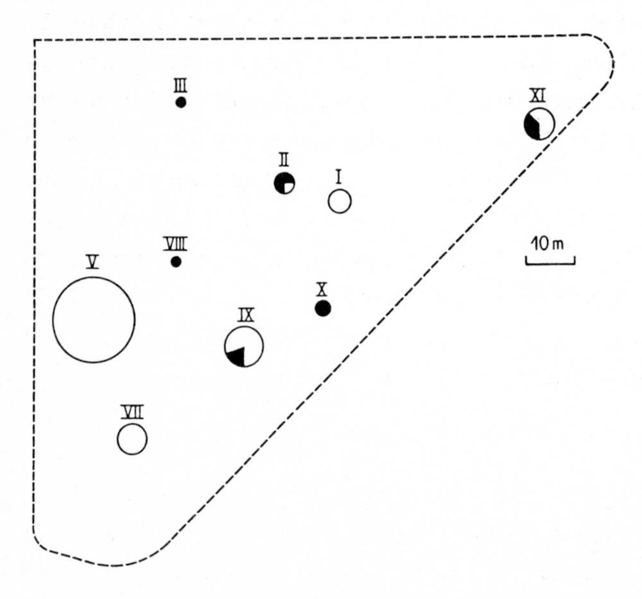

Fig. 5. Distribution of the patches of Thyme in the site 2. The scale is for the limits of the plots (interrupted line). The patches, the area of which is proportional to the number of individuals of Thyme, are represented by a very higher scale. In black : the proportion of ⚥ ; in white : the proportion of ♀ (see table 2) (Dommée and Guillerm, 1980).

Table 2. Numbers and corresponding proportion of female plants for subpopulations of site 2.

Code of the patches	III	VIII	X	II	I	XI	VII	IX	V
Numbers in the subpopulation	1	1	2	4	6	10	11	20	37
Proportion of ♀	0	0	0	0.25	1.00	0.60	1.00	0.80	1.00

Among those "islets" containing both forms and for which measures of fecundity are available (islets of 4 and 20 individuals), the ratio of ♀/⚥ fecundity is greater than 1. The islets containing only hermaphrodites (1 and 2 individuals) and only females (37 individuals), for which values are available, have fecundities close to 1, but with very wide individual variations.In these findings nothing suggest that hermaphroditism confers a superiority compared to the female state in isolated situations ; on the contrary a sex ratio to the advantage of the ♀ is, in "crowded" situations, observed. This suggests that a quite good pollen tranfer occurs between "islets" in the pioneer stages.

Germination of seeds of the 2 sexual forms. Seeds from females showed superior germination in comparison to hermaphrodites (Fig. 6).This result is wholly comparable with that found by Assouad et al. (1978). There is a trend toward slight negative relationship between the germination rates of seeds from hermaphrodites and from females in the population (r =-0.53, P # 0.25) ; this relation seems biologically significant for populations 3, 5 and 7.

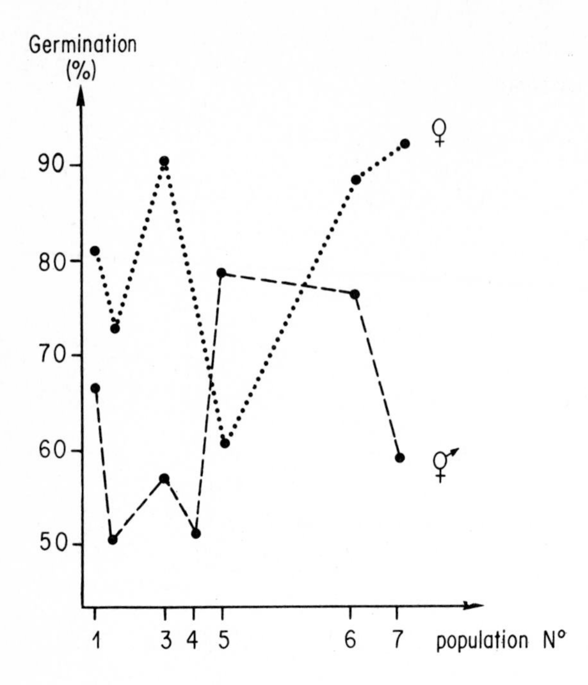

Fig. 6. Proportion of germination of seeds produced by female mother plants and by hermaphrodite mother plants according to the successional stage (population number).

The Offspring of the 2 Sexual Forms

Progeny plants, from 35 mother plants, 3 months old,were measured and then planted into an experimental plot.

Size of offspring.The size of the offspring from several of the mother plants per site is compared in Fig. 7.The sowings carried out in 1981 and 1982 involved different mother plants. For the sites early in the succession the average heights of the female's offspring (except for one) are above those of the hermaphrodites . Later on, the sizes of the descendants of the two sexual forms are comparable. The relationship ♀/⚥ is positively correlated with the proportion of ♀ ($r = 0.69$, $P < 0.05$) (Fig. 8). The advantage for the offspring of ♀ parents compared to those of ⚥ parents becomes more evident with the increasing number of ♀ plants

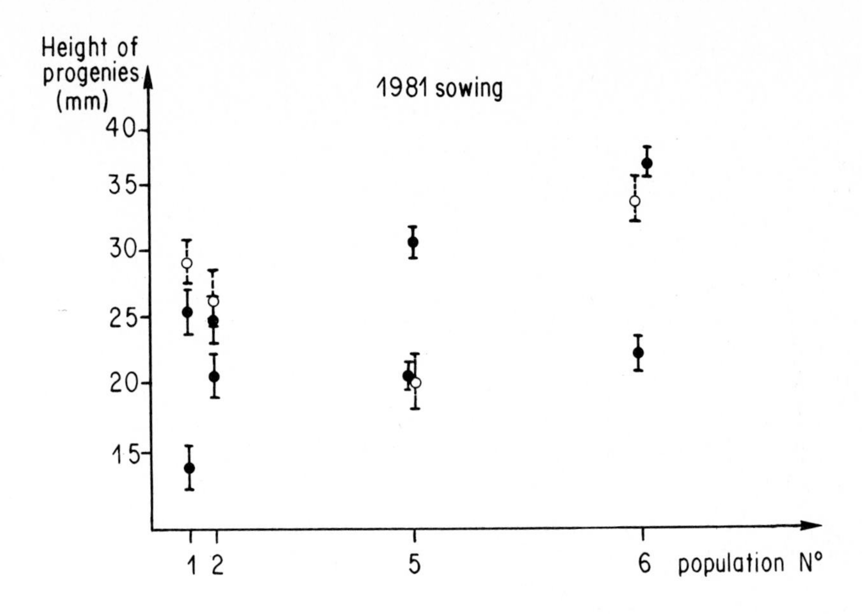

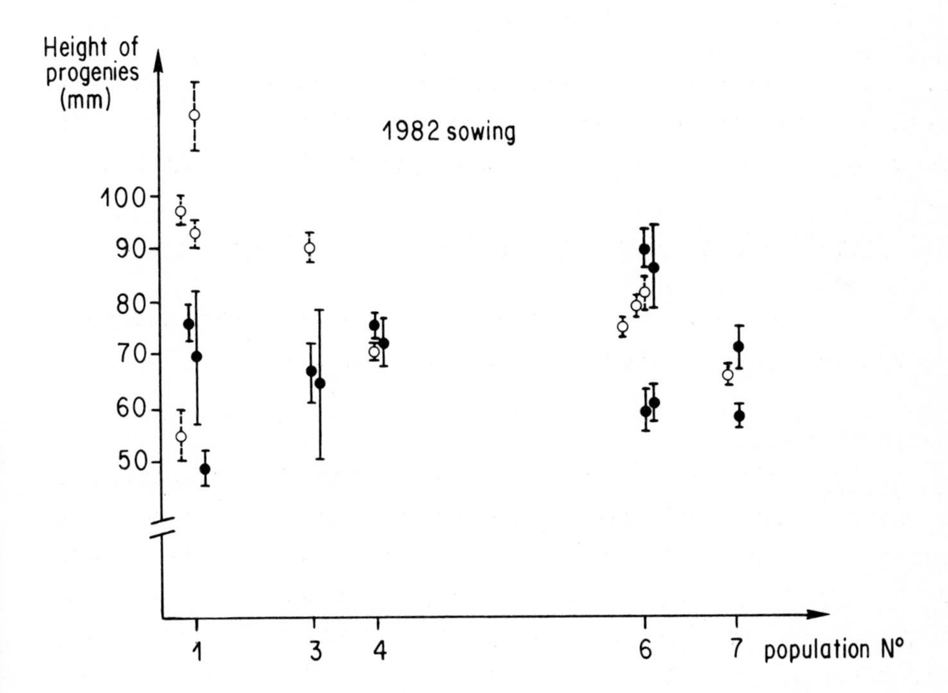

Fig. 7 Comparison of the size of the offspring from several mother plants per site ; ● ⚥ , o ♀ ($\bar{x} \pm s$).

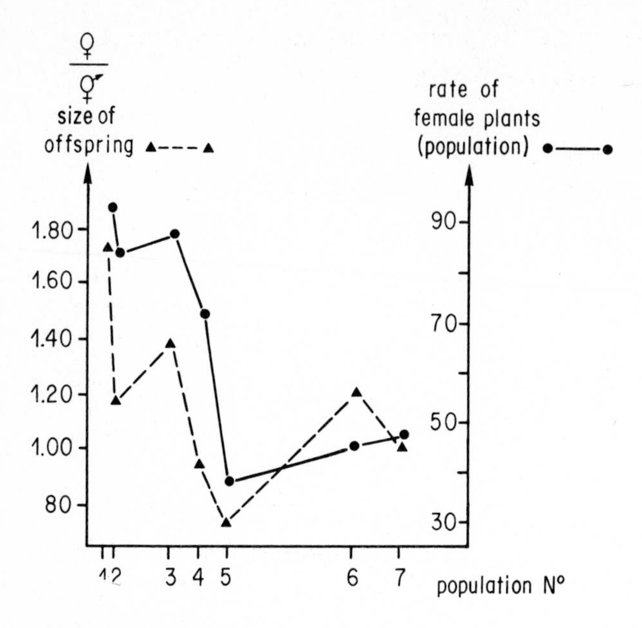

Fig. 8. Relationship between size of offspring of ♀ / size of offspring of ⚥ and proportion of ♀ according to sites.

in the site.It is worth noting that height gives a good estimation of the biomass ; more over a positive correlation is found (Bonnemaison et al. 1979) between vegetative biomass and reproductive output.

Sex of offspring. The ⚥ : ♀ ratio of those offspring whose height was measured (Fig. 7) are indicated in Table 3 and expressed as means in Fig. 9. In general the female mother plants produce more ♀ offspring than the ⚥ mother plants, but the individual results are very heterogeneous. Thus the segregations of the offspring from ⚥ and ♀ are similar for sites 1 and 5 respectively, whereas the differences between the segregations of the offspring of the two sexes are always great and show a tendency towards a very strong maternal heredity, for the sites 2, 3, 4, 6, 7.

Except in site 1 the hermaphrodite mother plants have produced a large majority of ⚥ offspring. The ratios observed in the offspring of the ♀ mother plants of sites

Table 3. Segregation obtained in situ from mother plants in the succession sites (⚥:♀)

Sites							
			Mother plants ⚥				
1	11:30	15:55	2:3	7:43	11:36		(46 :167)
2	20:0	29:0					(49 :0)
3	12:0	4:0					(16 :0)
4	17;23	10:0					(27 :23)
5	83:0	68:8					(151:8)
6	50:33	72:1	35:2	17:0	28:12	7:0	(209:48)
7	55:0	9:0					(64 :0)
			Mother plants ♀				
1	14:30	18:55	5:15	14:62	2:16		(53 :178)a
2	13:23						(13 :23) a
3	18:48						(18 :48) a
4	15:31						(15 :31) a
5	24:4						(24 :4)
6	30:19	18:28	36:46	38:43			(122:136)
7	0:43						(0 :43)

a All the pooled segregations from the ♀ individuals of the 4 sites are homogeneous (chi square test).

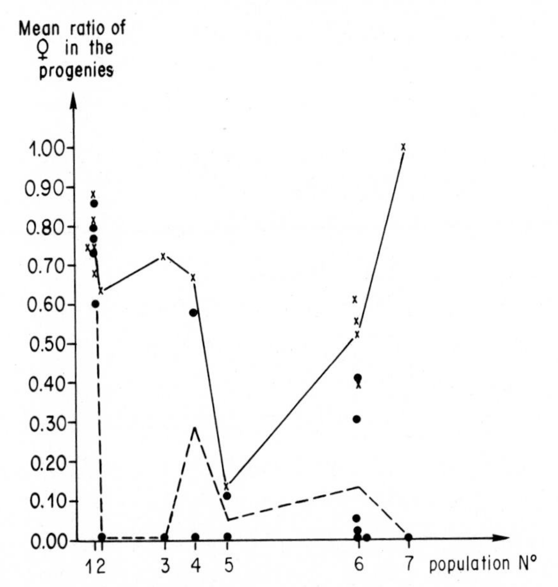

Fig. 9. Average proportion of ♀ in the offspring of ♀ mother plants (x) and ⚥ mother plants (●).

1,2,3, and 4 to the advantage of the ♀ reflect the situation in situ. The large variations in the segregations of the ⚥ mother plants at these sites do not seem to reflect upon the proportion of ♀ observed in situ.

In the later stages of the succession an analogous but symetrically opposite effect takes place meaning that a greater importance may be accorded to the ⚥ mother plants within the dynamics of reproduction (they rarely segregate ♀ types). Furthermore the large variations in the segregations of offspring from ♀ mother plants seem to have a minor influence on the sex ratio of these populations. However, the proportion of ♀ plants in situ remaining at a frequency of about 40 % indicates the important role of the female plants in reproduction.

DISCUSSION AND GENERAL CONCLUSION

The gynodioecy of *Thymus vulgaris* constitutes a good example of a relationship between a breeding system and the environment. In the succession studied, the abundance of ♀ plants was not related to the ♀/⚥ relative fecundity in a site. In the two richest populations in ♀ plants the fecundity of ♀ and ⚥ was comparable and other populations containing a much smaller number of ♀ plants showed a double or threefold advantage of ♀ in comparison to the ⚥ plants. Fecundity expressed here as the average number of seeds per fruit does not explain the observed sex ratios.

On the other hand, the viability and vigour of the offspring of the two forms vary from one site to the other and these variations are correlated with the sex ratio ; the values of these parameters are as an average lower for the ⚥ , more so when ♀'s are abundant in the site. This could indicate the existence of a strong effect of inbreeding depression of self-fertilization in the young populations : offspring from self-fertilization often carry deleterious (non viable seeds) or deficient characters (weak biomass, late flowering, slow growth, drop in chlorophyll level, reduced competitive fitness, Bonnemaison *et al.* 1979, Perrot *et al.* 1982). So theoretically, the number of gametes lost per individual would be higher in the hermaphrodites than in the ♀ ; under these conditions it may be assumed that the reproductive yield is brought to a maximum by a sex ratio in favour of ♀. But the knowledge that in Thyme there is a high coefficient of sexual maternal heredity is not enough to evidence that Thyme can adjust its sex-ratio. Thus one could propose an explanation for the maintaining of an allogamous system : maximum heterozygosity confers a wider genetic homeostasis. The genetic variance of the population is correlated to a high ecological variance.

Concerning evolution towards a tolerance of self-fertilization the present research seems to show that self-fertilization is much better tolerated in populations rich in ⚥ than in those containing few, this could be a sufficient reason for the decreasing of the ♀ proportion in the older populations in so far as the advantage linked with allogamy gradually becomes weaker. For example, if we compare the vigour of ♀ and ⚥, from the sowing of offspring, we can deduce that in older populations the vigours of the two forms are equivalent ; but this does not take into account the fact that ♀ forms have no investment in the production of stamens, and so would be more vigorous. Once the proportion of hermaphrodites increases the latter jointly carry out male functions (production of pollen) and female functions (production of seeds) with an increase in self -fertilization. Current experimental results seem to show that in the case of populations rich in ⚥ , the offspring produced by self -fertilization have a viability and vigour comparable to those from allofecundation, from which results an ability to accept and tolerate self -fertilization. The breeding system is thus close to autogamy.

The fact that many ♀ individuals are found in the young stages of the postagricultural succession implies either that the genes for restoration are rare in the populations or that they are actually there but inactive in one or several types of special cytoplasms. There are reasons for thinking that in the pioneering and young stages of Thyme growth the founding individuals grown from seeds of more or less diverse origin (transported by wind, sheeps or run off water) are cross-fertilized. These engender new nucleo-cytoplasmic combinations which are not likely to restore male fertility. The abundance of ♀ forms may be explained in this way. Male sterility is the loss of a function ; it happens that this loss has positive consequences by imposing cross-fertilization which favours recombinations and heterozygosity. This would suggest that the evidence of a high level of male sterility in the population of founders is a direct result of a new population of gametes instead of the consequence of a specific selective pressure in favour of females (Dommée and Guillerm 1980 ; Couvet 1984).

If this process acts in the same way as a selective differential, it seems that a high level of females is not only resulting from that phenomenon. In fact, if we look at long term disturbed environments (ranges for sheep grazing, slopes with intense run-off), they are occupied by populations with a stable high level of females plants (up to fifteen years). On so long periods a genetic balance has been established without change towards a decreasing of the female forms. This gives an evidence of a selective advantage in favour of females. The results in Table 3 seem to agree with the hypothesis and the model of the spreading of restorer genes for male fertility during the ageing of the population (Couvet 1984, Gouyon and Couvet 1984). In the course of succession, there is a trend towards decreasing the level of females in the offspring of ♀ mother plants, and rather rare segregations for ⚥ from old populations. It seems very coherent that there are more efficient restorer genes in a population dominated by ⚥. The model allows to describe the mathematical and predictable trajectory of the decreasing proportion of ♀. A nucleo- cytoplasmic equilibrium is reached for a low frequency of females, determined only by the nuclear and cytoplasmic interactions, independently from the environment. The result is an endogene balance of the population.

The hypothesis of an environmental selection of female plants is not contradictory if we consider that the environment neutrality of the genes of male sterility is only transient and relative. In that case, with time, the frequency of females is more and more a consequence of the selective pressure with a resulting diversification (high levels of females) or homogeneisation (low levels of females). The combination of the two hypothesis results in a balance between a mechanistic phase with no phenotypic selection at the founding of a colony, immediately followed by a selectionist phase. By chance, the variations of the female proportions are conditioned by endogene forces (nucleo-cytoplasmic model) and exogene forces (phenotypic selection) acting in the same direction. It seems that during the demographic decrease of the colony, a trend towards a new spreading of female forms occurs (Table 4).

On the whole the situation described for _T. vulgaris_ L. incorporates heterozygous advantage and the advantage of the female form which produced them, in the most unstable environments, the pioneer stages. Schaal & Levin (1976) have shown the superiority of heterozygotes in _Liatris cylindracea_, for their longevity, their reproductive capacity and their earliness in flowering etc.. Apart from that, variations in heredity of the ability for self- fertilization are described by Antonovics (1968) for _Anthoxanthum odoratum_ and _Agrotis tenuis_ and by Kahler et al. (1975) for _Hordeum vulgare._ Jain & Allard (1960) have also shown the advantage of heterozygotes in the latter species. Biémont (1984) shows that a reduction of the population size of _Drosophila melanogaster_ induced by a change of environmental conditions

Table 4 : Dynamic variations in the regime of reproduction and in the genetic structure of Thyme populations.

Populations (Sites)	1	2	3	4	5	6	7
Environment : composition and structure of the community	Fast modifications			Slow modifications			
"Age" of the population (years)		Young				Old	
	9	10	20	26	30	52	60
Size of the deme (plant/m2)	11	(.01)	13	6	3	4	2
% of ♀ plants (x)	.93	.83	.87	.79	.38	.45	48
♀ fecundity(1)(w) / ⚥ fecundity	.75	1.26	7.40	2.20	2.26	2.64	2.90
♀ germination / ⚥ germination	1.20	1.46	1.58	1.47	.77	1.13	1.55
HMS (2)	.77	.64	.73	.67	.14	.53	1.00
HMF	.22	1.00	1.00	.54	.95	.81	1.00
Expected proportion of ♀ at next generation (3)	↘ .77	↘ .55	↘ .72	↘ .65	↘ .10	↘ (dotted) .42	↗ .73
Breeding system	Allogamy		→		Autogamy		
Genetic structure (Index of heterozygosity)	.98		1.04	.87	.69	.50	.70
	Heterozygosity		→		Homozygosity		
Inbreeding depression on the basis of height of offspring {♀/⚥}		High			Low		
	1.74	1.18	1.39	.95	.74	1.22	1.03
Deleterious genes	Numerous				Eliminated		

(1) On the basis of average production of viable seed/fruit

(2) HMS = proportion of females in the maternal progenies of ♀

HMF = proportion of hermaphrodites in the maternal progenies of ⚥

(3) According to the model from Gouyon & Couvet (1984) :

$$P_{n+1} = [x.\ w.\ HMS + (1-x).\ (1-HMF)] / [x.\ w + (1-x)]$$

is not necessarily followed by a strong inbreeding depression; there are, in these conditions, homeostatic heterozygous individuals which have been selected for absence of deleterious genes.

In the hypothesis of the determining role of the depression of self- fertilization in maintaining the rate of females in the Thyme populations at a high level, it is necessary to specify the interactions between consanguinity and genetic determinism of male sterility. Ross and Gregorius (pers. comm.) suggest that it is a question of two completely different phenomena and that the maintenance of male sterility has nothing to do with an inbreeding effect. It is true that in Plantago lanceolata the combination of male sterility and self-incompatibility supports this view. But each species has it own biology and experience shows that general hypotheses are often ill based. Biémont et al. (1974) establish a direct relationship between the nuclear-cytoplamic interactions and the inbreeding depression. So, it seems very plausible that male sterility in Thyme interfers with inbreeding depression.

We consider that in Thyme, taking into account the present results, the species has a genetic system which enables it to easily maintain high rates of male sterility. Whatever the mechanism for this may be, we think that it results in guaranteeing allogamy. When in certain situations, mainly the destabilization of the environment, a vigorous selection against self-fertilization takes place it is reasonable to consider that the hermaphrodite phenotype is a disadvantaged type and only maintained at a high enough frequency to act as a pollen producer.

It is possible also that the forces resulting in inbreeding depression and male sterility may interfer with each other. Knowing that controlled crossings (Assouad 1972, Dommée 1973) show a nucleo-cytoplasmic determinism of male sterility and that the pioneer stages favour an intense genetic mixing (founders of diverse origin associated with pollen transport over great distances), the resulting new nucleo-cytoplasmic combinations have every reason not to contain restorers of male fertility.

One, may be observing, here, on an intraspecific scale, the phenomenon, well known by agronomists, of male sterility induced by hybridizations of different but related taxa. Thus the particularly high level of allogamy in these situations reinforces a tendency already favoured by the deleterious effects of self- fertilization. Moreover it is understood that this situation is not stable and that the modifications of the patterns of a population related to a subsequent genetic and ecological homogeneization reestablish a nucleo-cytoplasmic equilibrium by a genetic "purifying" of the deleterious genes allowing a restoration of male fertility.

REFERENCES

Antonovics J, (1968) Evolution in closely adjacent plant populations. V. Evolution of self- fertility. Heredity 23 : 219-238.

Assouad MW, (1972) Recherches sur la génétique écologique de Thymus vulgaris L. Etude expérimentale du polymorphisme sexuel,Thèse d'Etat, Univ. Sc. et Techn. du Languedoc, 224 p. C.N.R.S., A. O.7505.

Assouad MW, Dommée B, Lumaret R, & Valdeyron G,(1978) Reproductive capacities in the sexual forms of the gynodioecious species Thymus vulgaris L. Bot J. Lin. Soc. 77 : 29-39.

Biémont C, (1985) Effects of winter on genetic structure of a natural population of Drosophila melanogaster. Genet. Select. Evol. 17:27-42.

Biémont C, Bouffette AR & Bouffette J,(1974) Théorie chromosomique de l'inbreeding : Modèle probabiliste. Bull. Mathematical Biol. 36 : 417-434.

Bonnemaison F, (1980) Etude stationnelle de la dynamique du maintien d'un polymorphisme génétique :cas de quatre populations naturelles de Thymus vulgaris L. Thèse de 3e cycle, Univ. Sci. et Tech. du Languedoc, 82 p.+ Annexe.

Bonnemaison F, Dommée B, & Jacquard P, (1979) Etude expérimentale de la concurrence entre formes sexuelles chez le thym, Thymus vulgaris L. Oecol. Plant. 14. 1 : 85-101.

Chadefaud M, & Emberger L, (1960) Les Végétaux vasculaires. in : Traité de Botanique, t 2, Masson éd. 1539 p

Couvet D, (1984) Ecologie évolutive des populations de Thymus vulgaris L. : pourquoi des femelles ? Thèse de 3e cycle, Univ. Sc. et Techn. du Languedoc, 33 p. + Annexes.

Dommée B, (1973) Recherches sur la génétique écologique de Thymus vulgaris L. Déterminisme génétique et répartition écologique des formes sexuelles, Thèse d'Et., Univ. Sc. et Techn. du Languedoc. C.N.R.S. A.O. 8635, 129 p.

Dommée B, (1976) La stérilité mâle chez Thymus vulgaris L. : répartition écologique dans la région méditerranéenne française. CR. Acad. Sc. Paris 282, D : 65-68.

Dommée B, (1981) Rôle du milieu et du génotype dans le régime la reproduction de Thymus vulgaris L. Acta. Oecol., Oecol. Plant.2 (16) : 137-147.

Dommée B, & Guillerm JL, (1980). Etudes des relations entre la stérilité-mâle chez Thymus vulgaris L. et sa distribution dans les différents stades d'évolution des milieux perturbés. Rapp. Conv. de Recherche, Ministère de l'Environnement et du Cadre de Vie-CNRS, 100 p, 4 pl. phot.

Dommée B, Assouad MW, & Valdeyron G, (1978) Natural selection and gynodioecy in Thymus vulgaris L., Bot. J. Lin. Soc.77 : 17-28.

Dommée B, Guillerm JL, & Valdeyron G, (1983) Régime de reproduction et hétérozygotie des populations de Thym Thymus vulgaris L., dans une succession postculturale. CR. Acad. des Sciences t. 296 : 111-114, Sér. III.

Escarré J, (1979) Etude des successions post-culturales dans les hautes garrigues du Montpelliérais. Thèse de 3e cycle, Univ. Sci. et Techn. du Languedoc, 134 p.

Gouyon PH, & Couvet D, (1984) Selfish cytoplasm and adaptation : variations in the reproductive system of Thyme. In : Phenotypic and Genotypic Variation Within and Between Plant Populations, 2d Int. Symp. on the Structure and Functioning of Plant Populations (To be published).

Gregorius HR, Ross MD, & Gillet E, (1982) Selection in plant populations of effectively infinite size : III. The maintenance of females among hermaphrodites for a biallelic model. Heredity 48 : 329-343.

Jain SK, & Allard RW, (1960) Population studies in predominantly self-pollinated species.I. Evidence for heterozygote advantage in a closed population of barley. Proc. Nat. Acad. Sci. U.S.A. 46 : 1371-1377.

Kahler AL, Clegg MT & Allard RW,(1975) Evolutionary changes in the mating system of an experimental population of barley (Hordeum vulgare L.). Proc. Nat. Acad. Sci. U.S.A. 72 : 943-946.

Perrot V, Dommée B, Jacquard P, (1982) Etude expérimentale de la concurrence entre individus issus d'autofécondation et d'allofécondation chez le thym (Thymus vulgaris L.) Act. Oecol., Oecol. Plant. Vol. 3 (17) 2 : 171-184.

Ross MD, (1978) The evolution of gynodioecy and subdioecy. Evolution 32 : 174-188.

Schaal BA, & Levin DA, (1976) The demographic genetics of Liatris cylindracea Michx (Compositae). Am. Nat. 110 (972): 191-206.

Valdeyron G, Dommée B, & Valdeyron A, (1973) Gynodioecy : Another computer simulation model. Am. Nat. 107 (955) : 454-459.

Valdeyron G, Dommée B, & Vernet Ph, (1977) Self-fertilisation in male fertile plants of a gynodioecious species : Thymus vulgaris L. Heredity 39 : 243-249.

Van Damme JMM, (1984) Why are so many Plantago species gynodioecious ? . In : Phenotypic and Genotypic Variation Within and Between Plant Populations, 2nd Int. Symp. on the Structure and Functioning of Plant Populations (To be published).

Pines, Plantains and Thyme: three examples of frequency-dependent selection.

M.D. Ross

Department of Forest Genetics and Forest Tree Breeding, University of Göttingen, Göttingen, West Germany.

ABSTRACT

A method for obtaining fitness values in seed plants based on the concept of successful gametes is presented, where a gamete is regarded as successful if it takes part in fertilization. A previous study is cited to show that there are at least three causes of frequency-dependency in such fitness values, namely: (1) non-constant pollen:ovule ratios among individuals (sexual asymmetry); (2) differential individual selfing rates for ovules; and (3) cytoplasmic fertility effects. Experimental results are cited to show that the first cause operates in *Pinus sylvestris, Plantago lanceolata* and *Thymus vulgaris,* the second in *Thymus vulgaris*, and the third in *Plantago lanceolata* and *Thymus vulgaris*. Models of gene-cytoplasm controlled sexual asymmetry (including gynodioecy) are presented, and it is shown for recessive gynodioecy that despite the occurrence of complete dominance in each cytoplasm there may be overdominance or underdominance for fitness of nuclear genotypes averaged over all cytoplasms, in equilibrium populations. Hermaphrodite populations may show cyclic behaviour, with overdominance, underdominance or intermediate heterozygote fitness, averaged over all cytoplasms for a dominant model.

NATO ASI Series, Vol. G5
Genetic Differentiation and Dispersal in Plants
Edited by P. Jacquard et al.

1. INTRODUCTION

Studies of sex-polymorphic plant populations have shown that fitness values in such populations were frequency dependent, even when selection values (male and female fertilities) were constant (Ross and Shaw, 1971; Lloyd, 1974). Further studies showed that such frequency-dependent fitnesses were also characteristic of hermaphrodite populations, under simple but realistic assumptions (Ross, 1977a; 1977b), and a detailed study of such fitnesses showed that they directly determined gametic frequencies in the next generation (Gregorius and Ross, 1981). It was later shown that there were at least three different causes of frequency-dependent selection in populations of seed plants, namely: (1) non-constant pollen to ovule ratios among individuals (called "sexual asymmetry" by Ziehe and Gregorius, 1981); (2) different ovule selfing rates among individuals; and (3) cytoplasmic fertility effects (Ross, 1984). The present paper gives a fitness concept for seed plants, and cites experimental results to show that all three causes of frequency-dependent selection are present in real populations. The paper then deals with cytoplasmic fertility effects in more detail by presenting models of gene-cytoplasm controlled fertility variation (including gynodioecy) and considering fitness in such models.

2. DERIVATION OF FITNESS VALUES

Fitness values and related quantities, such as functional sex and selfing rate, are based upon the concept of successful gametes, which are defined as gametes which take part in fertilization (Gregorius and Ross, 1981). We now obtain fitness values for a simple monogenic model, noting that the method is not confined to this or other genetic models, but may be applied to phenotypes or individuals without making genetic assumptions. We assume a gene locus A which has alleles A_1 and A_2 and controls male and female fertility. Genotype A_1A_1 has ϕ_{11} ovules or seeds, μ_{11} pollen grains, ovule selfing rate σ_{11} and frequency P_{11}, and so on (Table 1). Non-selfed ovules are assumed to be fertilized at random.

TABLE 1 A one-locus two-allele model of fertility variation

	Genotype		
	A_1A_1	A_1A_2	A_2A_2
Female fertility	ϕ_{11}	ϕ_{12}	ϕ_{22}
Male fertility	μ_{11}	μ_{12}	μ_{22}
Ovule selfing rate	σ_{11}	σ_{12}	σ_{22}
Frequency	P_{11}	P_{12}	P_{22}

Genotype A_1A_1 has σ_{11} self-fertilized ovules, $(1-\sigma_{11})$ cross-fertilized ovules, $\sigma_{11}\phi_{11}$ selfing pollen grains and $\bar{\phi}_{(1-\sigma)}\mu_{11}/\bar{\mu}$ crossing pollen grains, where $\bar{\phi}_{(1-\sigma)}$ is the mean number of cross-fertilized ovules per individual and equals

$$(1-\sigma_{11})\phi_{11}P_{11} + (1-\sigma_{12})\phi_{12}P_{12} + (1-\sigma_{22})\phi_{22}P_{22}$$

and where $\bar{\mu}$ is the mean number of pollen grains per individual and equals

$$\mu_{11}P_{11} + \mu_{12}P_{12} + \mu_{22}P_{22}$$

The total fitness is the sum of these four quantities and equals

$$w_{11} = (1+\sigma_{11})\phi_{11} + \bar{\phi}(1-\sigma)\mu_{11}/\bar{\mu}$$

The degree to which an individual functions as female is called its "functional sex" (Ross, 1982), and is defined as its number of fertilized ovules divided by the total number of its successful gametes (i.e. its fitness value). Functional sex thus equals ϕ_{11}/w_{11}. The "combined selfing rate" is defined for all gametes and is therefore the total number of ovules and pollen grains which take part in selfing divided by the fitness value, or $2\sigma_{11}\phi_{11}/w_{11}$. Thus the combined selfing rate is 2σ times the functional sex (Ross and Gregorius, 1983), which illustrates the close connection between the two quantities. Corresponding values are given for these measures are given to the other genotypes.

We see that the fitnesses and related quantities are frequency-dependent since the mean numbers of pollen grains and crossed ovules per individual are frequency-dependent. We may see that this frequency-dependency results from the sexual asymmetry, by substituting symmetry or constant genotypic ratios for pollen to ovule fertility in the fitness values. Since μ_{ij}/ϕ_{ij} now = K, we may write $\mu_{ij} = K\phi_{ij}$ in the fitness value of any genotype A_iA_j, obtaining the constant $2\phi_{ij}$. Similarly, we may easily see that frequency-dependency results from different genotypic ovule selfing rates by setting all ϕ_{ij}'s equal to ϕ and all μ_{ij}'s equal to μ, which yields

$$w_{11} = \phi(2 + \sigma_{11} - \sigma_{11}P_{11} - \sigma_{12}P_{12} - \sigma_{22}P_{22})$$

which reduces to 2ϕ when all σ's are set equal (Ross, 1984). The frequency-dependency may be negative or positive. An example of negative frequency-dependency is found for hermaphrodites in a gynodioecious population, where it reflects the greatly increased fitness of the hermaphrodites as they become rare, since they pollinate all of the females.

3. PINES, PLANTAINS AND THYME: OBSERVATIONS AND EXPERIMENTS

The previous section has shown that there are at least three possible causes of frequency-dependent selection in seed plants, namely sexual asymmetry, different ovule selfing rates among individuals, and fertility interactions between nucleus and cytoplasm. This section briefly gives some of the evidence for the actual existence of these three causes.

(a) Sexual Asymmetry: Pines, Plantains and Thyme

Ross (1984) has estimated the number of male and female cones per tree in a clonal population of Scots Pine (*Pinus sylvestris*) from Germany. The population is a seed plantation, made up of many clones, each of which is present in multiple copies (see Ross, 1984 for further details). Table 2 gives the results for the four most asymmetrical of the 25 clones studied. The Table shows the extreme cases for ratios of male/female cone number. Clone CH19 had the highest and CH10 the second highest ratio, whereas clones CH10 and RD8 were 24th and 25th, respectively, among a total of 25 clones. A full publication of these results, with statistical tests, will be made elsewhere, but this extract is sufficient to show that sexual asymmetry is present in this population. The frequency-dependency in fitness caused by the asymmetry is readily seen by imagining a population consisting of, say, clone CH 19 only. In the model previously considered, if we set the frequency of genotype $A_1 A_1$ equal to one then the fitness of this genotype reduces to 2ϕ. Similarly, in a pine plantation consisting of clone CH19 only, this clone would have a fitness of twice the number of female cones, namely 27.50. Its actual fitness in the pine population, however, is very much higher (192.2) because of its high male fertility, and the effect caused by the presence of the other clones in the population (the frequency-dependency) is very considerable.

TABLE 2 Male and female fertilities, viabilities and fitness in four clones of Scots Pine (*Pinus sylvestris*)

Clone	Number of trees: 1981	1959	Mean no. of female cones	Est. mean no. of male cones	μ/ϕ	Fitness
CH 19	22	23	13.75	99829	7260	192.2
CH 2	24	24	13.50	68852	5100	143.0
CH 10	20	23	500.33	48335	97	557.2
RD 8	23	24	223.50	13487	60	259.9

The results are for $\sigma = 0.101$. $\bar{\phi} = 92.47$, $\bar{\mu} = 44665$ and the population size is 562.

We have seen that pine populations may show frequency-dependent selection because of asymmetry, and reviews of the literature together with theoretical considerations give good reason to suppose that asymmetry is widespread in hermaphrodite and monoecious plants (Ross, 1977b; 1984; Ross and Gregorius, 1983). For example, a race of the annual hermaphrodite crucifer *Leavenworthia crassa* is polymorphic for flower colour, and plants with yellow-centred flowers had 0.88 as many ovules and 1.21 as many pollen grains as the recessives for yellow flowers (Lloyd, 1965, Tables 3, 6). In *Lupinus nanus* a blue-flowered wild type had 5.68 ovules and 12,460 pollen grains per flower, whereas a pink-flowered mutant had 6.46 ovules and 9,060 pollen grains (Horovitz and Harding, 1972). More extreme forms of asymmetry are found in populations which consist of unisexual and hermaphrodite individuals. For example, gynodioecious populations consist of females and hermaphrodites, and have been much studied recently. Among the more intensively studied gynodioecious species are the plantain *Plantago lanceolata* (Van Damme, 1983) and the garden thyme, *Thymus vulgaris* (Valdeyron et al., 1977; Dommée, 1981).

(b) Differential Ovule Selfing Rates: Thyme

Some of the evidence for differential genotypic ovule selfing rates within populations was considered previously (Ross,1984), so that here only the evidence for such variation in thyme will be briefly considered. Dommée (1981) showed that different plants sometimes had different crossing rates when grown on the same site, and that the same plant could have similar crossing rates when grown on different sites.

(c) Gene-cytoplasm Fertility Effects: Plantains and Thyme

Gynodioecious species often show cytoplasmic influences on the inheritance of male sterility (Correns, 1928; Ross, 1978). Among the best-studied cases is *Plantago lanceolata,* in which Van Damme (1983) has found two widespread types of male sterility which show complex interactions between nucleus and cytoplasm. *Thymus vulgaris* shows similar gene-cytoplasm effects (Dommée, 1973), which are theoretically possible in hermaphrodite

TABLE 3. Computer simulations of two-allele two-cytoplasm models of fertility variation, where ϕ = ovule fertility, μ = pollen fertility, σ = ovule selfing rate, P = frequency and w = fitness value

	Genotype					
	NA_1A_1	NA_1A_2	NA_2A_2	SA_1A_1	SA_1A_2	SA_2A_2
Population 1, gynodioecious						
ϕ	0.75	1.00	0.75	0.75	0.75	1.25
μ	0.75	1.00	0.75	0.75	0.75	0
σ	0.50	0.50	0.50	0.50	0.50	0
P	0.277	0.208	0.135	0.158	0.158	0.064
w	1.579	2.106	1.579	1.579	1.579	1.250
Population 2, gynodioecious						
ϕ	0.50	1.0	0.50	0.50	0.75	0.75
μ	0.50	1.0	0.50	0.50	0.75	0
σ	0.75	0.75	0.75	0.75	0.75	0
P	0.145	0.103	0.107	0.269	0.257	0.120
w	1.081	2.161	1.081	1.081	1.621	0.750
Population 3, hermaphrodite						
ϕ	80	80	95	90	90	80
μ	200000	200000	50000	100000	100000	200000
σ	0.50	0.50	0.50	0.50	0.50	0.50
P	0.963	0.010	0.003	0.022	0.001	0.001
w	160.7	160.7	152.7	155.4	155.4	160.7
Population 4, hermaphrodite, with the same ϕ, μ, and σ values as Population 3						
P	0.544	0.004	0.001	0.447	0.003	0.001
w	174.6	174.6	156.1	162.3	162.3	174.6
Population 5, hermaphrodite, with the same ϕ, μ, and σ values as Population 3						
P	0.000	0.001	0.024	0.003	0.009	0.963
w	161.2	161.2	152.8	155.6	155.6	161.2

species also (Gregorius and Ross, 1984). When a nuclear genotype has different fertilities in different cytoplasm types, its fitness taken as a mean over all cytoplasms is clearly dependent on the cytoplasm frequency.

4. MODELS OF GENE-CYTOPLASM FERTILITY INTERACTIONS

The purpose of this section is to consider the effects of cytoplasm frequencies on fitness values, and to give some results of computer simulations of gene-cytoplasm fertility variation, both of hermaphrodite and gynodioecious populations. We now extend the model given in Table 1 by considering two cytoplasms, called N and S, where a cytoplasm type may affect the fertility or selfing rate of any nuclear genotype. Thus nuclear genotype A_1A_1, for example, may have different male or female fertilities in N and in S cytoplasms. We now have genotypes NA_1A_1 (with female fertility ϕ_{N11}, male fertility μ_{N11}, ovule selfing rate σ_{N11} and frequency P_{N11}), SA_1A_1 (with corresponding model parameters), and so on, giving six genotypes in all.

We now apply this model by considering a two-allele two-cytoplasm model of gynodioecy, where only genotype SA_2A_2 is female with ovule selfing rate zero. All other genotypes (including NA_2A_2) are hermaphrodite and have the same selfing rate for the ovules. Table 3 gives the equilibrium results for two gynodioecious populations, and non-equilibrium results for three hermaphrodite populations, where in all cases all six genotypes and therefore both cytoplasm types are present. The first two populations are gynodioecious, as is seen from the second and seventh lines, where the μ

values for type SA_2A_2 are zero.

In the first population there is overdominance in N cytoplasm, since each homozygote is three-quarters as fertile as the heterozygote for both ovules and pollen. In S cytoplasm, however, allele A_2 is recessive for increased ovule but zero pollen fertility. The two-thirds increase in ovule fertility of the female over the hermaphrodite in S cytoplasm would not be sufficient to maintain females if only S cytoplasm were present (Lewis, 1941). The fitness values (the w's, line 5) show the expected overdominance in N cytoplasm, and lower female than hermaphrodite fitness in S cytoplasm. When we obtain the mean fitness of any nuclear genotype over both cytoplasms, however, we obtain values of 1.579 for A_1A_1, 1.878 for A_1A_2, and 1.474 for A_2A_2, so that there is overdominance for fitness over both cytoplasms.

The second gynodioecious population in the Table shows overdominance in both cytoplasms. This result is given both to show that overdominance can maintain gene-cytoplasm gynodioecy and to indicate the effects of selfing. An equilibrium with all six genotypes present was obtained for the given fertility values only for the ovule selfing rate of 0.75. Selfing rates of 0, 0.25, and 0.5 gave all-hermaphrodite equilibria polymorphic for the three nuclear genotypes in N cytoplasm only, whereas a selfing rate of 1 gave a gynodioecious polymorphism in S cytoplasm only. Other examples of the dependence of a polymorphism on intermediate selfing rates have been found (e.g. Ross and Gregorius, 1983).

A more striking example of cytoplasmic effects on nuclear genotypic fitnesses is given in Table 3 for hermaphrodite populations. The Table shows that nuclear gene A_1 is dominant for ovule and pollen fertility in both cytoplasms. The Table then gives frequencies and fitnesses for three non-equilibrium populations, since the populations had apparently not reached equilibrium even after several thousand generations. Similar situations were found by Gregorius and Ross (1984). As expected, the Table shows that A_1A_1 always has the same fitness as A_1A_2 in each cytoplasm, but mean fitnesses over both cytoplasms for A_1A_1, A_1A_2 and A_2A_2, respectively, are: 160.6, 160.1 and 153.9 for the first population; 169.0, 169.8 and 162.0 for the second; and 156.1, 156.0 and 161.0 for the third. Thus the results may show intermediate heterozygote fitness, overdominance, or underdominance, according to the nuclear and cytoplasmic frequencies. Similar results were found for a similar model of hermaphrodite populations (Gregorius and Ross 1984). Fitnesses over both cytoplasms showed underdominance, or equal fitnesses of A_1A_2 and A_2A_2, or intermediate fitness of the heterozygote. Some of the populations of Gregorius and Ross (1984) apparently do not show equilibrium frequencies, but instead show spiral or cyclic behaviour. One population, however, showed a spiral approach to an interior equilibrium point, and all the populations studied which were near this point showed underdominance for the nuclear genotypes taken over both cytoplasms, despite dominance in each cytoplasm.

4. DISCUSSION

We have seen by means of a survey of theoretical and observational work that there are at least three causes of frequency-dependent selection in seed plants, in addition to those usually considered, e.g. those resulting from pollinator behaviour. It seems probable that all three causes operate frequently in nature. Sexual asymmetry, for example, probably occurs frequently within and among plant populations (Cruden, 1977; Ross, 1984), and selfing rates probably differ frequently among genotypes, although further studies are needed here. Cytoplasmic fertility effects occur in natural populations of gynodioecious plants, but are also often found after hybridization within and among species (Edwardson, 1970) suggesting that migration and introgression may often cause such effects in natural populations.

REFERENCES

Correns C (1928) Bestimmung, Vererbung und Verteilung des Geschlechtes bei den höheren Pflanzen. Borntraeger, Berlin

Cruden RW (1977) Pollen-ovule ratios: a conservative indicator of breeding systems in flowering plants. Evolution 31: 32-46

Domméé B (1973) Recherches sur la génétique écologique de *Thymus vulgaris* L. Déterminisme génétique et répartition écologique des formes sexuelles. Thèse d'État Univ Sc et Techn du Languedoc

Dommée B (1981) Roles du milieu et du génotype dans le régime de la reproduction de *Thymus vulgaris* L. Acta Oecologica/Oecol Plant 2: 137-147

Edwardson JR (1970) Cytoplasmic male sterility. Bot Rev 36: 341-420

Gregorius H-R, Ross MD (1981) Selection in plant populations of effectively infinite size: I. Realized genotypic fitnesses. Math Biosci 54: 291-307

Gregorius H-R, Ross MD (1984) Selection with gene-cytoplasm interactions. I. Maintenance of cytoplasm polymorphisms. Genetics 107:165-178

Lewis D (1941) Male sterility in natural populations of hermaphrodite plants. New Phytol 40: 56-63

Lloyd DG (1974) The genetic contribution of individual males and females in dioecious and gynodioecious Angiosperms. Heredity 32: 45-51

Ross MD (1977a) Frequency-dependent fitness and differential outcrossing in hermaphrodite populations. Amer Nat 111: 200-202

Ross MD (1977b) Behaviour of a sex-differential fertility gene in hermaphrodite populations. Heredity 38: 279-290

Ross MD (1978) The evolution of gynodioecy and subdioecy. Evolution 32: 174-188

Ross MD (1984) Frequency-dependent selection in hermaphrodites: the rule rather than the exception. Biol J Linn Soc (in press)

Ross MD, Gregorius H-R (1983) Outcrossing and sex function in hermaphrodites: a resource-allocation model. Am Nat 121: 204-222

Ross MD, Shaw RF (1971) Maintenance of male sterility in plant populations. Heredity 26: 1-8

Valdeyron G, Dommée B, Vernet P (1977) Self-fertilisation in male fertile plants of a gynodioecious species: *Thymus vulgaris* L. Heredity 39: 243-249

Van Damme JMM (1983) Gynodioecy in *Plantago lanceolata* L. II. Inheritance of three male sterility types. Heredity 50: 253-273

Ziehe M, Gregorius H-R (1981) Deviations of genotypic structures from Hardy-Weinberg proportions under random mating and differential selection between the sexes. Genetics 98: 215-230

Habitat, Genetic Structure and Dynamics of Perennial and Annual Populations of the Asian Wild Rice *Oryza perennis*

H. Morishima
National Institute of Genetics
Misima, 411 Japan

ABSTRACT

Populations of the Asian common wild rice, *Oryza perennis*, show a perennial-annual continnum. The perennial type growing in deep swamp is characterized by low reproductive effort, high outcrossing rate, late flowering, and many other traits, as compared with the annual type growing in shallower, temporary swamp, indicating the differentiation in adaptive strategy. From the study of hybrid progenies between the perennial and annual types, it was suggested that a gene for flowering time and a key factor governing seed productivity are linked and located on a chromosomal segment marked by an isozyme locus (*Pox*-1). In nature, the annual populations are fixed with a particular allele at this locus which is polymorphic in the perennial populations. Selection and fixation of this chromosomal block could bring about effective seed productivity which meets the requirement for the annuals. Observations of the permanent study-sites suggested that the annual populations are fugitive, while the perennial populations persist in stable habitat. The evolutionary potentiality of the intermediate perennial-annual type which can propagate both by seeds and by ratoons was discussed in relation to the evolution of the cultivated type.

DIFFERENTIATION IN ADAPTIVE STRATEGY AND HABITAT CONDITIONS

The common wild rice, *Oryza perennis* Moench is a complex species distributed throughout the humid tropics. The Asian forms (= *O*.

NATO ASI Series, Vol. G5
Genetic Differentiation and Dispersal in Plants
Edited by P. Jacquard et al.

Table 1. Comparison between perennial and annual types of Asian *O. perennis*.

Attribute	Perennial	Annual
Habitat:		
Water condition	Deep	Shallow
Disturbance	Low	High
% plant cover	High	Low
Companion species	Perennial	Annual
Propagation:		
Regenerating ability	High	Low
Seed productivity	Low	High
Seed dormancy	Weak	Strong
Awn development	Low	High
Buried seeds	Few	Many
Resource allocation:		
Seed/whole plant	Small	Large
Awn/seed	Small	Large
Pollen/seed	Large	Small
Outcrossing rate:	High$_{o}$	Low
Photoperiod sensitivity:	High	Low
Tolerance to:		
Deep water	High	Low
Drought (seedling)	Low	High
Submergence (seedling)	Low	High
Morphology:		
Plant stature	Tall	Short
Tillers	Few	Many
Population structure:		
Between population diversity	Small	Large
Within population diversity	Large	Small
Heterozygosity	High	Low
Sterile plants	Many	Few

rufipogon Griff.) of this taxon show a wide variation in the breeding system ranging from perennial to annual habit, and differ in many life history traits. Correlation study of these various traits revealed that they were intercorrelated and divisible into two groups of intra-positive and inter-negative correlations. This is due to the differentiation of Asian wild rice populations into two types having contrasting character assortment, i.e., the annual type characterized by high seed productivity, high reproductive allocation, high seed dispersing ability, early flowering, high selfing rate, short stature, tolerance to various adverse conditions, and the perennial type having opposite characteristics (Sano and Morishima 1982).

O. perennis is distributed in swampy lowlands, ponds, depressions, roadside ditches, etc. In Asia, perennial populations are generally found in deep swamps which are relatively stable, while annual populations are in shallower, temporary swamps which are parched in the dry season and frequently disturbed by man and cattle. Thus, the habitats of the two types differ in water regimes and in the grade of disturbance. Further, perennial wild rices tended to coexist with perennial plant species and annual wild-rices with annual species. These observations suggest that the two types have different niches. The association found between life history traits and habitat conditions can be explained as the outcome of r- and K-selection (MacArthur and Wilson 1967). K-selected perennial types show greater vegetative allocation which help them to maintain themselves in stable and crowded habitats. In contrast, r-selected annual types have higher reproductive allocation and produce many seeds. The habitat conditions, life-history traits, and population structure found in the perennial and annual types of Asian wild rice are summarized in Table 1.

A perennial-annual continuum is found among populations. Most characters discriminating between the perennial and annual types show a continuous variation with many intergrades. Out of the characters so far examined, two showed a discontinuity of variation between the perennial and annual types. Annual types generally have earlier heading dates and accordingly longer critical day-length for flower initiation than the perennial types growing in the same latitude. The critical day-length of different wild-rice strains showed the two parallel latitudinal clines represented by the perennial and annual types, respectively, the intermediate types being allocated to either of them (Morishima et al. 1961). A trend to discontinuous variation was also observed in reproductive allocation as shown by the proportion of seed weight to total plant weight, although perenniality as shown by regenerating ability of stem segments showed a continuous variation. Accordingly, most of morphologically intermediate populations were classified into two types, one allocating more energy to seed production (annual type) and the other allocating more energy to vegetative organs (perennial type) (Sano et al. 1982, Fig.1; Morishima et al. 1984, Fig.1)

Low frequency of intermediate types in reproductive allocation suggests that those populations tend to shift to either the perennial or the annual type unless disruptive or diversifying selection is strong. The dominance of wild rice in the community was lower in the intermediate type than in the perennial and annual types. This might indicate ecological instability of the intermediate type in natural state. The intermediate populations, however, are considered to have a high evolutionary potentiality as discussed later.

CONSIDERATION ON THE GENETIC BASIS OF DIFFERENTIATION INTO PERENNIAL AND ANNUAL TYPES

Recently, there has been many studies on the evolution of life history traits. Much of the interest in this field was focused on the relationship of combination of life history traits with habitat condition and its significance in determining the amount of genetic variability in natural populations. However, the genetic control of life history traits and the coherence of relevant genes remained almost unexploited since the pioneering work by Clausen, Hiesey and coworkers (1958, 1960). In the Asian wild rice, various traits characterizing the perennial and annual types showed a continuous variation among natural populations, as well as in the F_2 populations between the two types. Thus, these traits seem to be governed by multiple genes, which might be distributed on the different chromosomes. Without precise experiment, it is difficult to identify major genes controlling these characters, if any. A selection experiment was carried out in an attempt to obtain an insight into the genetic basis of life-history strategy, as follows.

Two hybrid populations, one between an Asian wild-annual and a wild-perennial type (W106 X W1294, Cross I) and the other between an Asian cultivated and a wild-perennial type (T65 X W120, Cross II), were propagated in bulk in our experimental field. In F_3, the seeds were collected by two different methods; one from naturally shed seeds on the ground and the other from the seeds remaining on the panicles. The same procedure was repeated in each group in subsequent generations until F_6. This resulted in selection for early and late flowering groups in Cross I in which the progeny uniformly had a high degree of seed shedding but segregated for flowering time. In Cross II where degree of seed shedding segregated in a wide range, selection for shedding and non-shedding plants resulted. In F_3 and F_6 generations, various metric characters and frequencies of alleles at two independent isozyme loci, *Pox*-1 and *Acp*-1, were examined.

In F_3, we failed to find out significant correlations between characters and allozymes, except for a loose association of earliness with *Pox*-1 in Cross I. In F_6, some correlated responses to selection and significant associations between characters and allozymes were detected. As shown in Table 2, in Cross I, selection for late flowering was effective in decreasing flowering plants under a short-day (ca. 12 hours) treatment and correlated responses were found in many other characters. Namely, the early flowering group (seeds from the ground) had shorter awns, taller plant height, greater values of seed productivity components, higher reproductive effort, lower regenerating ability, shorter anthers, less pronounced seed dormancy than the late-flowering group (seeds from the panicles). This pattern of character association is consistent with that found between the perennial and annual types except for awn length, plant height and seed dormancy. In Cross II where selection was directed to shedding *vs*. non-shedding group, correlated response was significant only in awn length and seed dormancy.

Table 2. Comparisons of character measurements between differently selected groups (S, R)[a], and between groups homozygous for Pox-1 (2A,4A) of F_6 derived from two hybrid populations

Character	Cross I[b]		Cross II[b]		Cross I		Cross II	
	S	R	S	R	2A	4A	2A	4A
	(44)	(53)	(83)	(89)	(59)	(30)	(76)	(45)
% flowering plants	99	41	99	99	86	59	100	100
Shedding degree (index)	4.5	4.7	4.4	2.4	4.5	4.7	2.5	2.9
Awn length (cm)	5.9	7.2	5.1	4.2	6.5	6.7	4.5	4.8
Culm length (cm)	102	84	108	107	89	87	108	107
Panicle length (cm)	10.6	8.6	12.0	12.1	8.7	10.3	11.9	13.5
Panicle No.	7.5	7.7	6.6	6.1	7.9	6.5	6.3	6.2
Spikelet No./unit panicle length	.33	.23	.32	.36	.29	.24	.33	.33
Spikelet No./plant	291	163	223	252	244	201	248	207
100 grain weight (g)	1.98	1.68	1.90	1.82	1.87	1.64	1.88	1.71
Panicle weight / plant (g)	56	22	44	45	43	19	48	37
Reproductive allocation (%)	23.2	8.9	20.9	20.6	18.3	10.0	23.8	17.4
Regenerating ability (index)	0.9	1.7	1.5	1.3	1.2	1.8	1.4	1.3
Anther length (mm)	3.2	3.9	3.5	3.6	3.3	3.9	3.5	3.8
Dormancy index	150	224	129	85	169	197	97	113

a S: Naturally shed seeds were collected.
R: Seeds remaining on panicles were collected.
b Cross I: W106 x W1294, Cross II: T65 x W120
c Number of plants observed.
[dashed box]: Directly selected; [double box] [single box] ; Difference between two groups is significant at 1% and 5% level, respectively.

The parents of these crosses carried different codominant alleles at Pox-1 (2A and 4A) and Acp-1 (-4 or +4 and +9). At the Pox-1 locus, the frequency of 2A increased during F_3 to F_6 in each selection group except for late-selected group (Cross I). The F_6 plants homozygous for 2A/2A and 4A/4A are compared with regard to various character measurements in Table 2. In both crosses, 2A/2A plants tended to have greater values of seed productivity components, higher reproductive allocation, shorter anther length, less pronounced seed dormancy, and in Cross I, they had higher proportion of flowering plants and lower regenerating ability of stem segments than 4A/4A plants, although not all the differences were statistically significant. This is consistent with the trend of character association found between the perennial and annual types except for seed dormancy. At the Acp-1 locus, allelic frequencies did not change in a certain direction. No particular trend of association of Acp-1 alleles with characters was found in Cross I; in Cross II awn length and spikelet number per unit length of panicle significantly differed between plants having alleles +4 and +9.

From these observations, the following points can be suggested: a) One of major genes controlling flowering time or photoperiod sensitivity differing between the perennial and annual types is linked with Pox-1 locus. b) There could be a key factor governing seed productivity and reproductive effort which is located on a chromosomal segment marked by Pox-1. c) Seed shedding is partly correlated with awn length and seed dormancy, developmentally or by linkage, but independent from seed productivity genetically.

In a multilocus comparison of isozymes, the perennial and annual types of the Asian wild rice could not be distinctly classified (Second 1984). Pox-1 alleles, however, were found to be differently distributed between the two types. As shown in Table 3, perennial populations were highly polymorphic with alleles, OC, 2A, and 4A, while annual as well as intermediate perennial-annual populations were fixed for 2A (OC in a few cases). At Acp-1, another locus extensively assayed, perennial populations were polymorphic similarly as at Pox-1, while annual populations were fixed for any of the multiple alleles. Differential distribution of 2A allele was also found between adjacent sub-populations within population observed at two sites in India and Thailand, where sub-populations tended to be differentiated into the perennial and annual types in response to environmental heterogeneity within the site (Morishima *et al*. 1984). Plants of the sub-populations showing a high reproductive allocation were fixed for 2A, while those showing lower reproductive allocation were polymorphic (Table 4).

Table 3. Frequencies of *Pox*-1^{2A} in natural population of Asian wild rice

Type	Frequency of *Pox*-1^{2A} (%) 10	20	30	40	50	60	70	80	90	100	No. of popul.
Perennial	9	1	2	1	3	2	1	2		4	25
Intermed.	1			1	1				1	6	10
Annual						1		1		13	15

Table 4. Comparison of frequency of *Pox*-1^{2A} between perennial and annual sub-populations growing on the same site

Population	Sub-population	No. of plants	Reproductive allocation(%)	Freq.of *Pox*-1^{2A}
Bhubaneswar, India (Pond)	Periphery	32	51.5 + 6.0	1.00
	Center	39	19.5 + 13.3	0.51
Chiangrai, Thailand (Ditch)	A	11	14.7 + 13.0	0.95
	B	30	43.8 + 7.9	1.00
	C	22	20.4 + 8.1	0.77

It can be inferred that the primitive type of *O. perennis* was a perennial type. The differentiation of annual types might have followed a gradual process as in many other species, which moulded the combination of genes to adapt the plants to drier and more disturbed habitats. In general, non-random associations found among many traits and gene loci can be attributed to coadaptation through selection, chromosomal linkage, coherence by inbreeding, pleiotropy, etc. (Hedrick *et al*. 1978). Our present observations suggest that the linkage between gene for earliness and that for seed productivity must have played a functional role forming an adaptive gene block (Grant 1975) in the differentiation of the Asian wild rice. Selection and fixation of this chromosomal segment can bring about effective and constant seed productivity which meets the requirement for the annuals. Further, many other genes and regulatiang systems would have accumulated to establish successful annuals.

STRUCTURE AND DYNAMICS OF THE PERENNIAL AND ANNUAL POPULATIONS

Precise demographic data have not been available so far for natural populations of wild rice. It was confirmed, however, that typical annual populations propagate only by seeds, perennial populations only by ratoons, and the intermediate populations by both seeds and ratoons (Sano *et al*. 1980). This difference in propagating habit associated with a difference in pollinating system (perennials have higher outcrossing rate than annuals) brings about distinct difference in the genetic structure of populations.

Genetic diversity was assessed on the basis of metric characters and allozymes using the first generation plants propagated from seeds collected from natural populations. As given in Table 5, perennial populations were highly polymorphic as compared with annual populations, and had higher frequency of heterozygote and lower fixation index. On the other hand, inter-populational diversity was much greater among annual than among perennial populations. Intermediate populations were in an intermediate situation in overall variability. According to our field observation, they seemed to be heterogeneous and consisted of various types rather than a uniform population of intermediate plants. In such populations, intra-populational differentiation into the perennial and annual types can occur due to a difference in water regime and degree of habitat disturbance within site (Morishima *et al*. 1984). The habitats of the intermediate perennial-annual types are characterized by stronger disturbance, higher species diversity and lower dominance of wild rice than either of the typical perennial or annual types.

Table 5. Genetic diversity between and within populations and other parameters estimated for Asian wild rice

Type	Average for *Pox*-1 and *Acp*-1					Average for 6 characters		
	No.of popul.	H_S	G_{ST}	Fixation index	Hetero-zygote(%)	No.of popul.	C_S	C_{ST}
Perennial	24	.290	.532	.083	.230	9	114	33
Intermediate	12	.117	.688	.427	.091	7	7	63
Annual	14	.092	.670	.914	.014	10	48	81

a: H_S and G_{ST} stand for average gene diversity within population, and relative magnitude of gene diferentiation among populations, respectively (Nei 1975).
b: C_S and C_{ST} stand for within-population genetic variance shown by $\overline{\ln X}$ x 1000, and relative magnitude of between-populational genetic variance to total genetic variance (%), respectively.

Since the perennial and annual populations markedly differ in genetic structure, they respond to environments differently. The annual types are *r*-strategist and seem to be fugitive. We observed that at Saraburi in Thailand a huge area previously planted with cultivated rice was occupied by an annual type of wild rice within three years since abandoned. The area is now used as the army camp, and destruction of vegetation due to burning and digging by army caterpillars must have provided a suitable habitat for the annual wild-rice. Another roadside population we are monitoring in the same district was a nearly pure stand of annual wild rice when first observed in 1974, but declined quickly year after year, losing its genetic variability (Fig. 1). It was inferred that the colonizers invaded a bare site made by road construction, and with relaxation of disturbance they were replaced by perennial competitors. According to our recent observation, the remaining plants are not typical wild rice but seemingly hybrids which absorbed genes from cultivars grown nearby. Those plants showed more vigorous growth than the earlier generation plants when tested under a uniform condition.

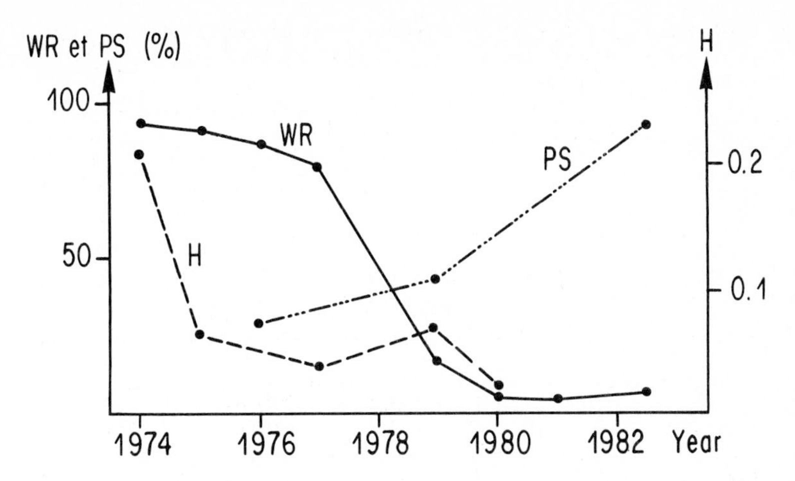

Fig. 1 Changes observed in an annual-type population of wild rice growing at Saraburi, Thailand.
WR: Wild rice cover; PS: Proportion of perennial species in the accompanying plants; H: Average gene diversity based on six isozyme loci.

Perennial populations are usually found in water-logged sites which would be a stable habitat. As long as the conditions remain unchanged, they may persist in the habitat. Since they propagate mainly asexually, new genets are not many, although they preserve a large amount of genetic variability mainly in heterozygous state (Table 5). Nowadays, original habitats of the perennial types are threatened by various unfavorable conditions, such as urbanization, improved agricultural management and disturbance. It is becoming difficult to find a true perennial-type wild rice not introgressed by cultivars.

Threatened by environmental destruction, many perennial populations are disappearing or shifting to an intermediate type which has an adaptability to cope with the new environments. Although most of intermediate populations prevailing at present seem to be introgressed by cultivars, transitional situations from perennial to annual type could have existed in the habitats which were subjected to drought and disturbance. Such intermediate populations could have released a large amount of genetic variability by seed propagation, and some plants could have been domesticated by man.

REFERENCES

Clausen J, Hiesey WM (1958) Experimental studies on the nature of the species IV. Genetic structure of ecological races. Carnegie Inst Washington Publ 615, Washington DC

Clausen J, Hiesey WM (1960) The balance between coherence and variation in evolution. Proc Nat Acad Sci 46: 494-506.

Grant V (1975) Genetics of flowering plants. Columbia Univ Press, New York

Hedrick P, Jain S, Holden L (1978) Multilocus systems in evolution. Evolutionary Biology 11:101-184

MacArthur RH, Wilson EO (1967) The theory of island biogeography. Princeton Univ Press, Princeton

Morishima H, Oka HI, Chang WT (1961) Directions of differentiation in populations of wild rice, *Oryza perennis* and *O. sativa* f. *spontanea*. Evolution 15:326-339

Morishima H, Sano Y, Oka HI (1984) Differentiation of perennial and annual types due to habitat conditions in the wild rice *Oryza perennis*. Pl Syst Evol 114:119-135.

Nei M (1975) Molecular population genetics and evolution. North-Holland Publ Comp, Amsterdam Oxford

Sano Y, Morishima H, Oka HI (1980) Intermediate perennial-annual populations of *Oryza perennis* found in Thailand and their evolutionary significance. Bot Mag Tokyo 193:291-305

Sano Y, Morishima H (1982) Variations in resource allocation and adaptive strategy of a wild rice, *Oryza perennis* Moench. Bot Gaz 143:518-523

Second G (1984) Evolutionary relationships in the Sativa group of *Oryza* based on isozyme data. Genetique, Selection, Evolution (in press)

Genetic and environmental constraints to selection response for juvenile characters in *Geranium carolinianum*

Deborah Ann Roach

Department of Botany
Duke University
Durham, N.C. 27706 USA

ABSTRACT

This paper summarizes studies on components to variation at the juvenile stage of the life cycle, and the phenotypic and genetic correlations between juvenile and adult characters in *Geranium carolinianum*. These factors are considered with respect to their effect on the potential response to natural selection for juvenile characters. Although differential survival and growth rates of individuals at the early stages of the life cycle are an important determinant of species distributions and of general life history patterns, these studies with *G. carolinianum* suggest that selection for juvenile characters may be slow, despite genetic variation for fitness characters. Juvenile traits are subject to large environmental effects, and juvenile performance (e.g. size) often shows a negative genetic correlation with adult fitness components.

INTRODUCTION

Understanding the factors which contribute to the differential success of individuals at early stages of the life cycle is important for both ecological and evolutionary reasons. Ecologically, the differences in survival and growth rates between individuals at these early life stages are important in determining the dynamics of adult populations (Sagar and Harper 1961, Harper and White 1974, Grubb 1977). And, the dynamics of these early stages must be studied directly because the events that determine whether or not a seedling becomes established are usually not evident after the plant has matured (Harper 1977). From an evolutionary perspective, one of the important issues is to explain the evolution and diversity of life history patterns. Theoretical studies suggest that mortality patterns during the pre-reproductive

NATO ASI Series, Vol. G5
Genetic Differentiation and Dispersal in Plants
Edited by P. Jacquard et al.

phase of the life cycle are an important determinant of life history patterns (Bell 1976, Stearns 1976), but there have been no detailed studies on genetic variation at the juvenile stage in natural populations. In order to fully understand life history evolution, we first need to know how much genetic variation is present at these early life stages upon which selection can act. Then, we can ask whether there are any inherent constraints to selection for traits at these early stages and thus, to life history evolution in general. The present paper summarizes the results from studies on the early life stages of _Geranium carolinianum_ (L.). These studies were designed to evaluate the factors which contribute to phenotypic variation at the early stages of the life cycle, and to consider the potential response to selection for traits expressed during this time.

STUDY ORGANISM

G. carolinianum is an annual plant which grows in fields and waste places throughout much of the United States. In North Carolina, it germinates in October and persists as a rosette, close to the ground, throughout the winter. In the present paper the rosette morphological type will be referred to as the "juvenile", or pre-reproductive, phase. In April, the stem of _G. carolinianum_ elongates, and the plant "bolts" to an erect "adult" form approximately 20 cm in height. It then flowers in May and sets seed and dies by early June.

SELECTION AND JUVENILE CHARACTERS

The questions concerning selection for juvenile characters have been derived primarily from descriptive demographic studies (for a review see Cook 1979). The general conclusion has been that the juvenile stage is a time of high vulnerability, often characterized by very high mortality rates. Given this consistently high rate of mortality, one can ask the question: Why has selection not acted on traits which would decrease the probability of mortality at the early life stages? Or, why has it not acted to shorten the vulnerable juvenile stage? (Williams 1966, Orians and Janzen 1974, Cook 1979). Even for species which do not have high mortality during the juvenile stage, selection on juvenile traits would be expected if there is a relationship between early juvenile size and adult survival or reproductive output (see e.g. Harper 1977). G. carolinianum has, for an annual species, a relatively high survival rate of 67% after germination (Fig. 1). There is a positive phenotypic correlation between juvenile leaf area and adult size (+0.56, $p<0.001$), and between juvenile leaf area and adult fecundity (+0.44, $p<0.001$; Roach 1984). Given this positive relationship, selection for increased juvenile size would be expected. However, some constraints to selection at the early life stages may limit the increase in plant size. In order to understand the potential response to selection for juvenile

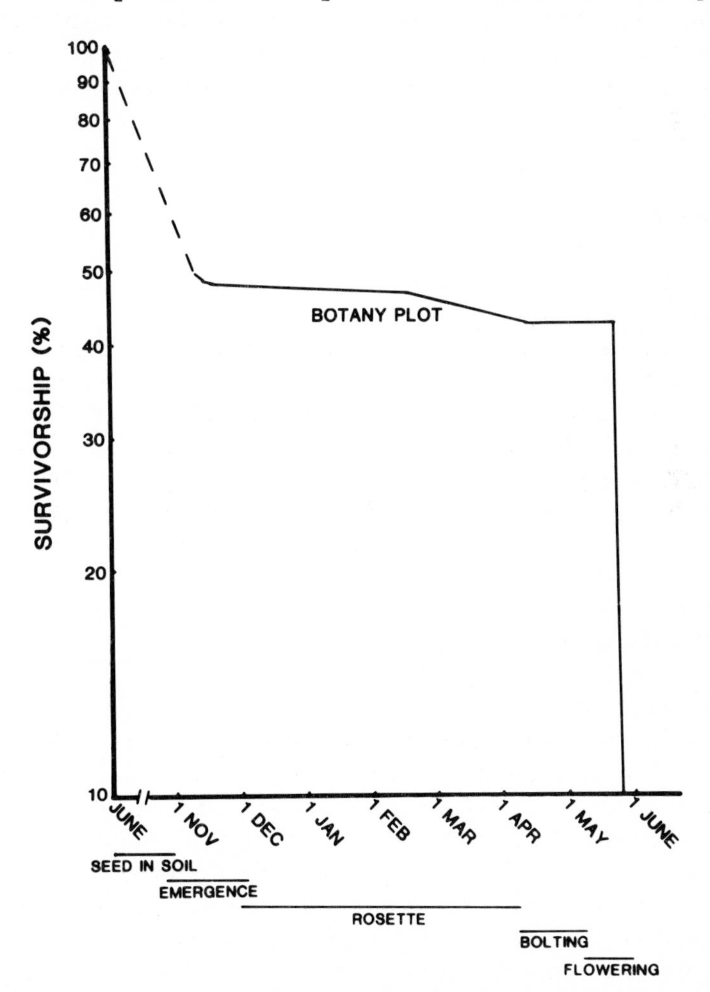

Fig. 1. Survivorship curve of G. carolinianum. The juvenile stage is defined as the time from seedling emergence through the end of the rosette stage, the adult stage is from bolting through flowering and seed dispersal. From Roach (1984).

size, it is important to examine not only the genetic and environmental components to variation at this stage, but also the genetic correlations between juvenile and adult traits (Lande 1982).

ENVIRONMENTAL COMPONENTS TO VARIATION

Phenotypic variation at the early stages of the life cycle may be due to several sources. The first, and most obvious source is the environment. Experimental studies have demonstrated that the environment immediately surrounding a seed or seedling is a critical determinant in the probability of seed germination as well as the subsequent growth rate of a seedling (c.f. Harper et al 1965). The present study revealed that the performance of G. carolinianum is dependent on the specific location in which an individual finds itself. In an experiment in which seeds were planted at random with respect to genetic family, a small area, 7 x 9 m, was very heterogeneous for favorable germination sites (Fig. 2).

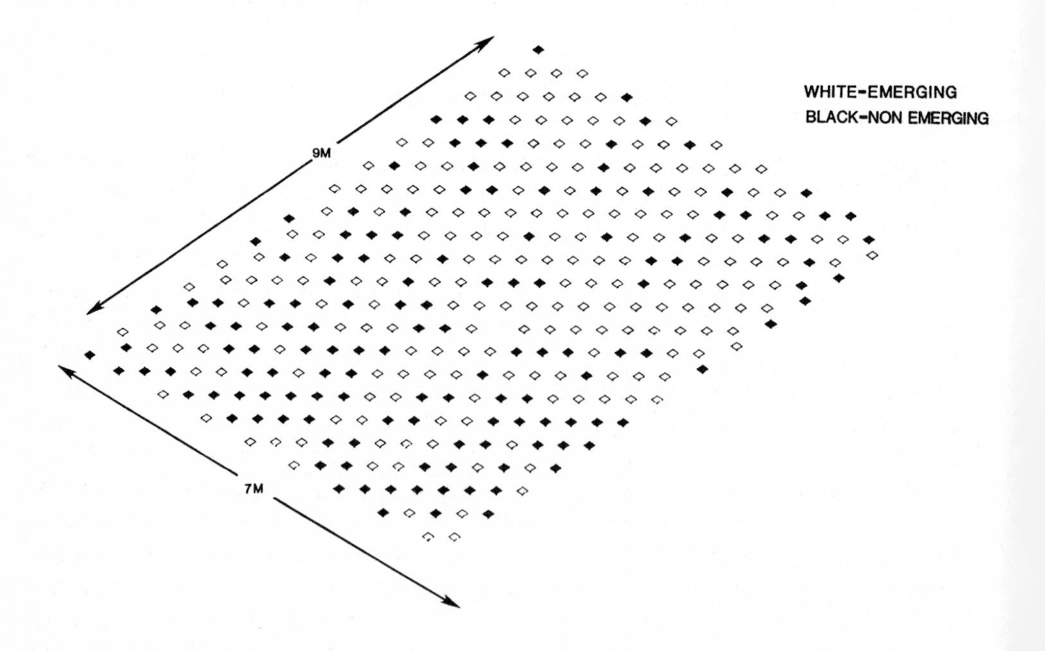

Fig. 2. Locations of all planted seed within a 7 x 9 m plot showing heterogeneity for favorable emergence sites. From Roach (1984).

(χ^2=22.04, df=8, p<0.01, for heterogeneity test of numbers of emerging vs. non-emerging in 9 subplots of the 7 x 9 m area). In a second experiment, there was a genetic response to environmental heterogeneity with time of emergence (Roach 1984). In other words, the time of emergence for individuals in a family was dependent on where they were located within the site. This type of genetic response to environmental heterogeneity could result in multiple niche selection where different genotypes are favored in different areas of the field (Hamrick and Allard 1972). Such specialization however would be difficult in the face of gene flow. There could also be a plastic response to multiple niche selection and the creation of generalized genotypes. This could result in a constraint to selection in two ways. First, the production of a generalized genotype may be at a cost of performance in any given habitat. And secondly, non-environmental differences between families may become obscured. The differences between families would also be reduced if there were no multiple niche selection. A large environmental component to the phenotypic variation will reduce the heritability for these characters and dictate that the response to selection would be very slow.

GENETIC AND EXTRA PARENTAL CONTRIBUTIONS TO VARIATION

In addition to the direct effect of the immediate environment on the juvenile individuals, the environment of the maternal parent may influence the juvenile phenotype of the offspring. Maternal effects in plants have their major effect through the quality and quantity of seed reserves, and they have been shown to have a significant influence on early seedling growth rates in agricultural species (eg. Edwards and Emara 1969, Hayward and Nsowah 1969). In natural populations, small changes in seed weight have been shown to have a significant effect on rates of germination and seedling characters (Black 1958, Schaal 1980). Studies with G. carolinianum are one of the first direct examinations of maternal effects in natural populations (Roach 1984). In that study nuclear and extra-nuclear maternal and paternal contributions were distinguished for four juvenile characters. This was done using a chain-block mating design, with the resulting seeds then planted back into the population from which the seeds had originally been collected. The results showed that while seed weight did have a genetic component, there was also a strong extra maternal contribution to this variation. This extra maternal contribution could be either environmental or genetic. If the maternal effect is environmental, then like other environmental effects, it would act as a factor slowing the response to selection on seed size. Such environmental effects have been shown in a number of species under controlled environmental conditions (Gutterman 1980). If during the time of seed maturation *G. carolinianum* is sensitive to the environment on the micro-scale observed in the previously mentioned study (Fig. 2), then clearly, seed size would have a large environmental component, and selection for this trait would at best be very slow. If on the other hand, this maternal effect is genetic, then it would increase the amount of genetic variation upon which selection can act. There is a positive phenotypic correlation between seed size and cotyledon size (+0.36, $p<0.001$), which suggests that selection would favor increased seed size. There may however be an additional constraint to selection for this character: selection for increased seed size may be at the cost of decreasing seed number. The existence of such a trade-off has not been measured in *G. carolinianum*, but it has been found frequently in between species studies (e.g. Werner and Platt 1976, Wilbur 1976).

Further analysis of the components to variation at the juvenile stage suggested an extra paternal contribution to cotyledon size, but no other direct genetic effects (Roach 1984). Extra-paternal effects are caused by cytoplasmic inheritance of organelles. *Geranium* spp. and *Pelargonium* (Geraniaceae) are the classic examples of biparental inheritance of plastid characteristics (Tilney-Bassett 1967). Thus, any selection for increased cotyledon size in *G. carolinianum* would correspond to indirect selection for paternal cytoplasmic effects, which may or may not be entirely genetic.

At the late juvenile stage, just prior to the change in morphology to the erect adult form, the components to variation in leaf area and rosette area were estimated. For both of these traits there was evidence for a small amount of genetic variation but little maternal or paternal contributions. Similar results have been found in other studies with agricultural species where the extra parental effects were restricted to only the very earliest stages of the life cycle (e.g. Lewis and Garcia 1979, Kotecha 1981).

GENETIC CORRELATIONS

Selection for juvenile characters may also be limited by genetic correlations between juvenile and adult characteristics. If there is a genetic correlation between two characters, caused either by linkage disequilibrium or pleiotropy, then selection for one of the two characters will result in an indirect response in the second character (Lande 1980). A classic example of negatively correlated characters is seed size and seed number: selection for increased seed size is often at the expense of decreasing seed number. This concept of constraints has been expanded to include the correlations between traits expressed at different stages of the life cycle in a study with G. carolinianum (Roach 1984). The results of this study showed that there was a negative genetic correlation between juvenile leaf area and adult

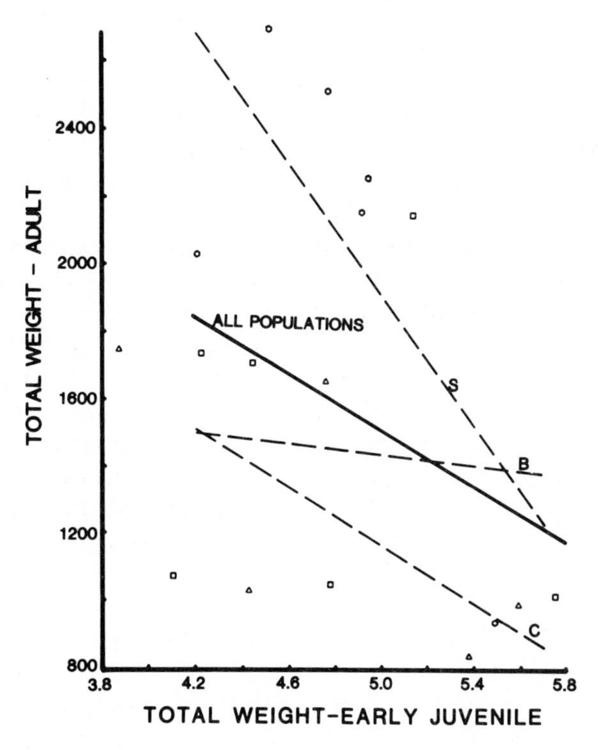

Fig. 3. Plot of family mean adult total weight vs. early juvenile total weight. The dotted lines represent the regressions on families from three populations (squares: population B, triangles: population C, and circles: population S). The solid line is the z-transformed average over the three populations.

fecundity (-0.49 ±0.08). The consequences of this are that selection for increased juvenile size will be at the expense of reducing fecundity. In other words, there will be balancing selection where individuals in one family will have a higher relative fitness at the juvenile stage, and individuals in another family will have a higher relative fitness at the adult stage. This is clearly seen in a plot of family mean adult vs. juvenile total weight (Fig. 3). Other studies have also shown that the intensity and the direction of selection may change during the life cycle (Ford 1975, Clegg, Kahler and Allard 1978, McGraw and Antonovics 1983), but the study with _G. carolinianum_ is the only one which directly measures genetic correlations in a natural population. The trade-offs between selection at different stages will result in little or no net progress in selection for these traits.

The correlation structure also offers further evidence for a large environmental component to the covariation between juvenile size and adult size and fecundity characters. A phenotypic correlation between two characters contains both a genetic and an environmental component. A positive phenotypic correlation and a negative genetic correlation suggest that the environmental correlation must be large and positive.

SUMMARY

Given the positive phenotypic correlation between juvenile size and adult size and fecundity in _G. carolinianum_, we would expect selection for increased juvenile size. The standardized regression of juvenile leaf size and fecundity has a slope of 0.53. This slope may be used as the selection differential (S), and if we assume a heritability (h^2) of only 10%, then we can calculate the expected response to selection (R) (Falconer 1981):

$$R=h^2S=(0.10)(0.53)=0.05$$

Thus, the predicted response to selection to increase juvenile leaf area by 1 mm would be a 0.05 mm increase per generation. Over several generations, juvenile size should be substantially increased. It seems, however, that there may be a number of different constraints to selection for juvenile characters in _G. carolinianum_. Segregating genetic variation is indispensable if selection is to operate. The results of the studies reported here showed some genetic variation for seed size, rosette size, and late juvenile leaf area. There must also be some genetic variation, albeit small, in early juvenile traits because there were measureable genetic correlations between leaf area and total dry weight and adult fecundity at this stage. The amount of genetic variation in these characters may in large measure be obscured however by environmental heterogeneity. G. carolinianum is a weedy annual plant, and annuals as a rule have a high level of phenotypic plasticity (Bradshaw 1965). This plasticity will only act to mask the differences between genotypes and to slow the response to selection for these characters.

Environmental heterogeneity is but one of the possible constraints to selection for juvenile characters. Negative pleiotropic effects were first considered by evolutionary biologists as a mechanism to explain the evolution of the deleterious phenomenon of senescence (Williams 1957). According to this theory, mutations with deleterious effects at the end of the life cycle will spread in the population because of their pleiotropic positive effects on the fitness of the organism earlier in the life cycle. A similar argument might be used for the evolution of juvenile traits. The negative genetic correlations between juvenile size and major fitness components at the adult stage observed in G. carolinianum suggest that traits which are relatively disadvantageous to juvenile individuals may be maintained in the population. A balance between components of fitness may result in little net progress of selection for increased size at the juvenile stage because this would be at a cost of reducing adult fecundity.

In conclusion, there are both environmental and genetic constraints to selection response for juvenile characters. The environmental constraints may be either a direct response to micro-site heterogeneity, or indirect maternal effects. The genetic constraints may be due to the genetic correlations between seed size and seed number, or between juvenile plant size and adult fitness components. The results of these studies with G. carolinianum demonstrate that no study of life history evolution can be complete without a clear partitioning of environmental and genetic effects at all stages of the life cycle, and an examination of the phenotypic and genetic correlations between traits expressed at the different stages.

REFERENCES

BELL, G., 1976. - On breeding more than once. *Amer. Nat.*, 110, 57-77.

BLACK, J.N., 1958. - Competition between plants of different initial seed sizes in swards of subterranean clover (*Trifolium subterraneum* L.) with particular reference to leaf area and the light microclimate. *Austr. Jour. Agric. Res.*, 9, 299-318.

BRADSHAW, A.D., 1965. - Evolutionary significance of phenotypic plasticity in plants. *Adv. Genet.*, 13, 115-153.

CLEGG, M.T., KAHLER, A.L. & ALLARD, R.W., 1978. - Estimation of life cycle selection in an experimental plant population. *Genet.*, 89, 765-792.

COOK, R.E., 1979. - Patterns of juvenile mortality and recruitment in plants. In *Topics in Plant Population Biology.*, (Solbrig, O.T., Jain, S., Johnson, G.B., & Raven, P.H., Eds.). Columbia Univ., New York, 2, 207-231.

EDWARDS, K.J.R. & EMARA, Y.A., 1969. - Variation in plant development within a population of *Lolium multiflorum*. *Hered.*, 25, 179-194.

FALCONER, D.S., 1981. - *Introduction To Quantitative Genetics*. Longman, New York.

FORD, E.B., 1975. - *Ecological Genetics*. Chapman and Hall, London, 442 pp.

GRUBB, P.J., 1977. - The maintenance of species-richness in plant communities: the importance of the regeneration niche. *Biol. Rev.*, 52, 107-145.

GUTTERMAN, Y., 1980. - Influences on seed germinability: phenotypic maternal effects during seed maturation. *Israel Jour. Bot.*, 29, 105-117.

HAMRICK. J.L. & ALLARD, R.W., 1972. - Microgeographical variation in allozyme frequencies in *Avena barbata*. *Proc. Nat. Acad. Sci., U.S.A.*, 69, 2100-2104.

HARPER, J.L., 1977. - *Population Biology of Plants.* Academic Press, London, 892 pp.

HARPER, J.L., WILLIAMS, J.T. & SAGAR, G.R., 1965. - The behavior of seeds in soil. I. The heterogenetiy of soil surfaces and its role in determining the establishment of plants from seed. *Jour. Ecol.*, 53, 273-286.

HARPER, J.L. & WHITE, J., 1974. - Demography of plants. *Ann. Rev. Ecol. and Syst.*, 5, 419-463.

HAYWARD, M.D. & NSOWAH, G.F., 1969. - The genetic organisation of natural populations of *Lolium perenne.* IV. Variation within populations. *Hered.* 24, 521-528.

KOTECHA, A., 1981. - Inheritance of seed yield and its components in Safflower. *Canad. Jour. Genet. Cytol.*, 23, 111-118.

LANDE, R., 1980. - The genetic covariance between characters maintained by pleiotropic mutations. *Genetics.* 94, 203-215.

LANDE, R., 1982. - A quantitative genetic theory of life history evolution. *Ecology*, 63, 607-615.

LEWIS, E.J. & GARCIA, J.A., 1979. - The effect of seed weight and coleoptile tiller development on seedling vigor in tall fescue *Festuca-Arundinacea*. *Euphytica*, 28,393-402.

McGRAW, J.B. & ANTONOVICS, J., 1983. - Experimental ecology of *Dryas octopetala* ecotypes. I. Ecotypic differentiation and life cycle stages of selection. *Jour. Ecol.*, 71, 879-898.

ORIANS, G.H. & JANZEN, D.H., 1974. - Why are embryos so tasty? *Amer. Nat.*, 108, 581-592.

ROACH, D.A., 1984. - Ecological genetics of life history characteristics in *Geranium carolinianum*. Ph.D. Thesis, Duke University.

SAGAR, G.R. & HARPER, J.L., 1961. - Controlled interference with natural populations of *Plantago lanceolata, P. major*, and *P. media. Weed Res.* 1, 163-176.

SCHAAL, B.A., 1980. - Reproductive capacity and seed size in *Lupinus texensis. Amer. Jour. Bot.* 67, 703-709.

STEARNS, S.C., 1976. - Life-history tactics: A review of the ideas. *Quart. Rev. Biol.* 51, 3-47.

TILNEY-BASSETT, R.A.E., 1967. - The inheritance and genetic autonomy of plastids. In *The Plastids: Their Chemistry, Structure, Growth and Inheritance.* (Kirk, J.T.O. and Tilney-Bassett, R.A.E., Eds.), W.H.Freeman, London.

WERNER, P.A. & PLATT, P.J., 1976. - Ecological relationships of co-occurring goldenrods (*Solidago: Compositae*). *Amer. Nat.*, 110, 959-971.

WILBUR, H.M., 1976. - Life history evolution of seven milkweeds of the genus *Asclepias*. *Jour. Ecol.*, 64, 223-240.

WILLIAMS, G.C., 1957. - Pleiotropy, natural selection, and evolution of senescence. *Evol.*, 11, 398-411.

WILLIAMS, G.C., 1966. - *Adaptation and Natural Selection: A critique of some current evolutionary thought.* Princeton, Princeton, N.J.. 307 pp.

The effects of neighbors as environments: characterisation of the competitive performance of *Danthonia spicata* genotypes

Steven E. Kelley

Department of Botany, Duke University, Durham, North Carolina, 27706

ABSTRACT

A novel phytometer technique is utilised to elucidate the nature of the interspecific competitive interactions between two co-occuring perennial grasses, *Danthonia spicata* and *Anthoxanthum odoratum*. In this paper the Finlay-Wilkinson method is employed to determine whether competing genotypes act as environments in terms of their effects on genotypic fitness. Nonconspecific competitors are found to affect genotypic fitness in a qualitatively different way than physical factors. The unpredictability inherent in competitive interactions may act as a selective force to maintain genetic variability and sexual reproduction within the population.

INTRODUCTION

Both genotypic performance and fitness are clearly context-dependent. The fact that genotypes may be *differentially* affected by the influence of varying habitats is of distinct evolutionary relevance. Genotype-environment interaction indicates the potential for population differention under selection in different habitats. More importantly, genotype-environment interaction may furnish a tenable mechanism for the preservation of genetic variation within populations and thus explain the so-called paradox of variation--one of the fundamental problems confronting evolutionary biology since the 1960's (Lewontin 1974).

The elucidation of the phenomenon of genotype-environment interaction has in part been accomplished through the use of a regression technique developed initially by Yates and Cochran (1938 cited in Finlay and Wilkinson), and later refined by

NATO ASI Series, Vol. G5
Genetic Differentiation and Dispersal in Plants
Edited by P. Jacquard et al.

TABLE 1.

SPECIES	R^2	REFERENCE
Avena sativa	.63 -.99	Langer, Frey, and Bader 1979
Cicer mollervetch	.74 -.97	Townsend et al 1979
Dactylis glommerata	.71 -.97	Gray 1982
Fescue	.89 -.99	Nguyen et al 1980
Phalaris arundinacae	.89 -.98	Barker et al 1981
Pisum sativum	.81 -.95	Hobbs and Mahon 1983
Secale cereale	.40 -.90	Kaltsikes 1969
Triticum aestivum	.71 -.99	Walton 1968
	.91 -.98	Keim and Koonstad 1979

Finlay and Wilkinson (1963). According to this method each environment receives a score based upon the performance of all genotypes planted in it. The performance of a particular genotype in each environment can then be plotted against all environments. Finlay and Wilkinson (1963) plotted the performance of different varieties of barley grown in three regions over a three year period. They found that (with a log transformation) the yield of a particular variety of barley was linearly related to the environmental scores. Subsequent utilisation of the technique has proceeded primarily amongst agronomists who are concerned with the yield stability of newly-developed crop varieties in the face of fluctuating field-to-field and year-to-year conditions.

Repeated application of the Finlay-Wilkinson method has led to an appreciation that three general properties seem to characterise genotype-environment interaction. First and perhaps most striking is the finding that genotypic or varietal performance appears to be linearly related to the environmental scale even in the absence of a logarithmic transformation. Numerous studies have documented the linear relationship between genotype or varietal performance and environments for a diverse array of untransformed performance measures including mean above-ground yield (vegetative and reproductive), root yield, percentage of seedling emergence, and resistance to nematodes (see Bilbro and Ray 1976; Breese

1969; Cooper 1981; Frey 1972; Fripp and Caten 1971; Grey 1982; Jika et al 1980; Johnson and Whittington 1977; Kaltsikes 1969; Langer, Frey, and Bailey 1979; Perkins and Jinks 1968; Saeed and Francis 1983; Walton 1968; Wilcox et al 1979 for examples). Utilising the coefficient of regression as a crude measure of linearity, R^2 values in the literature typically range from .75 to .99. (Table 1, references are cited therein). Only two studies failed to demonstrate a linear relationship between genotypic or varietal performance and the environment (DePauw, Farris, and Williams 1981; Imbrie et al 1981).

A second general conclusion has been that where significant genotype-environment interaction occurs, this interaction is due primarily to differences in the slopes of the lines representing genotypic performance plotted against environments. In theory genotype-environment interaction could be attributable either to differences in the slopes of the genotypic performance lines (termed linear instability) or to differences in the degree of dispersion about those lines (termed non-linear instability). Yet when environments represent different fields or different years the bulk of the difference between genotypes can be ascribed to heterogeneity among slopes of lines. Genotypes do not consistently differ in the degree of dispersion found about the linear regression. Numerous studies find no significant interaction variance attributable to dispersion about regression lines (Baihaki, Strucker, and Lambert 1976; Hill and Samuel 1971; Johnson and Whittington 1977; Langer, Frey and Bailey 1979; Phillips, Forest, and Hayter 1979). An equal number of studies do find statistically-significant variation about the regresion lines (Breese and Hill 1973; Eberhart and Russell 1966; Frey 1972; Paroda and Hayes 1971; Perkins and Hill 1968; Saeed and Francis 1983).

The third characteristic property of the genotype-environment data is the positive correlation between the mean yield of a particular variety or cultivar and the slope of the regression line; a trade-off exists between yield and stability (Baihaki, Strucker, and Lazmber 1976; Bilbro and Ray 1976; Cooper 1981; Eberhart and Russell 1966; Fatunla and Frey 1974; Frey 1972; Fripp and Caten 1971; Gray 1982; Hill and Perkins 1969; Hill and Shepard 1971; Hobbs and Mahon 1983; Johnson and Whittington 1977; Kaltsikes 1969; Laing and Fischer 1977; Langer, Frey, and Bailey 1979; Perkins and Jinks 1968;

Phillips, Forest, and Hayter 1979; Saeed and Francis 1983; Townsend et al 1979; Walton 1969.) Varieties or cultivars which exhibit the highest mean yield seem to be especially sensitive to the particular environmental conditions under which they are grown. Studies which have failed to produce such a positive correlation are few (Ford 1981; Langer et al 1979; Nguyen et al 1980; Wilcox et al 1979).

The phenomenology of genotype-environment interaction implies that such interactions could have important evolutionary consequences. Differences between genotypes in the slopes of performance against environment lines suggest that in some environments the lines will cross; the relative ranking of genotypes would thus be dependent upon the environmental context. Sufficient environmental change over a short spatial or temporal scale would promote the maintenance of genetic variation within a population. The existence of a trade-off between mean yield and stability of performance functions as constraint to evolution. Selection for high performance in a favorable environment (growth, fecundity, survivorship) would lead to a predomination of genotypes that perform especially poorly in less-favorable environments and vice versa. In the presence of such a constraint environmental variation prevents the domination of a population by a single superior genotype.

While genotype-environment interaction is a potentially-important evolutionary mechanism, its role is restricted by the scale over which the environment changes. In the vast majority of the studies conducted to date different environments are simply geographically widely-distributed fields planted in different years. Environmental variation here is occuring on a broad temporal and spatial scale. These environments differ specifically in physical parameters such as ambient temperature, soil nutrient content, and moisture-availability. It is not immediately clear that such parameters change sufficiently over a small scale to be relevant to evolutionary processes occuring within a single population.

The importance of genotype-environment interaction to evolution within the population would greatly be increased if neighboring individuals acted as different environments. Recent evidence attests to the paramount role neighbors play in individual performance (Mack and Harper 1977) and genotypic fitness has been shown to be sensitive to both the presence and identity of neighboring species (Turkington et al 1979; Turkington and Harper 1979). The suggestion is herein made that not only do different species constitute different environments but that different genotypes of another species constitute different environments for a particular species. Given that natural plant communities are usually mixtures of species--diversity is the rule and monotony the exception--and that most populations consist of a variety of genotypes, it becomes critical to ask whether the genotypes of competing species do act as different environments in a manner analagous to varying physical parameters. Is the performance of a specific genotype linearly related to the environments represented as different competing genotypes? Are genotype-genotype interactions present amongst naturally occuring genotypes? Is genotypic interaction when present primarily attributable to differences amongst the slopes of lines of performance against environment? And finally, is there a trade-off between yield and stability in the face of competing genotypes?

In order to make claims as to the importance of genotype-genotype interactions within a natural population it is essential to demonstrate that such interactions occur between naturally-occuring genotypes--not artificially-developed varieties or cultivars. Further, it is necessary to show that such interactions occur in the context of a natural plant community and not just within the confines of unrealistic and perhaps oversimplified ecosystems such as would be found in a common garden or planted field.

This paper undertakes to examine this question of whether competing nonconspecific genotypes do act as different environments for the genotypes of a particular species in a manner analagous to varying physical parameters. The experiment described utilises a novel phytometer technique which permits the elucidation of competitive interactions between specific genotypes of two species grown in a natural plant community. The regression technique of Finlay and Wilkinson is utilised to determine whether the general phenomena discovered for crop plants grown in different fields are duplicated by competing naturally co-occuring genotypes in a natural population.

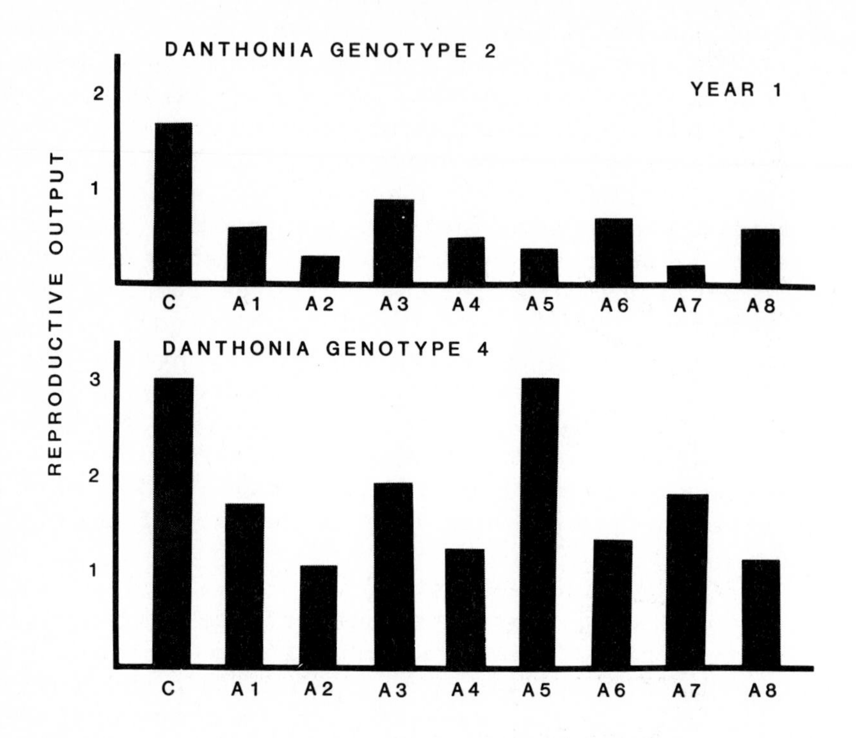

Figure 1. Reproductive output of Danthonia genotypes two and four when in competition with itself (C) and the eight Anthoxanthum genotypes (A1 - A8).

MATERIALS AND METHODS

The two species of grass utilized in this study, *Danthonia spicata* and *Anthoxanthum odoratum*, commonly co-occur in a variety of habitats in North Carolina. Both are short-lived perennial, caespitose grasses that are important components of old fields, lawns, roadsides, and woodland margines thorughout North Carolina and much of the eastern United States. Both species grow actively in early spring, flower in late spring, go dormant during the hot summer months, and both have another period of active growth in the fall when temperatures are

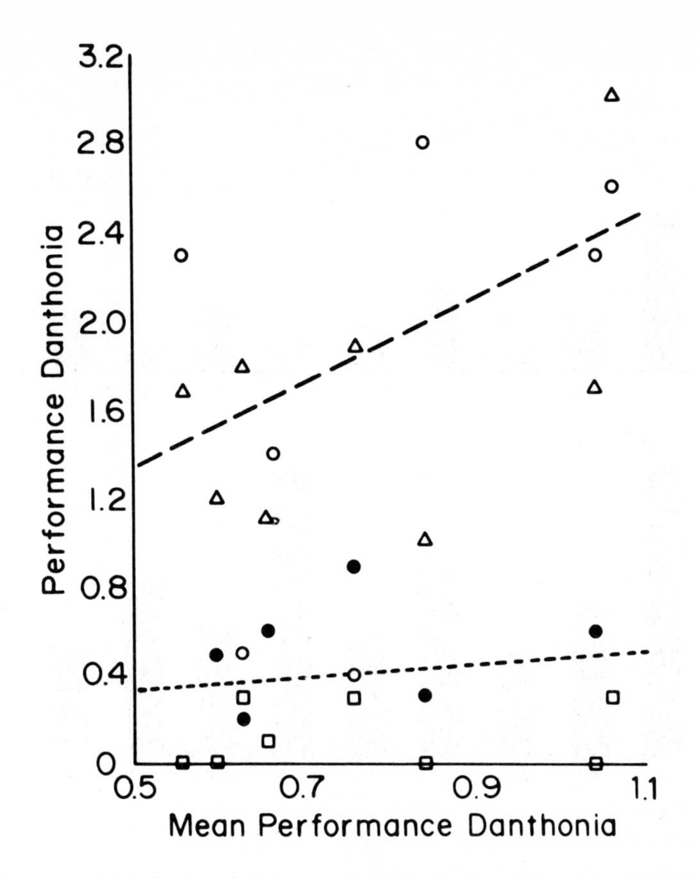

Figure 2. Reproductive output of particular Danthonia genotypes when in competition with particular Anthoxanthum genotypes plotted against the mean performance of all Danthonia genotypes competed against particular Anthoxanthum genotypes. Different symbols represent different Danthonia genotypes.

cooler. They are the two dominant spring grasses in an old field located on the Duke University campus, Durham, North Carolina, the site of this study (Fowler and Antonovics 1981).In the summer of 1981 eight individuals of each species were collected from an area of species overlap in the field. Presumably in this area, each respective population had been subjected to selective pressures through competition with the other species. Each individual was assumed to be a distinct genotype and this was later verified through comparison of electrophoretic banding patterns. Plants were grown in the greenhouse to a sufficient size and then divided into ramets each one to three tillers in size. At the same time soil plugs were taken from the area where the plants were collected. Soil plugs were individually placed in 7 x 7 cm plastic bottomless plastic pots. The pots and soil plugs were then placed and treated in a steam soil sterilizer for one hour in order to kill the living vegetation without undue disturbance of the soil structure.

Ramets of _Anthoxanthum odoratum_ and _Danthonia spicata_ were planted together in the pots in an incomplete diallel design. Eight genotypes of _A. odoratum_ were planted with each of the _D. spicata_ genotypes and each of the 64 combinations were replicated five times. Two ramets of each of the genotypes planted together in the same pot served as controls. The controls were also replicated five times for a grand total of 400 pots each containing two ramets. After a brief acclimation period of ten days the pots were brought back into the field and firmly sunk into the same holes the soil plugs originated from. The plants were watered once but no further treatments were applied. A more detailed description of the technique is presented elsewhere (Kelley and Clay 1985).

The following spring (May 1982) the plants were resurveyed. Survival, size as assessed by the number of vegetative tillers, and the number of reproductive tillers was noted. In order to control for initial size differences, two corrected response variables were analyzed. Growth was defined as the final number of tillers divided by the initial number of tillers. Reproductive output was defined as the number of inflorescences produced in the spring per initial tiller in size. A more exhaustive reportage of the data is presented elsewhere (Kelley and Clay 1985). This paper will focus exclusively on the some of the growth and fecundity data obtained for _Danthonia spicata_.

RESULTS

Danthonia spicata exhibited strong competitive interactions as assessed by both growth and fecundity. Overall significant differences were found between _Danthonia_ genotypes (Kelley and Clay 1985). Genotypic performance was dependent on the _Danthonia_ genotype, the _Anthoxanthum_ genotype, and an interaction between the two. This is illustrated in Figure 1 which shows fecundity per initial tiller in size for two Danthonia genotypes competed against the _Anthoxanthum_

Table 2.
Linear Regression of Danthonia Genotypes on Environment. Regression coefficient, F value and its associated probablility, and R-square are shown.

Growth

Genotype	R C	F Value	Pr>F	R-Square
1	.233	0.16	.70	.026
2	-.277	0.10	.76	.017
3	1.012	1.32	.29	.180
4	1.429	2.57	.16	.300
5	1.678	2.55	.16	.298
6	.346	0.13	.73	.021
7	.407	0.14	.72	.023
8	3.143	5.80	.05	.492

Fecundity

Genotype	R C	F Value	Pr>F	R-Square
1	-.097	0.09	.78	.014
2	-.115	0.06	.81	.010
3	.689	1.18	.32	.165
4	2.048	3.66	.10	.379
5	.690	1.90	.22	.240
6	.043	0.02	.89	.003
7	.850	1.20	.32	.166
8	3.884	8.17	.03	.577

genotypes and against the control. It is readily apparent that the mean performance of genotype four across all *Anthoxanthum* genotypes exceeds that of genotype two. Further, the parallelism in shape between the two curves indicates that certain *Anthoxanthum* genotypes (e.g. 3) consistently seem to have a less detrimental effect on the performance of both *Danthonia* genotypes whereas others (e.g. 2) have a more detrimental effect. Finally it is clear that particular combinations interact in rather different ways suggesting the presence of an interaction. *Anthoxanthum* genotype 5 has a very negative impact upon *Danthonia* genotype 2, whereas

Table 3.
Analysis of variance of growth and fecundity.
Data have been transformed with square root transformation.

Growth

Source	D F	Type I S S	F	Pr>F
Danthonia gen.	7	13.49	3.09	.0040
Anthoxanthum gen.	7	5.33	1.22	ns
Interaction				
Regression	7	7.13	1.63	.1259
Deviations	42	39.70	1.51	.0291

Fecundity

Source	D F	Type I S S	F	Pr>F
Danthonia gen.	7	31.18	3.18	.0001
Anthoxanthum gen.	7	3.09	1.76	.0959
Interaction				
Regression	7	0.98	0.56	ns
Deviations	42	15.14	1.43	.0502

Danthonia genotype 4 has its highest fecundity in combination with the same genotype.

Both growth and fecundity data can be plotted according to the Finlay-Wilkinson regression technique. The performance of *Danthonia spicata* genotypes provides a measure of the different environments created by the *Anthoxanthum* genotypes. Accordingly, the horizontal axis (Figure 2.) represents the mean performance (fecundity per initial tiller in size) of all the *Danthonia* genotypes against each associated *Anthoxanthum* genotype. (Growth data are not shown.) The vertical axis is the performance of each genotype against each associate. In the Finlay-Wilkinson schema a genotype whose yield was very sensitive to the nature of its associates would give a slope of greater than one and a genotype with very stable behavior a slope of less than one. Genotypes with generally high yields will have regression lines lying towards the top of the figure, and vice versa.

As can be seen from Figure 2. there are two apparent groupings of *Danthonia* genotypes. Those with generally high yield (on the upper part of the graph) seem to be more sensitive to the genotypic identity of their competitor, while those *Danthonia* which are generally low-yielding are much more insensitive. The lines which are shown are hand-drawn approximations for separate regressions for the two groupings of genotypes. The best linear fit to the data can be determined for each genotype through regression analysis. The slopes, intercepts, and R^2 values are displayed in Table 2. Slopes of the genotypes range in value from -.277 to 3.143 for growth, and -.096 to 3.880 for fecundity. Three of the slopes are negative (*Danthonia* genotype two for growth, and one and two for fecundity) suggesting that these genotypes do slighly better against *Anthoxanthum* genotypes which generally have larger detrimental effects on all *Danthonia* and slightly worse against *Anthoxanthum* genotypes which have generally less detrimental effects on all *Danthonia*. However, only one of the individual regressions is significant at the .05 level for both growth or reproductive output (genotype eight). Log transformation did not improve the linearity.

An analysis of variance was performed on the growth and reproductive output data to determine if significant genotype-genotype interaction was present. Data were transformed by square root transformation to achieve homogeneity of variance according to Cochran's test (Dixon and Massey 1957). The results are shown in Table 3. Transformed growth data show a significant *Danthonia* main ($P<.004$) and a significant interaction ($P<.0291$). The interaction variance is attributable not to heterogeneity of regression lines but due to a signficant differential dispersal around the regression lines (Table 3); interaction variance due to regressions approaches significance ($P<.1259$). Transformed fecundity data show a similar pattern; significant *Danthonia* genotype main effect and interaction are present. Once again the interaction is due to differential dispersal about regression lines ($P<.0502$) and not to heterogeneity of slopes of such lines.

In accordance with the Finlay-Wilkinson technique two numerical values can be extracted from the plot of performance against environment: the mean yield of each *Danthonia* genotype

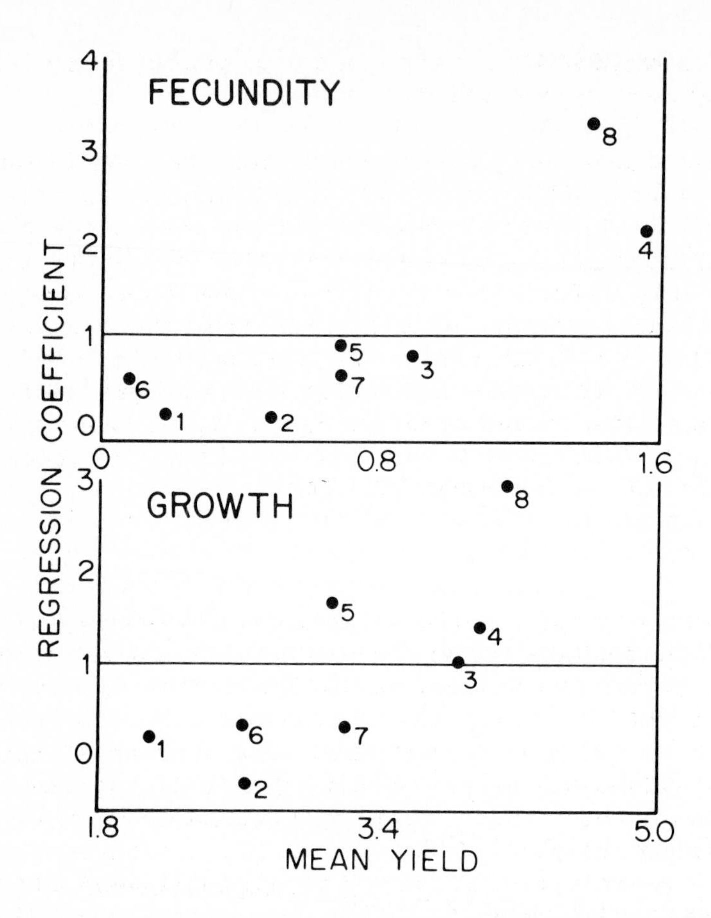

Figure 3. Regression coefficient (stability) of fecundity and growth of <u>Danthonia</u> genotypes when competed against all <u>Anthoxanthum</u> genotypes plotted against overall mean performance of that same <u>Danthonia</u> genotype.

in the face of all associates and the slope of the regression calculated for each genotype. The results are displayed in Figure 3 for both growth and reproductive output. Note that genotypes seem somewhat evenly distributed across the horizontal axis. A statistically-significant positive correlation exists between the regression coefficient and the mean yield (Pearson correlation coefficient .761, $P<.05$ for growth, .841, $P<.05$ for reproductive output). Although higher yielding genotypes do display increased sensitivity such that they do proportionally worse against good competitors than low yielding genotypes, these high yielding genotypes still outperform low-yielding genotypes in practically every genotype-genotype combination.

There was one confounding aspect to this study. At flowering it was discovered that three of the Danthonia genotypes, numbers 4, 5, and 8, were infected by the systemic, parasitic fungus, Atkinsonella hypoxylon. The infected genotypes did not consistently differ from the other noninfected genotypes. When these three genotypes were deleted from the analyses of variance, the results remained qualitatively the same. Nevertheless, it is clear that infected genotypes generally were more sensitive to the genotypic identity of the competing Anthoxanthum. Whether such sensitivity is attributable to infection has not been determined.

DISCUSSION

Use of a novel technique has permitted the characterisation of the competitive performance of naturally-occuring genotypes of one species grown in combination with a second species with which it co-occurs both spatially and temporally. The results were somewhat surprising. Given the broad overlap of the two species in the field it might have been naively expected that genotypes would have been subject to selective pressures operating through competitive interactions such that all would have equal competitive ability (no Danthonia main effect). This same selective pressure could have been expected to homogenise both the respective populations with regards to competitive ability such that genotypic performance would be insensitive to the genotypic identity of the competing Anthoxanthum genotypes (no Anthoxanthum main effect or interaction.)

While genotypes were sensitive to the identity of their competitors, use of the Finlay-Wilkinson technique was not especially well-suited to the elucidation of interspecific interaction. Genotypes grown in the presence of different competitors did not act like crop varieties grown in different fields. In particular, performance of most Danthonia genotypes was not linearly related to the environments. Further, unlike genotype-environment interaction, genotype-genotype interaction was attributable not to heterogeneity among the slopes of the regression lines but rather to the differential dispersal of genotypic performance about the regression lines. Although such slopes were not statistically different from each other, a plot of the regression coefficient against mean yield did show the characteristic trade-off between yield and stability. The correlation was quite strong and highly significant.

Few authors have applied the Finlay-Wilkinson method to data from naturally-occuring genotypes grown in their own environments. Ford (1981) who studied eight agamospecies of *Taraxacum* transplanted into three natural sites found that regression of performance on environment did generate statistically-significant differences between genotypes in performance (the appearance of linearity was inevitable with only three environments). There was no apparent correlation, however, between mean yield and the regression coefficient.

Equally-few authors have considered the applicability of the Finlay-Wilkinson method to interspecific interactions between naturally-occuring genotypes. Jacquard and Caputa (1970) in an analysis of the performance of nine grassland species grown in combination with each other found satisfactory linear fit for the relationship between performance and the environmental scale. They further observed that low yielding species tended to be more insensitive to the identity of the competing species, but there were many exceptions to the trend. Rousvoal and Gallais (1973) in a study of interactions between five component species of a natural grassland under two cutting regimes found similar results. Breese and Hill (1973) reanalysed data from two sets of experiments designed to examine species-species interaction. For both transformed and untransformed data they noted that the interaction term was comprised of significant variation due to linear regression and significant variation due to deviation from regression.

No previous study has applied the Finlay-Wilkinson method to genotype-specific interspecific competitive interactions occuring within a natural population. The data presented here suggest that such interactions may be qualitatively-different from interactions occuring at the level of the species. Unlike different species, genotypes do not mimic environments in terms of their effects on the genotypic performance of a competing species.

These data suggest that genotype-genotype interactions could potentially play a tremendous role in evolution occuring within the community. Genotype-genotype interaction does result in alteration of the relative rankings of genotypes in response to the interspecific genotypic context. The fitness of a particular genotype is sensitive to the genetic identity of neighboring species. Further, the tradeoff between stability and yield does appear to exist at the genotypic level within the population. Whether such a tradeoff between high yield (or fitness in a particular environment) and stability (or fitness in other environments) is physiological or a constraint imposed by negative genetic correlations between fitness traits (or pleiotropic effects) remains to be determined. In either case such a trade-off functions as a constraint to the evolutionary process; high fitness in a particular microenvironment (in this case represented by a specific competing genotype) necessitates low fitness in a second microenvironment (represented by another competing genotype.) Genotypic diversity within one species could thus promote genetic diversity within a second species.

The fact that genotypic performance can be linearly related to environments (scored according to the performance of all genotypes grown in them) suggests that different genotypes tend to perceive the physical environment in similar ways. As the data from this experiment indicate, however, when competing genotypes constitute the different environments, genotypic performance is not predictable. Models for the adaptive significance of sexual reproduction and outcrossing often rely on environmental unpredictability to explain how such traits are maintained within the population (Maynard Smith 1978; Williams 1975). Genotype-specific competitive interactions result in unpredictability of genotypic fitness and may thus provide the major selective force within the population acting to maintain sexual recombination and outcrossing.

Evidence from this study indicates the existence of different sorts of genotypes within the population. In our small sample of *Danthonia spicata* genotypes were two (not highly distinct) groups--one set of genotypes with high fecundity and growth was more sensitive to the genetic identity of the competitor; the other set was less fit (at least for the fitness components assayed) but more stable. In fact, the data suggests these genotypes might be slightly more fit in the context of more aggressive *Anthoxathum* genotypes. The sorts of population genetic processes that give rise to and maintain such distinct types of genotypes need to be explored.

Clearly the present work has limitations. This study concerns itself only with two component species in a community comprised of well over forty species. Further, it focusses on a particular stage of the life cycle (the adult) and ignores such critical stages as the seed and establishment phase. Finally, it considers only two components of fitness--adult fecundity and vegetative size. Nevertheless, important insight was gained through the use of a novel phytometer technique. Importantly, this study presents the first evidence that genotypic performance is sensitive to the genotypic identity of competing species (Kelley and Clay 1984). Genotype-genotype interaction was found to be primarily attributable to nonlinear deviations from regression. Further, the study represents the first indication that a tradeoff between mean performance and stability of performance at the level of genotypes exists within a natural population of plants. Such a constraint may operate to explain the plethora of variation evidenced in natural plant communities. With further work along these lines, the generality of the phenomena observed can be verified.

REFERENCES

Baihaki A, Stucker RE, Lambert JW (1976) Association of genotype x environment interactions with performance level of soybean lines in preliminary yield tests. Crop Sci 16:718-21

Barker RE, Hovin AW, Carson IT, Drolson PN, Slepes DA, Ross JC Caster MD (1981) Genotype-environment interactions for forage yield of reed canary grass clones. Crop Sci 21: 567-71

Bilbro JD, Ray LL (1976) Environmental stability and adaptation of several cotton cultivars. Crop Sci 16:821-824

Breese EL (1969) The measurement and significance of genotype-environment interactions in grasses. Hered 24:27-44

Breese EL, Hill J (1973) Regression analysis of interactions between competing species. Hered 31:181-200

Campbell LG and Kern JJ (1981) Cutlivar x environment interactions in sugarbeet yield trials. Crop Sci 22:932-935

Cooper RL (1981) Development of short-statured soybean cultivars. Crop Sci 21:127-131

DePauw RM, Faris DG, Williams CJ. (1981) Genotype-environment interactions of yield in cereal crops in northwestern Canada. Can J Plant Sci 61:255-63

Dixon WJ, Massey FJ Jr (1957) Introduction to statistical analysis. McGraw Hill, New York Toronto London

Eberhart SA, Russell WA (1966) Stability parameters for comparing varieties Crop Sci 6:36-40

Fatunla T, Frey KJ (1974) Stability indices of radiated and nonradiated oat genotypes propagated in bulk populations. Crop Sci 14:719-724

Finlay KW, Wilkinson GW (1963) The analysis of adaptation in a plant-breeding programme. Aust. J. Agric. Res. 14:742-754

Ford H (1981) Competitive relationships amongst apomictic dandelions. Biol J Linn Soc 15:355-368

Fowler NL, Antonovics J (1981) Competition and coexistence in a North Carolina grassland. I. Patterns in undisturbed vegetation. J Ecol 69:825-841

Frey KJ (1972) Stability indices for isolines of oats (Avena sativa L.). Crop Sci 12:809-812

Fripp YJ, Caten CE (1971) Genotype-environmental interactions in Schizophyllum commune. I. Analysis and character. Hered 27:393-407

Gray E (1982) Genotype x environment interactions and stability analysis for forage yield of orchard grass clones. Crop Sci 22:19-23

Harper JL (1977) Population biology of plants. Academic Press, London New York San Francisco

Heinrich GM, Francis GA, and Eastin JD (1983) Stability of grain sorghum yield components across diverse environments. Crop Sci 23:209-212

Hill J, Samuel CJA (1971) Measurement and inheritance of environmental response amongst selected material of Lolium perenne. Hered 27:265-276

Hobbs SLA, Mahon JD (1983) Variability and interaction in the Pisum sativum (L.)-Rhizobium leguminosarum symbiosis. Can J Plant Sci 63:591-9

Imbrie BC, Drake DW, DeLacy IH, Byth DE (1981) Analysis of genotypic and environment variation in international mungbean trials. Euphyt 30:301-311

Jacquard P, Caputa J (1970) Manifestation et nature des relations sociales chez les vegetaux superierieurs. Oecol Plant 3:137-168

Jika NI, St-Pierre CA, Denis JC (1980) Adaptations der cultivars de Sorgho-grain a differents regimes hydriques.

Can J Plant Sci 60:233-239
Johnson, GF, Whittington WJ (1977) Genotype-environment interaction effects in F_1 barley hybrids. Euphyt 26:67-73
Kaltsikes PJ (1969) Study of the general adaptation in fall rye by means of regression analysis. Can J Plant Sci 49:761- 764
Keim DL, Koonstad WE (1979) Drough resistance and dryland adaptation in winter wheat. Crop Sci 19:574-576
Kelley SE, Clay K (1985) Submitted
Laing DR and Fischer RA (1977) Adaptation of semidwarf wheat cultivars to rainfed conditions. Euphyt 26:129-139
Langer I, Frey KJ, Bailey T (1979) Association among productivity production responses and stability indexes in oat varieties. Euphyt 28: 17-24
Lewontin RC (1974) The genetic basis of evolutionary change. Columbia, New York London
Mack R, Harper JL (1977) Interference in dune annuals: spatial pattern and neighbourhood effects. J Ecol 65:345-363
Maynard Smith J (1970) The evolution of sex. Oxford, New York London
Moll RH, Cockerham CC, Stuber CW, Williams WP (1978) Selection responses, genotype-environmental interactions, and heterosis with recurrent selection for yield in maize. Crop Sci 18:641-645
Nguyen HT, Sleper DA, Hunt KL (1980) Genotype-environment interactions and stability analysis for herbage yield of tall fescue synthetics. Crop Sci 20: 221-224
Paroda RE, Hayes JD (1971) An investigation of genotype-environment interactions for rate of ear emergence in spring barley. Heredity 26:157-175
Perkins JM, Jinks JL (1968) Environmental and genotypic-environmental components of variability. III. Multiple lines and crosses. Hered 23:339-356
Phillips MS, Forrest JMS, Hayter AM (1979) Genotype x environment interaction for resistance to the white potato cyct nematode (Globodera pallida, pathotype E) in Solanum vermei x S tuberosum hybrids Euphyt 28:515-519
Rousvoal D, Gallais A (1973) Comportement en association binaire de cinq espedes d'une prairie permanente. Oecol Plant 8:279-300
Saeed M, Francis CA (1983) Yield stability in relation to maturity in grain sorghum (Sorghum bicolor (L) Moench). Crop Sci 23:682-687
Schaller CW, Qualset CO, Rutger JM (1972) Isogenic analysis of

the effects of the awn on productivity of barley. Crop Sci 12:531-535

Tan WK, Geok-Yong T, Walton PD (1979) Regression analysis of genotype-environment interaction in smooth bromegrass. Crop Sci 19:393-396

Turkington R, Cahn MA, Vardy A, Harper JL (1979) The growth, distribution, and neighbour relationships of Trifolium repens in a permanent pasture. III. The establishment and growth of Trifolium repens in natural and perturbed sites. J Ecol 67:231-243

Turkington R, Harper JL (1979) The growth, distribution, and neighbor relationships of Trifolium repens in a permanent pasture. IV. Fine-scale biotic differentation. J Ecol 67:245-254

Walton PD (1968) Spring wheat variety trials in the prairie provinces. Can J Plant Sci 48:601-609

Wilcox JR, Schapaugh WT Jr, Bernard RL, Cooper RL, Fehr WR, Niehaus MH (1979) Genetic improvemnts in soybeans of the midwest. Crop Sci 19:803-805

Williams GC (1975) Sex and evolution. Princeton, Princeton

Genotypic and phenotypic variation within *Rumex acetosella* L. populations along a secondary succession

J. Escarré, C. Houssard and J.P. Briane

Laboratoire Associé 121 (CNRS), Univ. Paris XI, Bât. 362, 91405 Orsay Cedex

ABSTRACT

Energy allocation patterns to component organs of the sexes of *Rumex acetosella* L. a dioecious weed, were investigated in three natural old field populations abandoned since 9 months, 2 years and 15 years, and in controlled conditions. Males and females were recolted separately at two sampling dates. Ramets of three populations were planted in pots containing four different soil types, taken from three old fields and from a neighbouring wood, in all combinations between populations and soils.

Results from field studies and in controlled conditions show significant differences in mean dry weigths of plant parts among populations and between sexes. Young populations contained more total energy in inflorescences and seeds than the old one. The latter allocated more energy to vegetative structures, and had a narrower niche than the other two populations. These results are discussed in relation to demographic strategies and selective forces acting along successional gradients.

INTRODUCTION

Patterns of resource allocation in natural populations along a secondary succession were the subjet of much research in recent years, since the Mac Arthur and Wilson (1967) and Odum (1969) papers.

After Odum (1969) as a plant community becomes more mature in a successional sense, plants would tend to allocate a greater proportion of their total biomass to vegetative structures, in contrast with communities of highly disturbed habitats in which species allocate more to reproduction. Thus the average proportion of a plant resource biomass allocated to reproduction (reproductive effort) would be higher in less mature than in more mature communities.

NATO ASI Series, Vol. G5
Genetic Differentiation and Dispersal in Plants
Edited by P. Jacquard et al.

The results of field studies are rather contradictory (see Soule and Werner, 1983). Also, evidence of genetically programmed reproductive strategies of plant populations by controlled garden experiments (Thompson and Stewart, 1981) do not show general agreement with theory. Some results have shown a genetic component (Gadgil and Solbrig, 1972; Solbrig and Simpson, 1974; Law et al., 1977; Grace and Wetzel, 1981), but others found the differences to be plastic responses of plants (Hickman, 1975; Roos and Quinn, 1977; Holler and Abrahamson, 1977; Raynal, 1979; Jacquard et al., 1983).

The objectives of the present investigation were to study the effects of changing environmental conditions along an old field succession on fecundity and biomass allocation of *Rumex acetosella* L. populations and to test, by controlled garden experiments, whetter observed differences were genetically fixed or the result of phenotypic plasticity.

MATERIAL

Rumex acetosella L., red sorrel, is a low dioecious perennial herb, with vigorous vegetative spread by root-buds. In the Paris region populations studied belong to *R. acetosella sensu stricto* with 2n= 42 chromosomes (Gardou and Bigot, 1976).

a) Field studies. Data were taken from three adjacent populations in old fields on tertiary acid soils which had been abandoned respectively for 6 months (hereafter named P_0), 2 years (P_1) and aproximately 15 years (P_2).

Percentage total vegetation cover increases with age. Fields 6 month and 2 year old are compound of herbaceous vegetation. The constitution of the early successionnal community was made up essentialy of annual species. *Rumex* density was low (0,60 shoots/m^2) in the 6 month old field, and very high in the two year old field (50 shoots/m^2). In the latter, observations have been made on the "hummocks" (P_1b) and "hollows" (P_1c) which remain from an ancient ploughing. The third old field, 15 year old, is colonized by shrubs, essentially *Sarothamnus scoparius*, and the *Rumex* density is intermediate (15 shoots/m^2). Twice in mid-August and late September, ten randomly selected individual plants of each sex were harvested at ground level. The shoot biomass for each plant was ovendried at 70° C and separated into component parts (leaves, stems,shoot inflorescences and seeds or flowers).

b) Controlled garden experiments. Root segments of P_1 and P_2 populations of known sex were collected during summer 1982 and grown under uniform conditions of soil and light for 6 months prior to experiments. In April 1983 four root pieces 2-3 cm. long of the resulting individuals were transplanted in 12 cm. pots filled with the soil of the three former fields, representing every possible combination of sites and soils. Males and females were grown in separate pots. Any particular combination was represented by 16 individuals: four replicates with four individuals each (8 males and 8 females).

P_o population was raised from seeds collected in the field, so as to simulate the annual regeneration processes of populations from the seed bank in the soil after harvesting.

Soil samples used in the experiment were collected in each of the three old fields (named hereafter S_o, S_1 and S_2) from a depth of 0-20 cm. A fourth complementary soil (S_3) was used, coming from an adjacent oak coppice, from which *Rumex* is absent (Table 1).

After 90 days cultivation, the above ground biomass of each individual was processed in the same manner as for the field plants.

Below ground biomass which consisted of roots was calculated on a per pot basis, because of the difficulty in separating exactly the contribution of each one of the four individuals.

TABLE 1. Chemical and physical characteristics of soils of the population sites: So, 6 month old field; S_1, two years old field; S_2, 15 years old field; S_3, oak coppice.

SOIL	CLAY	SILT	SAND	C/N	pH	P_2O_5	Ca	Mg	K
	(	o / o	)			(o/oo)	(M.	EQ. P.	100)
S_o	0	17,7	82,3	13,91	5,0	0,13	0,9	0,06	0,182
S_1	0,1	16,8	83,1	11,6	4,8	0,13	1,0	0,10	0,322
S_2	0	18,0	82,0	13,33	5,6	0,03	1,6	0,14	0,136
S_3	0,6	12,8	86,6	14,05	4,5	0,03	2,4	0,33	0,234

During the experiment, males of the P_2 population (15 years old) did not succeed in S_3 soil and only in part in S_1 soil where they were partially replaced by females of the same population.

METHODS

Data analyse were performed using a multiple discriminant analysis (Fisher, 1936). The variables used were: shoot height, vegetative height, and dry weights of plant parts for each sampled individual.

The method summarizes the differences between groups of individuals (in this case the differences between populations and sexes) by the use of discriminant functions. The latter are linear combinations of the variables which maximize the intergroup variance and minimizes the intragroup one. Essentially the results are a series of non correlated functions with a decreasing discriminant power.

Discriminant function scores for each sex and population are presented in Fig. 1. A X^2 transformation of Wilk's lambda which shows the group differences has been calculated for the two analyses.

Inter-group comparisons were performed using variance analysis and the Student-Newman-Keuls test; sex differences were assessed by a t test (Sokal and Rohlf, 1969).

Niche breadth (B) for each population and sex was calculated using Levins' formula (Levins, 1968): $B = 1/(p_i^2) \cdot s$; when p_i is the proportional response (in biomass units) of a species in the ith state and s is the number of resource states. The four soils used were considered as four edaphic states each characterized by a nutrient level (Table 1). The value of B can range from 0 (narrow niche) to 1.

RESULTS

a) Field observations. The multiple discriminant analysis shows significant differences between field populations (X^2 = 396,7 ddf. = 96; p<0,001).

The analysis of variance for component parts of the plants also showed highly significant effects of harvest dates, populations, sexes and their interactions (Table 2).

Male and female individuals of P_0 (6 month old) had the highest above ground biomass values (Table 3). This result appears on the discriminant analysis diagram (Fig. 1a), P_0 individuals are clearly isolated from the other three populations.

Individuals from the P_2 old field (15 year old) maintained the lowest shoo biomass for both sexes, and on two harvesting dates, but their average values

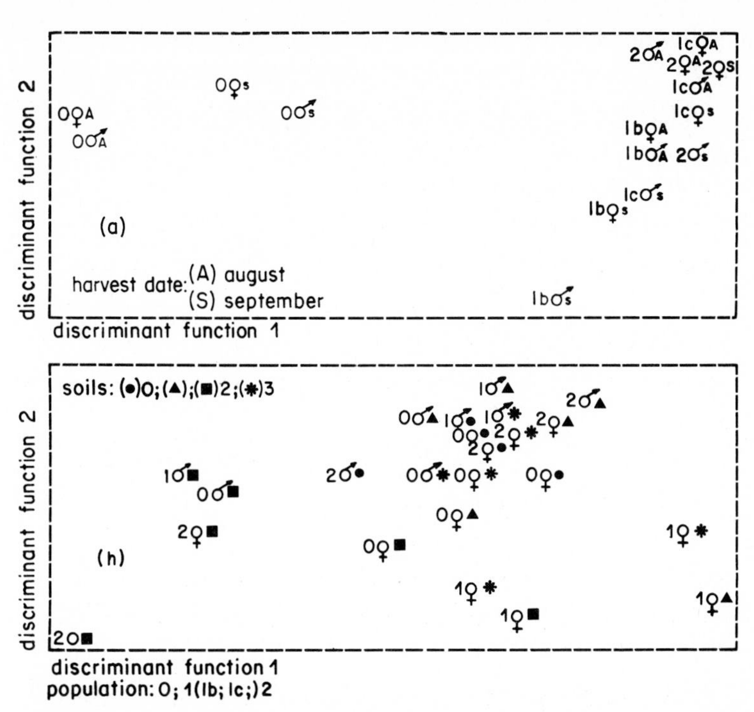

FIGURE 1. Discriminant function scores for each sex of three populations of *Rumex acetosella* (0,1 and 2 before sex symbols) collected in three old fields abandoned for 6 months (0), two years (1) and fifteen years (2).
a) field studies at two harvest dates (A) August; (S) September; (1c) and (1b) populations from the 2 year old field collected in the hollows (1c) and in the hummocks (1b) of an old ploughing.

b) garden experiments: populations 0,1 and 2 were transplanted in the soils of the old fields and in a fourth soil comming from an adjacent oak coppice.

TABLE 2. Comparisons by mean of an analysis of variance between populations and sexes of *Rumex acetosella*, collected in three old fields at two different dates (August and September).
n.s. not significant; (+), (*) and (**) are significant at 0,10; 0,05 and 0,01 respectively.

	SHOOTS	LEAVES	FLOWERING SHOOTS	FLOWERS, SEEDS	TOTAL BIOMASS	REPRODUCTIVE EFFORT
POPULATIONS	**	**	**	**	**	**
DATES	**	*	**	**	**	**
SEXES	**	*	**	**	**	**
POP. x SEXES	n.s.	n.s.	n.s.	+	n.s.	n.s.
POP. x DATES	n.s.	n.s.	n.s.	*	n.s.	*

TABLE 3. Mean values for characters in each sex of four populations of *Rumex acetosella* collected in three old fields at two different dates (August and September). Means within a row containing the same letter are not significantly different at the 0,05 level with Student-Newman-Keuls test.

p(♂≠♀) probability that differences between ♂ and ♀ occurred by chance.
n.s. not significant; (+), (*) and (**) are significant at 0,10; 0,05 and 0,01 respectively.

HARVEST DATA		AUGUST				SEPTEMBER			
POPULATION		P_0	P_1c	P_1b	P_2	P_0	P_1c	P_1b	P_2
	MALES	6,48a	1,16b	0,72bc	0,57c	5,39a	0,62b	0,37b	0,35b
TOTAL	FEMALES	6,58a	2,77b	1,19c	0,86c	6,09a	1,37b	0,72c	0,64c
BIOMASS	p(♂≠♀)	n.s.	**	*	n.s.	n.s.	*	**	*
(g/plant)	POPULATION	6,53a	1,96b	0,95c	0,71c	5,74a	0,99b	0,54c	0,49c
	MALES	0,20a	0,22a	0,32a	0,13b	0,17a	0,12a	0,15a	0,11a
REPRODUCTIVE	FEMALES	0,26a	0,44b	0,44b	0,19a	0,21a	0,21a	0,31a	0,25a
EFFORT	p(♂≠♀)	n.s.	*	n.s.	*	n.s.	**	**	n.s.
	POPULATION	0,23a	0,33b	0,38b	0,16a	0,19a	0,16a	0,23a	0,18a

were not significantly different for individuals from P_1b (hummocks). The P_1c populations collected in the hollows presented intermediate values between P_0 populations and the other two ones (P_1b and P_2).

Average shoot biomass and reproductive effort values were greater for females than for males, but the degree of significance varied between populations and harvesting dates (Table 3). There were no differences between sexes of P_0 individuals but there were for the other three populations.

Generally the mean biomass values for each population were greater in August than September, howewer the diminutions were only significants for P_1c females (hollows) $t=2,44$ $p< 0,05$; and P_1b males (hummocks) $t=2,30$ $p< 0,05$.

Male and female reproductive effort of the two P_1 populations was greater in August than for the corresponding P_0 and P_2 populations (Table 3). Differences between sites were not significant at the 5 % level for both sexes in September. This result is explained by phenological differences at the population level (specially P_1 and P_2):there was a significant decrease in reproductive effort of female individuals from the P_1 site (hollows $t=5,52$ $p< 0,01$;hummocks t= 3,21 p<0,01) and a increase (non-significant) of females of P_2 population from August to September.

b) Controlled garden experiments. In preliminary studies of P_1 populations (two year old), P_1c (hollows) and P_1b (hummocks) cultivated under uniform garden conditions, the equal total production among individuals suggested that the differences observed in the field habitats were the result of phenotypic plasticity, therefore the two populations were pooled for subsequent studies.

Multiple discriminant analysis showed significant differences between populations (X^2 = 69,1; df. = 24; p<0,001), and separated the sexes into two non overlapping sets (Fig. 1b). Male and female individuals from P_o and P_1 habitats were clearly separated by the second discriminant function, which was not the case for the P_2 populations.

A two-way analysis of variance (soils x populations) was realized with the eight heaviest flowering individuals, irrespectively of their sex, for to having equal sample size, because of the mortality of P_2 male individuals.

Results show significant differences between populations and soils for nearly all variables studied (Table 4).

Above ground biomass was generally maximum in P_1 populations (Table 5) and minimum in P_2 populations, though sometimes the differences were not statistically significant. The pattern $P_1 > P_o$ was the reverse of the field results (see Table 3).

In the three populations of *Rumex acetosella*, the individuals of P_2 had the lowest reproductive effort than either of the populations P_o and P_1 which were nearly identical in this regard.

TABLE 4. Comparisons by mean of an analysis of variance between populations of *Rumex acetosella* coming from three old fields and cultivated in four different soils: S_o, S_1, S_2 from the old fields, S_3 from an adjacent oak coppice. p is the probability of the null hypothesis. Symbols as in Table 2.

	SHOOTS	LEAVES	FLOWERING SHOOTS	FLOWERS, SEEDS	TOTAL BIOMASS	REPRODUCTIVE EFFORT
POPULATIONS	**	n.s.	**	**	**	**
SOILS	**	**	**	+	**	n.s.
POP. x SOILS	**	n.s.	n.s.	n.s.	**	n.s.

SOIL		S_o 6 months old			S_1 2 years old			S_2 15 years old			S_3 Oak coppice		
POPULATION		P_o	P_1	P_2	P_o	P_1	P_2	P_o	P_1	P_2	P_o	P_1	P_2
FLOWERING MALES		9	7	3	7	7	2	11	8	4	4	8	0
FLOWERING FEMALES		6	7	6	8	7	6	5	7	8	7	6	7
TOTAL POPULATION		16	16	16	16	16	12	16	16	16	16	16	8
TOTAL BIOMASS (g/plant)	MALES	1,01a	1,24a	1,32a	1,34a	0,95a	0,44a	2,27a	2,24a	2,99a	1,34a	1,01a	.
	FEMALES	1,35a	2,95b	1,35a	1,99a	2,60a	1;06b	2,88a	3,21a	2,56a	1,63a	2,54a	1,31a
	p(♂≠♀)	n.s.	**	n.s.	+	**	n.s.	n.s.	n.s.	n.s.	n.s.	n.s.	.
	POPULATION	1,08b	1,90a	0,89b	1,35a	1,62a	0,64b	2,46a	2,52a	2,65a	1,22a	1,49a	1,18a
REPRODUCTIVE EFFORT	MALES	0,16a	0,19a	0,12a	0,22a	0,16a	0,02b	0,24a	0,23a	0,14a	0,24a	0,20a	.
	FEMALES	0,39a	0,41a	0,26a	0,40a	0,49a	0,28b	0,34ab	0,43a	0,23b	0,28a	0,51b	0,20a
	p(♂≠♀)	**	*	n.s.	*	**	*	n.s.	*	n.s.	n.s.	**	.
	POPULATION	0,24a	0,26a	0,12b	0,24ab	0,30a	0,14b	0,27a	0,30a	0,18b	0,18a	0,30a	0,17a

TABLE 5. Mean values (g/plant) for characters in each sex of three populations of *Rumex acetosella* collected in three old fields and cultivated in four different soils (see Table 1). For each soil, means within a row containing the same letter are not significantly different at 0,05 level with a Student-Newman-Keuls test. Organization and symbols as in Table 3. (.) missing data (plants dead, or vegetative stages at harvest time).

Female individuals showed more between population differences than males (Table 5), notably when cultivated in S_0 and S_1 soils, the populations in S_2 and S_3 soils being no different. In the latter soil a big variation has been recorded between individuals irrespective of the populations.

TABLE 6. Comparisons by mean of an analysis of variance between responses of male and female individuals (above ground biomass) of each population of *Rumex acetosella* cultivated in four different soils (data in Table 5). Symbols as in Table 3

	SHOOTS			LEAVES			FLOWERING SHOOTS		
POPULATIONS	P_0	P_1	P_2	P_0	P_1	P_2	P_0	P_1	P_2
MALES	**	*	**	n.s.	n.s.	+	*	*	n.s.
FEMALES	n.s.	n.s.	*	n.s.	n.s.	**	n.s.	n.s.	n.s.
POPULATION	*	n.s.	*	n.s.	n.s.	**	*	n.s.	n.s.

	FLOWERS, SEEDS			TOTAL BIOMASS			REPRODUCTIVE EFFORT		
POPULATIONS	P_0	P_1	P_2	P_0	P_1	P_2	P_0	P_1	P_2
MALES	**	+	n.s.	*	*	*	n.s.	n.s.	n.s.
FEMALES	n.s.	n.s.	n.s.	n.s.	n.s.	*	n.s.	n.s.	n.s.
POPULATION	n.s.	n.s.	n.s.	*	n.s.	**	n.s.	n.s.	n.s.

TABLE 7. Niche breadth of populations of *Rumex acetosella* (P_0, P_1 and P_2) and of each sex, cultivated in four different soils. Niche breadth was calculated using Levin's (1968) indice.

	NICHE BREADTH		
POPULATION	P_0	P_1	P_2
MALES	0,90	0,87	0,69
FEMALES	0,92	0,99	0,87
POPULATION	0,88	0,95	0,74

Comparisons within each population showed different response patterns. Females from P_1 and P_o populations do not react significantly to four soils, the reverse is true for P_2 females and the males of three populations (Table 6).

Male individuals do not show differences, but they react more sharply than females to four soils (for ex. seven times aboveground biomass increase for the P_2 males between S_1 soil (0,44 g.) and S_2 soil (2,99 g.).

Females from P_o and P_1 grow larger than males under cultivation in S_o and S_1 soils. There were no statistical differences between the two sexes of P_2; however this result requires confirmation because of the heavy mortality and the small number of flowering male individuals (Table 5).

There were no significant differences on root biomass values between populations within each sex, but male individuals had below ground biomass values higher than females (3,05 g/pot for males, all populations pooled, and 2,48 g/pot for females; $t = 2,486$; $p<0,02$).

Female individuals from P_o and P_2 allocated more resources to below ground biomass (31 and 30 % of net production) than did females from P_1 (13 %; $p<0,01$), but the latter allocated more resources to seeds (Table 5).

Males and females from the late successional population P_2 (15 years old) had narrower response breadths than did individuals from two earlier populations P_o and P_1 (Table 7).

Females had broader niches than males, the latter showed a regular decrease in niche breadth from early to late populations.

DISCUSSION

Field populations of *Rumex acetosella* collected along a successional gradient show different patterns of biomass allocation, some of these differences persisted when individuals were grown under four different soils in the garden.

The 6 months old field P_o is usually a cultivated plot, vegetation cover is low and individuals of *Rumex acetosella* experience soil disturbances by ploughing and by fertilizer and herbicide application. The individuals from P_o had significantly higher values of above ground biomass allocation than either of the populations of P_1 and P_2 but their values of reproductive effort were intermediate.

Thus population P_o finds its optimal growth conditions in the 6 month old field, possibly in relation to the low vegetation cover. One explanation of these results may be found in the work of Putwain and Harper (1972) who report that suppression of species in herbaceous communities by herbicides favors the establishment of *Rumex acetosella* and induces an increase in its growth.

The vegetation cover of the 2 years old field (P_1) is intermediate between P_0 and P_2, but the density of *Rumex acetosella* is very high (see Material). Both in the field and in controlled conditions females of P_1 allocated more resources to seed production than the other two populations.

Female individuals from P_1 growing in the hollows, the moister environment, had consistenly greater size than females of hummocks. Male individuals were similar for all size measures.

In P_2, the old field abandonned for 15 years, vegetation cover is nearly 100 %, and male and female individuals had significantly lower reproductive effort and above ground biomass allocation than the other two populations both in the field and in the garden.

P_2 individuals show a narrow niche breadth and react to their original soil more strongly that the other two populations, these results would seem to express a great adaptation of these individuals to their substratum.

Werner and Platt (1976) have found similar results for a group of species of the genus *Solidago*, and Parrish and Bazzaz (1982) show that plant survival of early successional species was less affected by high nutrient concentrations than mid and late successional species, the latter had narrower response breadths than did early successional ones.

Population responses of 1 and 2 sites (P_1 and P_2) may be adaptative as they have been described in both the field and in controlled conditions. This is not the case for the P_0 population which does not produce as much above ground biomass in the garden as in the field, showing a plastic response of individuals.

Phenotypic plasticity of plants brings out a problem of field versus laboratory measurements (see Gadgil and Solbrig, 1972; Thompson and Stewart, 1981). According to the latter authors "genetic and environmental influences are not separable in the field...but...laboratory measurements are more likely to approach the genetically programmed reproductive strategies of the genotypes concerned..".

These variations in biomass allocation of a species along a successional gradient may be related to differences in regulation of the populations with respect to r and K selection theory (Mac Arthur and Wilson, 1967).

Individuals from the two early successional fields (P_o and P_1) where vegetation cover is moderately low, have the highest reproductive effort and biomass allocation values (r selected genotypes, or ruderal individuals, Grime (1979). In the old field (P_2) abandonned for 15 years, vegetation cover is very high and demographic regulation patterns are density dependent: under these conditions low fecundity genotypes are selected (K selection).

Male and female individuals of *Rumex acetosella* do not react in the same manner to the selective pressures along the successional gradient. Female populations present biomass allocation differences, and they exhibit more tolerant responses to soils, linked with a higher niche breadth than male individuals. The latter did not differ significantly in resource allocation patterns, but they presented strong biomass variations between the four soils; a high mortality rate in some of them, linked with a low niche breadth decreasing from the early to late populations. Males produce more below ground biomass than females an so, they may reproduce more abundantly by vegetative means. Putwain and Harper (1972) have noted this difference in root biomass between the sexes of *Rumex acetosella*, which has been connected by Zimmerman and Lechowicz (1982) to a greater drought tolerance of male plants.

With increasing maturity of the community, space becomes more and more overcrowded by vegetation and patterns of demographic regulation of populations are density dependent. It would be expected that natural selection would then favor the genotypes with the greatest propensity for clonal growth (Abrahamson, 1975, 1980; Newell and Tramer, 1978; Law et al. 1977), as the life histories selected under density dependent regulation involve very high rates of seedling mortality (Putwain et al., 1968; Sarukhan and Harper, 1973; Law et al. 1977). These characteristics may confer a competitive superiority on the males with increasing age or maturity of the community. Indeed we observe an increasing predominance of male individuals with stage in the succession (Escarré and Houssard, in preparation). Harris (1968) has already noticed that males of *Rumex acetosella* in New Zealand were more frequent than females in the old communities.

CONCLUSION

The evolution of the life history strategy of *Rumex acetosella* along a successional gradient lead us to describe separate regulation between sexes and between adjacent populations during a short period of time.

Our observations confirm previous results which showed - a rapid evolution of adjacent populations of a species under various environmental conditions (Snaydon and Davies, 1982; Kiang, 1982), - different environmental optima of the sexes of dioecious plants (Freeman et al., 1976; Wade et al. 1981; Zimmerman and Lechowicz, 1982) and - differences in biomass allocation between the sexes and of the predominance of one sex on the other associated with competition (Putwain and Harper, 1972; Wade, 1981 a & b; Meagher and Antonovics, 1982; Onyekwelu and Harper, 1979).

Lloyd and Webb (1977) propose that many of the observed differences between the sexes, are of direct selective value in relation to the distinct roles of males and females in sexual reproduction.

Accumulation of different growth phases and metabolic functions associated with pollen production and the provisioning of seeds may amplify these divergences with increasing maturity of the community, but may contribute to persistance of the species along a successional gradient.

ACKNOWLEDGEMENTS

We wish to thank J. Lacourt for his help in field work; M. Coudrin and J. Poncet for garden experiments; Drs. J.L. Harper, M. Hayward, A. Lacoste and B. Saugier for their comments on the manuscript; and Claudine Briane for linguistic help.

REFERENCES

Abrahamson WG (1975) Reproduction of *Rubus hispidus* L. in different habitats. Amer. Midl. Nat. 93: 471-478

Abrahamson WG (1980) Demography and vegetative reproduction. In Solbrig OT (ed) Demography and vegetative reproduction in plant populations, Blacwell p 89

Fisher RA (1936) The use of multiple measurements in taxonomic problems. Anns. of Eugenics 7: 179-188

Freeman DC, Klikoff LG, Harper KT (1976) Differential resource utilization by the sexes of dioecious plants. Science 193: 597-599

Gadgil M, Solbrig OT (1972) The concept of r and K selection: evidence from wild flowers and some theoretical considerations. Am. Nat. 106: 14-31

Gardou C, Bigot G (1976) Contribution à l'étude du *Rumex acetosella* L. S.L. (Polygonaceae) dans la région parisienne. Candollea 31: 257-271

Grace JB, Wetzel RG (1981) Phenotypic and genotypic components of growth and reproduction in *Typha latifolia*: experimental studies in marshes of differing successional maturity. Ecol. 62: 789-801

Grime JP (1979) Plant strategies and vegetation processes. Wiley & Sons. 222 p

Harris W (1968) Environmental effects on the sex ratio of *Rumex acetosella* L. Proc. N.Z. Ecol. Soc. 15: 51-54

Hickman JC (1975) Environmental unpredictability and plastic energy allocation strategies in the annual *Polygonum cascadense* (Polygonaceae). J. Ecol. 63: 689-701

Holler LC, Abrahamson WG (1977) Seed and vegetative reproduction in relation to density in *Fragaria virginiana* (Rosaceae). Am. J. Bot. 64: 1003-1007

Jacquard P, Foroughbakhch RF, Lumaret R (1983) Dynamique des relations de compétition entre trois graminées des successions post-culturales. Acta Oecol. Oecol. Plant., 4: 123-138

Kiang YT (1982) Local differenciation of *Anthoxantum odoratum* L. populations on roadsides. Amer. Midl. Nat. 107: 340-350

Law R, Bradshaw AD, Putwain PD (1977) Life-history variation in *Poa annua*. Evolution 31: 233-246

Levins R (1968) Evolution in changing environments. Princeton Univ. Press

Lloyd DG, Webb CJ (1977) Secondary sex characters in seed plants. Bot. Rev. 43: 177-216

Mac Arthur RH, Wilson ED (1967) The theory of island biogeography. Princeton Univ. Press

Meagher TR, Antonovics J (1982) The population biology of *Chamaelirium luteum* a dioecious member of the lily family: life history studies. Ecol. 63: 1690-1700

Newell SJ, Tramer EJ (1978) Reproductive strategies in herbaceous plant communities during succession. Ecol. 59: 228-234

Odum EP (1969) The strategy of ecosystem development. Science 164: 262-270

Onyekwelu SS, Harper JL (1979) Sex ratio and niche differenciation in spinach (*Spinacia oleracea* L.). Nature 282: 609-611

Parrish JAD, Bazzaz FA (1982) Competitive interactions in plant communities of different successional ages. Ecol. 63: 314-320

Putwain PD, Harper JL (1972) Studies in the dynamics of plant populations. V. Mechanisms governing the sex ratio in *Rumex acetosa* and *Rumex acetosella*. J. Ecol. 60: 113-129

Putwain PD, Machin D, Harper JL (1968) Studies in the dynamics of plant populations. II. Components and regulation of a natural population of *Rumex acetosella* L. J. Ecol. 56: 421-431

Raynal DJ (1979) Population ecology of *Hieracium florentinum* (Compositae) in a central New York limestone quarry. J. Appl. Ecol. 16: 287-298

Roos FH, Quinn JA (1977) Phenology and reproductive allocation in *Andropogon scoparius* (Graminae) populations in communities of different successional stages. Am. J. Bot. 64: 535-540

Sarukhan J, Harper JL (1973) Studies on plant demography: *Ranunculus repens* L. *R. bulbosus* L. and *R. acris* L. I. Population, flux and survivorship. J.Ecol. 61: 675-716

Snaydon RW, Davies TM (1982) Rapid divergence of plant populations in reponse to recent changes in soil conditions. Evolution 36: 289-297

Sokal RR, Rohlf FJ (1969) Biometry. W.H. Freeman and Co.

Solbrig OT, Simpson BB (1974) Components of regulation of a population of Dandelions in Michigan. J.Ecol. 62: 473-486

Soule JD, Werner PA (1981) Patterns of resource allocation in plants, with special reference to *Potentilla recta* L. Bull. Torrey Bot. Club 3: 311-319

Thompson K, Stewart AJA (1981) The measurement and meaning of reproductive effort in plants. Am. Nat. 117: 205-211

Wade KM (1981a) Experimental studies on the distribution of the sexes of *Mercurialis perennis* L. II Transplanted populations under different canopies in the field. New Phytol. 87: 439-446

Wade KM (1981b) Experimental studies on the distribution of the sexes of *Mercurialis perennis* L. III Transplanted populations under light screens. New Phytol. 87: 447-455

Wade KM, Armstrong RA, Woodwell SRJ (1981) Experimental studies on the distribution of the sexes of *Mercurialis perennis* L. I Field observations and canopy removal experiments. New Phytol. 87: 431-438

Werner PA, Platt WJ (1976) Ecological relationships of co-occuring goldenrods (*Solidago*: Compositae). Am. Nat. 110: 959-971

Zimmerman JK, Lechowicz MJ (1982) Response to moisture stress in male and female plants of *Rumex acetosella* (Polygonaceae). Oecol. 53: 305-309

Regulation of a Shortgrass Steppe Plant Community in Time and Space: Physiology and Genetics

George J. Williams III and M.R. Sackschewsky
Department of Botany
Washington State University
Pullman, WA., U.S.A. 99164-4230.

ABSTRACT

The shortgrass steppe plant community in the central grasslands of North America is dominated by a combination of species utilizing the C_3, C_4 and CAM carbon fixation pathways. Seasonal temperature and water availability create temporal gradients while the rolling hills topography creates spacial gradients of water and nutrients.

Photosynthetic response to growth temperature and *in situ* phenology suggests use of site resources by dominant species is offset. Temperature regulation of carbon fixation is limited by water availability. Species have a range of physiological adaptations to water stress.

Laboratory investigations of two grass species grown under controlled common environments suggests a role for genetic differentiation in the community. Patterns of leaf growth, leaf number, and flowering are different for populations from dry hilltop sites, intermediate mid-slope sites, and wet swale sites. Thus genetic differentiation is playing an important role in the spacial relations of this grassland community.

Environmental regulation of growth activity resulting in offset utilization of resources by dominant species has reduced potential for competition among them. These physiological adaptations coupled with genetic variation among intraspecific populations growing along topographically created environmental gradients provide insight into community dynamics in space and time.

NATO ASI Series, Vol. G5
Genetic Differentiation and Dispersal in Plants
Edited by P. Jacquard et al.

INTRODUCTION

The shortgrass steppe plant community is dominated by *Bouteloua gracilis*, C_4, *Agropyron smithii*, C_3, and *Opuntia polyacantha*, CAM. *Carex eleocharis*, C_3, is an important member of the community being one of the first species to grow and also it maintains some green leaves throughout the growth season. The air temperature at the research site near Nunn, Colorado, U.S.A. (40° 42'N, 104° 46'W) rises from a mean monthly maximum of 17C in April to a seasonal high of 30C in July. Soil water potentials as reflected by predawn plant water potentials become more negative as the seasonal rainfall decreases and seasonal warming drives evapotranspiration.

Periodic relief of water stress is realized with the onset of convectional rain storms that occur periodically during the summer months. Therefore the plant community is subjected to temporal changes in the temperature and water status resulting in a seasonal gradient. The topography of the local area produces another series of gradients. The surface of the land is characterized by rolling hills which result in spacial gradients of water and nutrients. The slopes of the hills have wet swales and progressively drier sites towards the tops. Therefore, the plants along these catenas can be subjected to various soil water regimes during the growth season. The populations can exist at any point in time with very different water status depending on the position of a population along the gradient. The water status can range from flooded conditions in the swales to very dry soils on the tops (-0.01 KPa to -3.5 KPa plant water potentials at predawn respectively).

The offset phenology patterns of the dominant species suggest that response of carbon gain and water status to temperature and moisture conditions would be a means of regulating the patterns. A series of experiments in the laboratory and the field were undertaken to examine this possibility.

During the experimental stage it became apparant that it was necessary to determine if genetic differences existed among the populations of A.smithii and B.gracilis distributed along the topographic gradient in the steppe. The following account is composed of two parts one describing the physiological examination of carbon gain regulation by temperature and water status and the other describes the examination of the possible occurance of genetic differences among populations along the topographic gradient. The physiological responses of steppe species are described by a synthesis of published works while the genetic aspects are described by presentation of unpublished results on common growth condition experiments.

MATERIALS AND METHODS

All plant material used in laboratory studies was obtained from the Central Plains Experimental Range (CPER) maintained by the United States Agriculture Research Service. All species were obtained from soil blocks dug from the CPER and transported to Washington State University in Pullman, Wa., U.S.A.. Plants were grown under two temperature regimes 20/15C, cool, and 35/15C, warm (day temperature/ night temperature). Photoperiods were 14h and with a maximum of 1,000 u mole quanta m^{-2} s^{-1} photon flux density. Gas exchange was measured according to Monson et al. (1982) or Gerwick and Williams (1978) for the cacti. Water potential was measured either by pressure bomb techniques or by use of a Wescor HR-33T microvoltmeter (modified for digital readout) and C-51 sample chambers.

RESULTS AND DISCUSSION

Seven perennial shortgrass steppe species were analyzed in laboratory experiments to determine their photosynthetic responses to growth and analysis temperature. The species included four graminoids, Agropyron smithii, Bouteloua gracilis, Buchloe dactyloides, and Carex eleocharis, and three cacti, Echinocereus viridiflorus, Mammillaria viviparia and Opuntia polyacantha. Therefore the species studied are representative of a range of photosynthetic pathways, phenological patterns and dominance types.

The dominants A. smithii, B.gracilis, and O.polyacantha exhibited rates of maximum photosynthesis at different analysis temperatures and these were dependent upon growth temperature. Maximum rates in the C_3, A.smithii, were by cool grown plants analyzed at 25C while B.gracilis, C_4, had greatest photosynthesis when warm grown plants were analyzed at 45C leaf temperatures (Monson et al., 1983). In A.smithii photosynthetic rates were lower in warm grown plants at all analyses temperatures except 45, 47, and 50C. The converse case was noted for B.gracilis with exceptions at 10 and 15C.

The dominant CAM species O.polyacantha exhibited maximum daily carbon gain when warm grown. Analysis also showed that warm grown plants had enhanced daily carbon gain when analyzed at 35/5C rather than 35/10 or 35/15 C (Gerwick and Williams, 1978).

The photosynthetic response to temperature of the two grasses suggests that maximum carbon accumulation and greatest growth is early in the cooler portion of the growth season for A.smithii and later for B.gracilis. The experiments also suggest that O.polyacantha would exhibit maximum carbon and growth during periods of the season characterized by warm days and cool nights in late August and early September.

Although C.eleocharis would not be characterized as a dominant based on biomass production, it is representative of other aspects of importance and variation in the steppe. Carex eleocharis is one the first species to became active in spring and maintains green leaf tissue into the warmer portions of the summer. The photosynthetic response of this species provides insights into the physiological basis of observed phenology. Plants grown under cool conditions have a lower photosynthetic temperature optimum, 25C, than warm grown ones, 38C (Monson et al., 1983). The time course for the acclimation of the temperature optimum requires five to six days based on experiments involving moving plants grown under a cool condition to a warm growth condition (Veres and Williams, 1984). Analysis of changes in the temperature optima show that leaves developed under a cool growth regime acclimate to a lower temperature optima, 28-30C, when moved to the warm regime than the optima of leaves developed under the warm regime, 35C, (Veres and Williams, 1984). Similar results were obtained by Monson (1984) upon examination of seasonal changes in the photosynthetic temperature optima of *C. eleocharis* in the field.

Buchloe dactyloides is a C_4 species increasing in importance as it appears to be replacing *B.gracilis* after disturbances to the prairie. Its photosynthetic temperature response is much like *B.gracilis* as both favor warm growth and analysis temperatures. *Buchloe dactyloides* favors slightly earlier growth activity in the field and its photosynthetic rates are slightly greater than those of *B.gracilis* when grown at cool temperatures and analyzed at cool temperatures (Monson *et al.*, 1983).

Analysis of the two subdominant CAM species, *E.viridiflorus* and *M.vivipara*, was initiated to determine if they differed from the dominant *O.polyacantha* in their carbon gain response to temperature. Each exhibited its greatest carbon gain under warm rather than cool conditions but neither had enhanced carbon gain due to lowering the analysis night temperature (Green and Williams, 1982). This difference between them and *O.polyacantha* suggested that they were most active at different times. The two subdominant cacti have a photosynthetic response to temperature that suggested they are active at the same time during the growth season as the two C_4 grasses.

Although the photosynthetic response to temperature by the species under study suggests the important role of temperature in this habitat; water has been found to exert a strong regulation of species activity. Gerwick and Williams (1978) have demonstrated that in *O.polyacantha* carbon gain enhancement by lower night temperature is lost when water stress is induced. They also showed carbon gain was reduced by 80% and after rewatering 60% of the original carbon gain ability was regained within 24h. Both of the other CAM species studied failed to exhibit the ability to recover from water stress exhibited by *O. polyacantha* (Green and Williams, 1982).

Agropyron smithii and *Bouteloua gracilis* both exhibited reduced photosynthesis during the growth season in response to increasingly more negative water potentials (Monson and Williams, unpublished). Seasonal temperature increases reduce photosynthesis in *A.smithii* and photosynthesis is drastically reduced by decreasing soil water content. However, artificial irrigation rapidly induces increases in daily photosynthesis (within 24h). Daily photosynthesis in *B.gracilis* also rapidly recovers from water stress upon irrigation. The main difference between the C_3 and C_4 species was that the C_4, *B.gracilis*, exhibited superior water use efficiency.

The distribution of populations of a species over environmental gradients sometimes suggests genetic differences among the populations. The distribution of *A.smithii* and *B.gracilis* over moisture gradients in the shortgrass steppe prompted our examination of the possibility of genetic differentiation among populations of these species. Using the common conditions of controlled growth chambers, samples from field sites along these moisture gradients were compared under cool and warm temperature conditions. Populations samples were obtained from hilltop, mid-slope, and swale sites and samples represent experimental populations of the species.

The results of the common growth experiments should however be interpreted with some caution. In this volume Hayward describes experiments with *Lolium perenne* showing that plants with both high and low rates of tillering can be selected from within a single clone. He further notes that this was possible only with young clones and that among young clonal material more perrenial clones produced a greater response than short-lived ones. Hayward cites comparable intances in other grass species (Pannicum, Pernes *et al*., 1970 and Phleum, Palenzona *et al*., 1973; see Hayward this volume for references). If the phenomena is similar in *A.smithii* and *B.gracilis* it seems of minor importance as experimental populations were generated from comparatively old field populations, however it is still better to be aware of the possibility of such responses.

TABLE 1 - Growth response, number of leaves per tiller, of three experimental populations of Agropyron smithii under A., a cool growth regime (20/15C) and B., a warm growth regime (35/15C).

A. 20/15C

significant interactions(*)	week	leaves per tiller of experimental populations		
		hilltop	mid-slope	swale
	1	2.7 ± 0.8*	2.6 ± 0.8	2.9 ± 0.5
	2	3.9 ± 0.5	3.5 ± 1.0	3.4 ± 0.7
	3	4.8 ± 0.9	3.9 ± 0.9	3.7 ± 0.9
	4	5.9 ± 1.6	4.7 ± 0.8	4.6 ± 1.0(13)
	5	7.3 ± 2.3	5.7 ± 0.9(13)	5.5 ± 1.1(13)
	7	9.4 ± 2.8(14)	7.5 ± 1.9(13)	7.2 ± 2.3(12)
	8	10.5 ± 3.0(14)	8.2 ± 2.0(12)	7.2 ± 1.8(12)

B. 35/15C

significant interactions(*)	week	hilltop	mid-slope	swale
	1	2.8 ± 0.6	2.5 ± 0.6	2.7 ± 0.6
	2	3.6 ± 0.9	3.3 ± 0.7	3.3 ± 0.8
	3	4.0 ± 0.8	4.1 ± 0.8	4.3 ± 0.7(14)
	4	4.8 ± 1.0	4.8 ± 1.3	5.7 ± 1.4(14)
	5	5.7 ± 1.1	5.7 ± 1.7	6.4 ± 1.8(14)
	7	7.7 ± 1.9	7.9 ± 2.7(14)	7.9 ± 1.9(14)
	8	8.2 ± 1.8	8.5 ± 2.7(14)	8.7 ± 2.0(14)

(*) significant interactions h = hilltop, m = mid-slope, s = swale levels of significance, pooled t(a) = $p<0.01$, (b) = $p<0.02$, (c) = $p<0.10$

* Numbers represent mean ± standard deviation, N = 15 unless indicated within parenthesis

TABLE 2 - Growth response, number of leaves per tiller, of three experimental populations of Agropyron smithii under A., a cool growth regime (20/15C) and B., a warm growth regime (35/15C).

A. 20/15C

sig. int.(*)	week	hilltop	mid-slope	swale
	1	95.7 ± 37.9	96.1 ± 36.9	92.1 ± 23.6
	2	194.0 ± 58.4	174.0 ± 63.3	159.7 ± 62.6
hxs=(a)	3	267.1 ± 79.5	228.4 ± 86.3	189.6 ± 79.7
hxs=(b)	4	384.5 ± 154.1	331.7 ± 108.7	274.3 ± 94.3(13)
	5	535.2 ± 285.3	432.1 ± 101.1(13)	373.5 ± 159.1(13)
hxs=(b)	7	962.1 ± 400.8(14)	805.8 ± 230.8(13)	622.8 ± 307.3(12)
	8	1094.2 ± 474.0(14)	912.3 ± 276.3(13)	672.4 ± 296.4(12)

(columns hilltop, mid-slope, swale: leaves per tiller of experimental populations)

B. 35/15C

sig. int.(*)	week	hilltop	mid-slope	swale
hxs=-(a)	1	140.0 ± 42.4	112.7 ± 43.3	100.8 ± 39.3
	2	176.7 ± 53.9	172.8 ± 52.5	130.7 ± 49.4
	3	234.6 ± 50.7	221.8 ± 64.2	191.6 ± 50.8(14)
	4	297.0 ± 57.0	297.3 ± 88.2	306.7 ± 58.7(14)
	5	405.0 ± 72.6	428.8 ± 113.4(14)	463.9 ± 103.2(14)
hxs=(b)	7	669.6 ± 134.1	704.3 ± 288.9(14)	795.4 ± 185.2(14)
hxs=(b)	8	754.5 ± 142.6	801.7 ± 359.3(14)	921.6 ± 235.7(14)

(*) significant interactions h = hilltop, m = mid-slope, s = swale levels of significance, pooled t(a) = $p<0.01$, (b) = $p<0.02$, (c) = $p<0.10$

* Numbers represent mean ± standard deviation, N = 15 unless indicated within parenthesis

The results show that experimental populations of A.smithii and B.gracilis differ in a number of traits and these differences are interpreted as being genetic. When grown under cool conditions experimental populations of A.smithii differed in leaf number per plant with hilltop having greater numbers than midslope and swale experiment populations (Table 1A). These differences were noted after four weeks of growth and persisted through eight weeks. Under warm growth conditions differences were only noted between hilltop and swale experimental populations with swale populations having greater leaf number per plant during weeks four and five (Table 1B).

TABLE 3 - Four week growth response, number of leaves per tiller, of three experimental populations of Bouteloua gracilis A. grown under a cool growth regime (20/15C) and B. grown under a warm growth regime (35/15C).

significant interactions *	leaves per tiller of experimental populations		
	hilltop	mid-slope	swale
A - 20/15C			
h x s $p < 0.02$	4.1 ± 1.8(*)	5.5 ± 1.5	5.8 ± 1.4
B - 35/15C			
NONE	5.8 ± 1.4	5.9 ± 1.0	6.7 ± 1.5

* Numbers represent mean ± standard deviation, N = 15
(*) h = hilltop, s = swale

The total leaf length per tiller (cm) was also a character that differed among experimental populations. Hilltop and swale experimental populations under cool growth conditions differed from week three through seven with the hillop sample having the greatest leaf length per tiller (Table 2A). Differences between hilltop and swale experimental populations occured under warm conditions and it should be noted that initially (first week due to sampling) hilltop samples had a greater leaf length per tiller than swale ones (Table 2B). However, this was reversed by week seven and persisted through week eight showing that this tendency is a very strong one.

Leaf number per plant among experimental populations of B.gracilis was very similar under both cool and warm growth conditions (Table 3 A and B). The only difference noted after four weeks of growth was between hilltop and swale experimental populations under the cool growth regime (Table 3A). Flowering patterns among experimental populations were different under both growth regimes with hilltop samples having greater numbers of flowering tillers

per tillers observed (Table 4). Flowering was greatest for all experimental populations under the warm growth condition which is the temperature most like the site temperature when flowering occurs in the steppe.

TABLE 4 - Flowering after four weeks growth of three experimental populations of Bouteloua gracilis under cool (20/15C) and warm (35/15C) growth regimes.

growth	Experimental population			
regime	hilltop	mid-slope	swale	x^2 (significance)
20/15C	4/ 9*	0/13	1/14	$x^2 = 9.67$ ($p < 0.01$)
35/15C	11/15	8/15	3/15	$x^2 = 8.72$ ($p < 0.05$)

* flower tiller per total number of tillers

The combination of traits examined for species in the shortgrass steppe provide a better understanding of how environment regulates the plant community via interaction with the different physiology of species and the within species genetics. Seasonal temperature gradients and species phenology suggest a role for temperature regulation of temporal vegetation patterns. Photosynthetic temperature response of steppe species reinforce this idea and further support the hypothesis that steppe species because of this response avoid coincident resource utilization. Thus physiological response to temperature reduces the potential for species competition especially among the dominant species.

The differences in genetics of experimental populations of *A.smithii* indicates that populations in the field on hilltops produce more leaves and these leaves grow faster while temperatures are cool than populations down the slopes. This will allow them rapid growth while water is available in those sites. Populations in the swales shows a greater ability than hilltops ones to grow their leaves under warm conditions. This indicates that when it is warm in the steppe and hilltops are dry and swales are wetter these swale populations can continue to grow. The flowering patterns of *B.gracilis* indicate that greater flowering more rapidly by hilltop populations than mid-slope and greater and more rapid flowering by mid-slope than swale populations could also be a genetic feature that is favorable. For example since the hilltop sites become drier more rapidly than the down slope sites it would be an advantage to flower more rapidly in these more rapidly drying sites. This is especially true since photosynthesis is drastically reduced by water stress.

The approach of intergrating the physiological and genetic studies of this plant community offer the promise of a better understanding not only of the temporal relationships among species but also their spatial ones.

ACKNOWLEDGMENTS

Thanks and recognition are extended to Paul Kemp, Cliff Gerwick, Janice Green, and John Veres whose hard work and initial insights have contributed so much to the study of the physiological ecology of the shortgrass steppe. Their efforts made this report possible along with the efforts of Russ Monson. Special thanks and recognition go to him as his effforts in research, his abundant curiosity and continual questioning have been invaluable.

We also thanks A. Berger and G. Valdeyron whose critical reading allowed us to improve the manuscript.

REFERENCES

GERWICK BC, WILLIAMS GJIII. (1978) Temperature and water regulation of gas exchange of *Opuntia polyacantha*. Oecologia (Berlin) 35 : 149-159

GREEN JM, WILLIAMS GJIII (1982) The subdominant status of *Echinocereus viridiflorus* and *Mammillaria viviparia* in the shortgrass prairie: The role of temperature and water effects on gas exchange. Oecologia (Berlin) 52 : 43-48

MONSON RK (1984) A field study of photosynthetic temperature acclimation in *Carex eleocharis* Bailey. Plant, Cell and Environment 7 : 301-308

MONSON RK, STIDHAM MA, WILLIAMS GJIII, Edwards GE, Uribe EG (1982) The temperature dependence of photosynthesis in *Agropyron smithii* Rydb. I. Factors affecting photosynthesis net CO_2 exchange and contribution from RuPB carboxylase measured *in vivo* and *in vitro*. Plant Physiol. 69 : 921-928.

MONSON RK, LITTLEJOHN RO, WILLIAMS GJIII (1983) Photosynthetic adaptation to temperature in four species from the Colorado shortgrass steppe; a physiological model for coexistence. Oecologia (Berlin) 58 : 43-51

VERES JS, WILLIAMS GJIII (1984) Time course of photosynthetic temperature acclimation in *Carex eleocharis* Bailey. Plant, Cell and Environment (in press).

Genetic Differentiation and Phenotypic Plasticity in Populations of *Plantago lanceolata* in response to nutrient levels.

Daan Kuiper
Department of Plant Physiology,
University of Groningen,
P.O. Box 14, 9750 AA Haren (Gn)
The Netherlands.

ABSTRACT

Three populations of *Plantago lanceolata* L. were studied with regard to their genetic differentiation and phenotypic plasticity. Eight random taken samples of each population were grown at two nutrient levels and subjected to an alteration in mineral supply. Growth, root respiration and root ATPase activity were followed during the experiment.

With respect to all measured characteristics genetic differentiation on population level was demonstrated. High growth rates correlated to higher respiration activities and higher Ca^{2+}- and Mg^{2+}-stimulated ATPase activities. Habitat characteristics, like fertility and water supply, were positively correlated with physiological properties of the tested plants.

Phenotypic plasticity of the measured characteristics was observed in all three populations. It is concluded that phenotypic plasticity plays a very important role in maintenance of *P. lanceolata*. In populations, in which growth is limited, plasticity is more directed to abiotic factors, while plasticity in the more productive populations strengthens competitive ability to other plants.

INTRODUCTION

The set of environmental conditions to which an organism is exposed, exerts considerable influence on its development. This is

NATO ASI Series, Vol. G5
Genetic Differentiation and Dispersal in Plants
Edited by P. Jacquard et al.

particularly evident in plants, which because of their immobility cannot escape the physical effects of the environment. As a result plants show morphological and physiological adjustments to their particular environment, a response referred to as phenotypic plasticity. The way in which an organism reacts to different environmental conditions is perhaps one of the most important characteristics for a succesful reproduction.

Survival of plant populations depends upon genetic composition of the population and phenotypic plasticity of the individual plants, the latter also being genetically determined. In this paper I define plasticity as all intragenotypic variation and in this way forming a part of phenotypic flexibility (Bradshaw, 1965). Phenotypic flexibility is the capacity of a genotype to function in a range of environments by plastic and/or stable responses. In botanical literature plastic responses in plants have generally been interpreted as having adaptive value.

Plasticity forms a short-term response to a changed environment. According to Marshall and Jain (1968):" Plasticity in morphological characteristics related to growth, besides plant differentiation or morphogenesis, is certainly of great adaptive value ". I prefer to consider plasticity and stability as being adaptive, when the responses will increase the fitness of the plant. In physiological experiments it is often hard to judge, whether a plastic response will benefit or will damage plants, since fitness measuring is unusual and difficult in those experiments. It is not important to discriminate between morphological and physiological plasticity (Bradshaw, 1965). Considering in essence all plasticity physiologically, I think, is the most direct approach, since morphology also is physiologically determined.

"Under varying environments individual phenotypic plasticity and genetic polymorphism could evolve as alternative strategies or in some varying combination such as to maximize fitness " (Lewontin, tin 1957). Correlation of genetic variation and phenotypic plasticity has often been discussed. Levins (1963) suggests a negative correlation between individual phenotypic plasticity and genetic variation on the population level. Results on <u>Bromus mollis</u> L. by Jain (1978) support this idea. On contrary, in agreement with their hypothesis Hume and Cavers (1982) demonstrated in <u>Ru-</u>

mex crispus L., that genetic variation and phenotypic plasticity were not complementary.

Genetic differentiation or genetic specialization has been found within (and between) species. Abundant literature is available on genetic differentiation, with respect to drought (Quarrie, 1980, 1981; Henson et al., 1981a, 1981b), to temperature or altitude (Berry and Raison, 1981; Somero, 1978; Graham and Patterson, 1982 ; Berry and Bjoerkman, 1980), to light intensity and/or light quality (Boardman, 1977; Wild, 1979 ; Bjoerkman, 1981), heavy metal tolerance (Antonovics et al. 1971; Ernst, 1982), mineral nutrition (Clarkson and Hanson, 1980; Chapin, 1980), to salt tolerance (Greenway and Munns, 1980; Albert, 1982) etc. Development of such genetic specialization depends upon the degree of selection pressure provided a minimum degree of available genetic variation. Pollard (1980) succeeded in selecting heavy metal tolerant genotypes from Plantago lanceolata L. after five generations in the genus Scenedesmus after relaxation of selection pressure (Stokes and Dreier, 1981).

Generalist plant species usually are species with a wide ecological tolerance, a high degree of genetic variation and heterozygosity (Nevo ,1978). The latter enables the plant to exploit more biochemical pathways for growth and reproduction (Adams and Shank , 1959). Price et al. (1980, 1981) suggested, that generalists possess a high DNA content, which may contribute to a high degree of heterozygosity and phenotypic plasticity. DNA content from annuals and ephemers was lower than that from perennials and non-ephemers respectively. This observation, a connection between perennial lifecycle, larger DNA content and more pronounced phenotypic plasticity, supported by results on Rumex acetosella L. (Farris and Schaal, 1983) and by a comparative study of genome size of 162 plants (Grime and Mowforth, 1982) showing a positive correlation between DNA content and generalist strategy. Thus, it might be concluded, that generalists have a higher degree of genetic variation, heterozygosity and phenotypic plasticity than specialists (Martinez and Armesto, 1983).

Phenotypic plasticity was studied in several Plantago species, since we are interested in the role of genetic variation and phenotypic plasticity in populations of Plantago species in main-

taining themselves in their habitats. This study is part of the central theme of the dutch Grassland species research group.
In this study I have chosen the cosmopolitan *Plantago lanceolata* L., a perennial rosette herb, which is obligately cross-pollinating (Sagar and Harper, 1969; Primack, 1978, 1980) and has a high degree of heterozygosity (H = 0.12, H Van Dijk unpublished results). Genetic variation within populations of *P. lanceolata* generally is higher than between populations (Fowler and Antonovics 1981; K Wolff, unpublished results). In the field genetic differentiation in *P. lanceolata* was demonstrated with respect to light by Teramura (1983), to the height of the vegetation (Warwick and Briggs, 1979; Teramura *et al*, 1981) and with respect to time of flowering (J van de Toorn and H ten Hove, unpublished results). However, in several reports on *P. lanceolata* plasticity of many characteristics is also emphasized, like leaf length and size of the rosette. Several characteristics of photosynthesis are highly plastic, resulting in a broad photosynthetic optimum response to light intensity (Teramura and Strain, 1979). Plasticity was also concluded from previous results on root respiration and shoot to root ratio (Lambers *et al*, 1981), on parameters of nitrogen metabolism (Stulen *et al.* 1981) and on root ATPase activity, enzymes involved in the response of uptake and transport processes (Kuiper and Kuiper, 1979; Kuiper, 1983b) to alterations in mineral supply and light.
These data suggest, that the ability of populations of *P. lanceolata* to occupy a wide range of local conditions is in large measure based on plastic responses of individual plants. The hypothesis that *P. lanceolata* was responsive to a change in the level of mineral nutrition, was tested in the present work by studying the responses of root Ca^{2+}- and Mg^{2+}-stimulated ATPases, root respiration and growth of plants of three populations.

MATERIAL AND METHODS

Populations of *P. lanceolata*

From each of the three chosen populations, Westduinen (=W),

Merrevliet (=M) and Hete-ren (=H), eight seed samples were taken at random. Each sample consisted of seeds collected from a single plant. *Westduinen* is an old coastal and mainly dry dune grassland with a species-rich vegetation, a relative high diversity in vegetation structure and many gradients in edaphic factors. The area is grazed extensively by cattle and lightly trampled (Blom *et al.*, 1979; Noe and Blom, 1980). *Merrevliet* is a silted up branch of a small river, formerly used as farm land and nowadays as a closed hay meadow. The watertable is a few centimeter below soil surface. The diversity of species is rather high (Van der Aart, 1984) *Heteren* is former farm land, now used as hay meadow. The vegetation is dominated by tall grass species and the diversity of species is rather low. Characteristics of the three populations are summarized in Table 1.

Table 1. Some characteristics of the three populations of *P. lanceolata*. W, Westduinen; M, Merrevliet; H, Heteren; P, phosphorus; N, nitrogen. (from Blom *et al.*, 1979 and from J. Mook and S. Troelstra, unpublished results)

Population	W	M	H
total P (ppm)	120	1230	800
total N (ppm)	1300	22800	2800
total organic matter (%)	2 - 20	86	33
moisture content soil (%)	4 - 30	69 - 86	18 - 44
soil	sand	peat	clay
height of vegetation (cm)	15	50	70
interception daylight at soil surface, minimum (%)	30	3	2

Growth Conditions

Seeds were placed on wet filter paper in Petri dishes. After about 5 days the seed coat of ungerminated seeds was cut open to provide a high germination percentage (c. 95 %). After day 14 the seedlings were transferred into a nutrient solution at half the strenght of that given by Smakman and Hofstra (1982), at pH 6.0. After 2 days half of the plants was transferred to a full nutrient solution (100%) and to a diluted solution (2% of the full nutrient solution). On day 32 after sowing reciprocal transfers were performed, resulting in four groups of plants, 100%, 2%-100%, 100%-2% and 2% plants. The experiment was done in a growth chamber at 20 ^{o}C, relative humidity of the air was about 60% and light intensity was about 350 micro E $m^{-2}s^{-1}$ visible radiation. Light period was 12 hours and the experiment took about 60 days.

Methods

<u>ATPase and protein determination</u> Differential centrifugation of the microsomal membrane fraction from roots has been described earlier (Kuiper, 1982). The ATPase activity was determined at $30^{o}C$ in 1 ml volume containing 3 mM ATP, 40 mM histidine-HCl buffer (pH 6.5), $MgCl_2$ or $CaCl_2$ at various concentrations (0 mM-10mM), and enzyme (0.1 ml)as a last addition. The reaction was stopped after 30 min by adding 33% trichloric acid. Inorganic phosphate was determined by the ammonium molybdate-$SnCl_2$ method (Lindeman, 1958). ATPase activity was calculated on protein basis. The protein content was determined according Lowry (1951). basis.

<u>Root respiration</u> Respiration of the roots was measured with a YSI (Yellow Springs Instruments) oxygen meter (model 53). Intact roots, excised from the shoots, were placed in a cuvette containing an aerated culture solution similar to that used for growth. The cuvette was carefully sealed, excluding any air bubbles in the solution, and the decrease of the oxygen concentration of the solution was measured by an electrode mounted in the cuvette. Thereafter, the culture solution was replaced by a solution (except Fe) and 25mM SHAM. Fe was omitted since it chelates with SHAM; pH was 6.0 and the temperature of the solution was kept at $20^{o}C$. The decrease of the oxygen concentration in this solution was measured, and the difference between the two measurements as-

cribed to the activity of the alternative respiratory pathway, which is sensitive to substituted hydroxamic acids, like SHAM (Kuiper, 1983a).

Growth Plants were harvested twice a week. The plant material was dried overnight at 90°C. The dry weights of shoot and root were determined seprately and from these data the shoot to root ratio ws calculated.

RESULTS

Growth

Shoot (Fig. 1) and root growth (not shown) of the eight samples of each population were averaged. In all treatments the growth rate of the shoot of H (=Heteren) plants was higher than that of the other two populations . Plants from W (=Westduinen) showed the lowest biomass development. Growth of transferred plants responded very quickly to an alteration of the nutrient solution. RGR (relative growth rate) of shoot and roots of the H plants was higher than that of the other two populations. The differences between these populations were significant ($P \leqslant 0.025$, Table 2).

Table 2. Relative growth rate of shoot and root.

Nutrition		Shoot	Root
100%	W	0.095 ± 0.041	0.091 ± 0.047
	M	0.128 ± 0.027	0.106 ± 0.020
	H	0.147 ± 0.021	0.129 ± 0.012
2%	W	0.091 ± 0.047	0.095 ± 0.042
	M	0.086 ± 0.015	0.084 ± 0.018
	H	0.122 ± 0.019	0.107 ± 0.021

Differences in growth among the populations of *P. lanceolata* were arranged from low to high (Table 3.). The dry weights of 100% and

2% plants increased in the order W, M, H. The difference between between the dry weights of the W and M populations were small, but significant in spite of the high population-specific variation in the W population. From these results I conclude growth capacity and RGR to be genetically differentiated on population level.

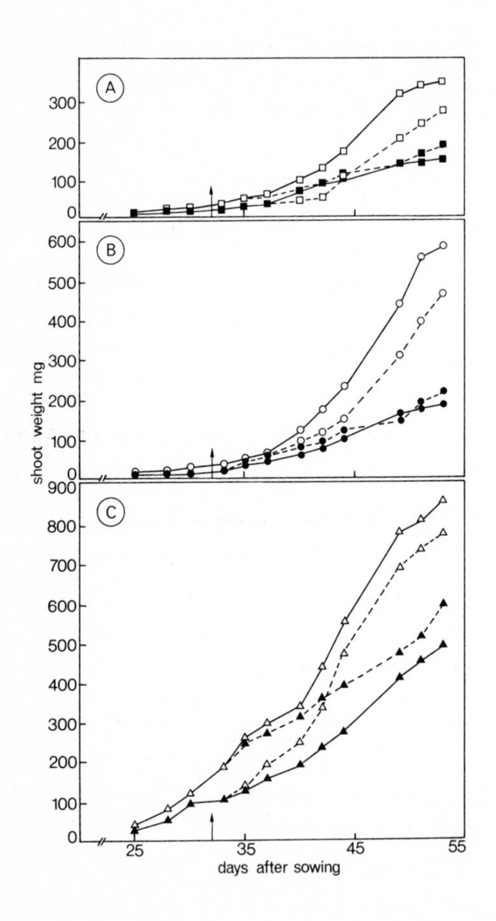

Fig. 1. Dry matter accumulation, shoot, in the plants of *P. lanceolata* of three populations, Westduinen (A; □—□, 100%; ■—■, 2%; ■---■, 100%-2%; □---□, 2%-100%), Merrevliet (B; ○—○, 100%; ●—●, 2%; ●---●, 100%-2%; ○---○, 2%-100%) and Heteren (C; △—△ 100%; ▲—▲, 2%; ▲---▲, 100%-2%; △---△, 2%-100%), ↑ time of transfer.

Shoot to root ratio was affected by mineral supply. The lower ratio in 2% plants is generally observed as a response to low mineral supply. The three populations could not be distinguished with aid of absolute values of the shoot to root ratio or behaviour after transfer. The shoot to root ratio appeared to be a plastic property (Table 4).

Table 3. Arrangement of dry weights of shoot and root in three populations of *P. lanceolata*, with an indication of significance level: xxx, $P \leq 0.005$; xx, $0.005 < P \leq 0.01$; x, $0.01 < P \leq 0.025$. (Time, 53 days after sowing).

Nutritional regime	Growth parameter	low ---------------------- high
100%	dry weight shoot	W --- xx --- M --- xxx --- H
	dry weight root	W --- xx --- M --- xxx --- H
2%	dry weight shoot	W --- xx --- M --- xxx --- H
	dry weight root	W ---- x --- M --- xxx --- H

Table 4. Shoot to root ratios of plants of three populations of *P. lanceolata*. (*n*=40)

Days after transfer		Treatment			
		100%	2% - 100%	100% - 2%	2%
W	1	3.5 ± 0.9	1.3 ± 0.7	3.5 ± 0.9	1.1 ± 0.4
	17	3.8 ± 1.1	3.7 ± 1.0	1.8 ± 0.8	1.4 ± 0.5
M	1	3.3 ± 0.4	2.0 ± 0.2	3.3 ± 0.6	2.4 ± 0.2
	17	3.5 ± 0.4	3.4 ± 0.4	2.1 ± 0.3	2.2 ± 0.3
H	1	3.5 ± 0.3	1.5 ± 0.1	3.5 ± 0.2	1.5 ± 0.1
	17	3.7 ± 0.2	2.9 ± 0.2	1.5 ± 0.1	1.3 ± 0.1

Total root respiration (Fig. 2) was affected by nutritional strength. The respiration of 2% plants was always lower than that of 100% plants. Total root respiration decreased in all plants during the experiment. The decrease in the 100% plants

Root Respiration

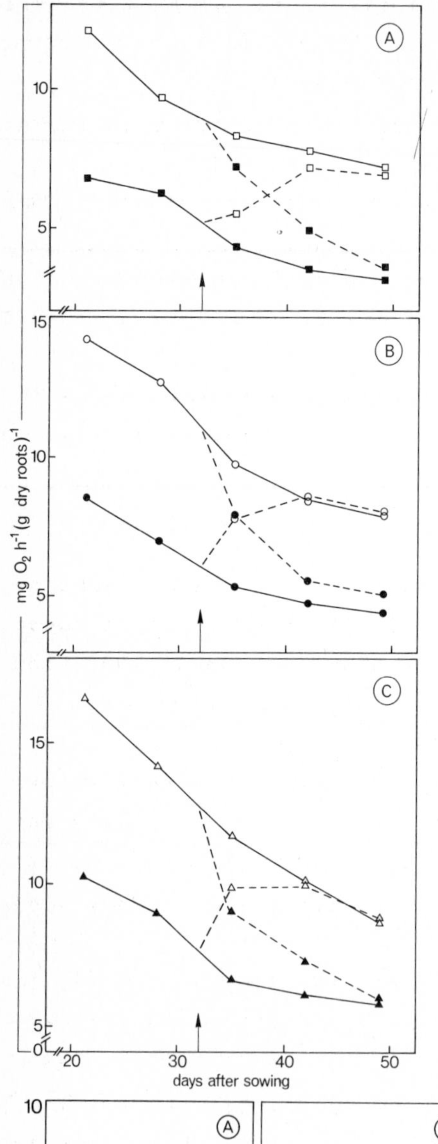

Fig. 2 Root respiration. Otherwise as for Fig. 1.

of the H population was larger than in W and M plants, although the proportional decrease was quite similar, 48%, 45% and 40% for H, M and W 100% plants respect. Cytochrome pathway, forming the ma-jor energy delivering part of respiration (Fig. 3) resembled the pattern of total respiration. The responses upon a transfer from one to the other mineral condition occurred very rapidly, especially in the M and H population. Thus the internal supply of energy-rich compounds is very responsive. The respiration results were arranged from low to high (Tab. 5). The H plants had the highest respiratory activity and W plants showed low activities corresponding with growth results.

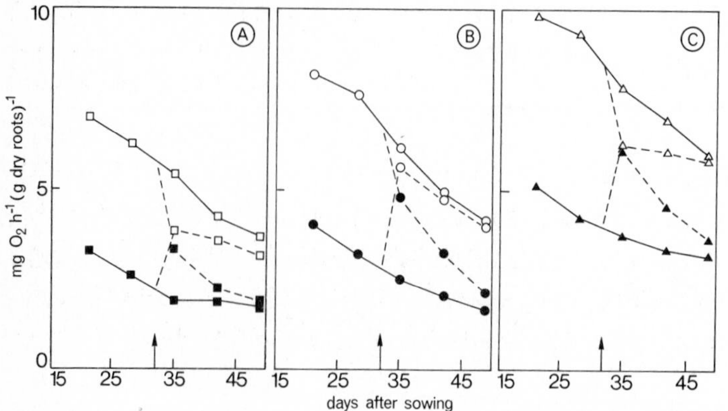

Fig. 3 Cytochromal activity. Otherwise as for Fig. 1.

Table 5. Arrangement of total root respiration and cytochromal activity in plants of three populations of P. lanceolata with indication of significance level. xxx, $P \leq 0.005$; xx, $0.005 < P \leq 0.01$; x, $0.01 < P \leq 0.025$. (time, from day 21 to 42).

Nutritional regime	Parameter	Range low ------------------- high
100%	total respiration	W -- x -- M ------ xx ----- H
	cytochromal activity	W -- x -- M ------ xx ----- H
2%	total respiration	W -- x -- M ------ xx ----- H
	cytochromal activity	W -- x -- M ------ xx ----- H

Root ATPase Activity

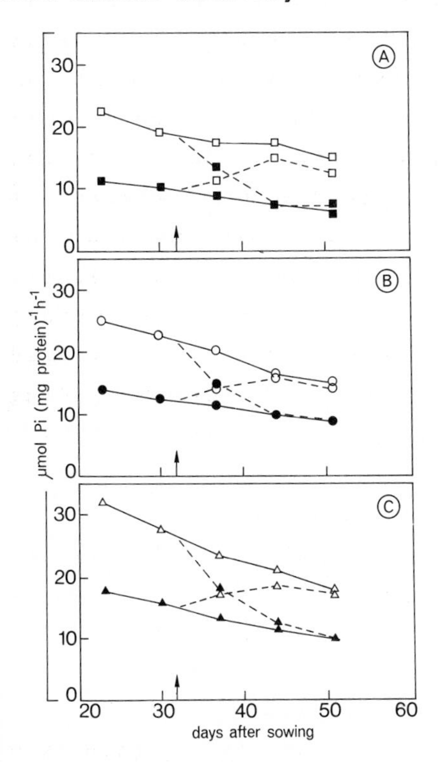

Fig. 4 Ca^{2+} activity. Otherwise like Fig. 1.

Vmax values of the Ca^{2+}-stimulated ATPase activity of roots of 100% plants was higher than that in 2% plants (Fig. 4). Plants of the H population showed the highest Vmax values and in the plants of the M population root ATPase activity generally was higher than that in the W population. These differences were at the 5% level. Transferred plants showed a rapid response upon a change in mineral supply. After 5 days the transferred plants showed intermediate Vmax values between those of 100% and 2% plants. In most cases the adaptive response was completed after 10 days. Mg^{2+} dependent ATPases resembled closely the Ca^{2+} stimulated one.

DISCUSSION

The present results confirm the hypothesis that *Plantago lanceolata* L. possesses a quick and adaptive response to mineral nutrition. In addition, transferred plants of each population responded similarly to an alteration of mineral supply, indicating that no large genetic differences for such an adaptive response were present. In conclusion, phenotypic plasticity for mineral nutrition is characteristic for *P. lanceolata*. In several aspects the plants of the H population showed the highest metabolic activity, i.e. in RGR, in respiration and in ATPase activity. Thus genetic differentiation was present in the studied populations. Recent information on allozyme variation about these populations do not support this genetic differentiation, based on physiological results; possibly genetic factors have been studied, which had not been subjected to selection pressure by mineral nutrition (unpublished results K Wolff). Our results on genetic differentiation in physiological parameters are supported by information on morphological properties. In the area of Westduinen the prostrate form of *P. lanceolata* is dominating during the whole growth period, in the M populations only in spring time; later it is replaced by the more erect habitus due to growth of vegetation. Leaf angle and leaf length are genetically differentiated among these populations (Mook *et al.*, 1981). This adaptive and morphological specialization (Warwick and Briggs, 1979; Teramura *et al.* ,1981) can be related to the management of vegetation: Westduinen has been grazed by cattle for centuries, while Heteren and Merrevliet are both hayfields and managed for that purpose for decennia. Vegetative reproduction (Teramura, 1983) and time of flowering (Van der Aart, 1984) of *P. lanceolata* seem to depend on light regime. Individuals of *P. lanceolata* from the W population (meadow) flower late, while plants from the H and M population (hayfield) flower earlier, in correlation with time of mowing. The present results on growth (Fig.1) and RGR (Table 2) give a consistent picture of a fast increment of the biomass in the H population. The slow growth in plants of the W population could not be distinguished significantly from that of the M population, even though differences in growth between M and W plants often were larger than those between M and H plants. Merrevliet and Heteren are both hayfields with a higher total phosphorus and nitrogen content than the Westduinen area (Table 1), although these values do not represent the available amounts of phosphorus and nitrogen. The persistence of an overal lower growth rate in W plants at growth chamber conditions indicates genetic background for slow growth, and may be interpreted as an adaptation to the lower nutrient availability and low water supply. Selection for a higher growth potential may be correlated with a larger vegetation height (longer leaves) and a higher content of organic matter and water. In conclusion, genetic differentiation between populations is present in P. lanceolata (Fowler and Antonovics, 1981; this paper).

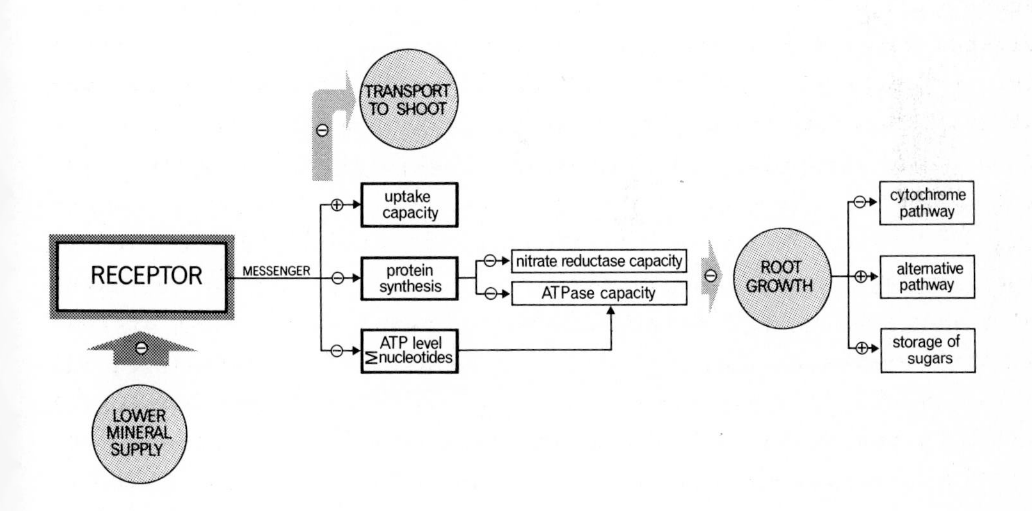

Fig. 5 Physiological model for phenotypic plasticity in roots.

Growth rate is positively correlated to physiological characteristics like root respiration, content of energy-rich compounds and activities of many enzymes (Kuiper, 1984). So I like to suggest a model for physiological plasticity (Fig. 5). An input site for information from the environment to the plant; sensitive to mineral nutrition. These signals have to be perceived and translated in plant characteristics. As far as I know no specific receptor, structural and/or functional for mineral nutrition is present in root tissues. Notwithstanding NO_3^- may act as an effector. A lower NO_3^- supply will imediately change the content of a substance, which plays the role of messenger and in this cytokinins are very promising agents (Kuiper, 1984). The changed cytokinin production will affect protein synthesis and the pool of nucleotides , both being highly correlated to growth rate. The growth rate of the roots will decrease following a lower NO_3 supply and will provoke a lower root respiration rate and a changed NADH oxidation between the two respiratory pathways (Fig. 5, Kuiper, 1982). Literature reports that cytokinins can respond very quickly to a changed nitrogen availability. Root growth and shoot growth, the latter being dependent on cytokinin supply from the root, both depending on the sensitivity of the receptormessenger system, respond quickly to an alteration in the mineral supply of the plant.Differences in phenotypic plasticity, sensitivity as well as amplitude, may be ascribed to genetically determined differences in the receptor system and the involved physiological processes.

Acknowledgements-I am grateful to Professors P.J.M. van der Aart, J. van Andel, W. van Delden and P.J.C. Kuiper for critical reading of the manuscript and Mr. E. Leeuwinga for drawing the pictures. These investigations were supported by the Foundation for Fundamental Biological Research (BION), which is subsidized by the Netherlands Organization for the Advancement of Pure Research (ZWO). Grassland Species Research Group Publ. No. 79.

REFERENCES

Adams, M.W. & Shank, D.B. (1959). The relationship of heterozygosity to homeostasis in maize hybrids. Genetics 44,777 - 786.

Albert, P. (1982). Halophyten. In Pflanzenoekologie und Mineralstoffwechsel. pp. 33 - 215, H. Kinzel (ed.). Ulmer Verlag.

Antonovics, J., Bradshaw, A.D. & Turner, R.G. (1971). Heavy metal tolerance in plants. Adv. Ecol. Res. 7, 2 - 85.

Berry, J. & Bjoerkman, O. (1980). Photosynthetic response and adaptation to temperature in higher plants. Annu. Rev. Plant Physiol. 31, 491 - 453.

Berry, J. & Raison, J.K. (1981). Responses of macrophytes to temperature. Plant-Ecol. Responses to the physical environment, (O.L. Lange et al.), Encyclop. Plant Physiol. vol. 12A, pp. 277 - 338.

Bjoerkman, O. (1981). Responses to different quantum flux density. Plant-Ecol. Responses to the physical environment, (O.L. Lange et al.), Encyclop. Plant Physiol. vol. 12A, pp. 57 - 107.

Blom, C.W.P.M., Husson, L.M.F. & Westhof, V. (1979). Effects of trampling and soil compaction on the occurrence of some *Plantago* species in coastal dunes IV: The vegetation of the two dune grasslands in relation to physical soil factors. Verh. Kon. Ned. Akad. Wetensch. series C. 82, 245 - 273.

Boardman, N.K. (1977). Comparative photosynthesis of sun and shade plants. Annu. Rev. Plant Physiol. 28, 355 - 377.

Chapin, F.S. (1980). The mineral nutrition of wild plants. Annu. Rev. Ecol. Syst. 11, 233 - 281.

Clarkson, D.T. & Hanson, J.B. (1980). The mineral nutrition of higher plants. Annu. Rev. Plant Physiol. 31, 239 - 298.

Ernst, W.H.O. (1982). Schwermetallpflanzen. In, Pflanzenoekologie und Mineralstoffwechsel. pp. 472 - 506. H. Kinzel (ed.) Ulmer Verlag.

Farris, M.A. & Schaal, B.A. (1983). Morphological and genetic variation in ecologically central and marginal populations of *Rumex acetosella* L. (Polygonaceae). Amer. J. Bot. 70, 246 255.

Fowler, N.L. & Antonovics, J. (1981). Small-scale variability in the demography of transplants of two herbaceous species. Ecology 62, 1450 - 1457.

Graham, D. & Patterson, B.D. (1982). Responses of plants to low, non freezing temperatures: proteins, metabolism and acclimation . Annu. Rev. Plant Physiol. 33, 347 - 372.

Greenway, H. & Munns, R. (1980). Mechanisms of salt tolerance in non halophytes. Annu Rev. Plant Physiol. 31, 149 - 190..

Grime, J.P. & Mowforth, M.A. (1982). Variation in genome size- and ecological adaptation. Nature 299, 151 - 153.

Henson, I.E., Mahalakshmi, V., Bidinger, F.R. & Alagarswamy, G. (1981a). Genotypic variation in Pearl millet (Pennisetum americanum) in the ability to accumulate abscisic acid in response to water stress. J. Exp. Bot. 32, 899 - 910.

Henson, I.E., Mahalakshmi, V.,Bidinger, F.R. & Alagarswamy, G. (1981b). Stomatal responses of Pearl millet (Pennisetum americanum) genotypes, in relation to abscisic acid and water stress . J. Exp. Bot. 32, 899 - 910.

Hume, L. & Cavers, P.B. (1982). Geographic variation in a widespread perennial weed, *Rumex crispus*. The relative amounts of genetic and environmentally induced variation among populations. Can. J. Bot. 60, 1928 - 1937.

Jain, S.K. (1978). Inheritance of phenotypic plasticity in soft chess,*Bromus mollis* L.(Gramineae). Experientia 34, 835 - 836.

Kuiper, D. (1982). Genetic differentiation in *Plantago major*: Ca^{2+}- and Mg^{2+}-stimulated ATPases from roots and their role in phenotypic adaptation. Physiol. Plant. 56, 436 - 443.

Kuiper, D. (1983a). Genetic differentiation in *Plantago major*: Growth and root respiration and their role in phenotypic adaptation. Physiol. Plant. 57, 222 - 230.

Kuiper, D. (1983b). Ca^{2+}- and Mg^{2+}-stimulated ATPases from roots of *Plantago* species: Responses to different light energy fluxes . New Phytol. 94, 39 - 45.

Kuiper, D. (1984). Genetic differentiation and phenotypic plasticity in *Plantago* species.- Thesis, Univ. of Groningen, The Netherlands.

Kuiper, D. & Kuiper, P.J.C. (1979). Ca^{2+}- and Mg^{2+}-stimulated ATPases from roots of *Plantago lanceolata*, *Plantago media* and *Plantago coronopus*: Response to alterations of the level of mineral nutrition and ecological significance. Physiol. Plant. 45, 240 - 244.

Lambers, H., Posthumus, F., Stulen, I., Lanting, L., Van de Dijk, S. & Hofstra, R. (1981). Energy metabolism of *Plantago lanceolata* as dependent on the supply of mineral nutrients. Physiol. Plant. 51, 85 - 92.

Levins, R. (1963). Theory of fitness in a heterogenous environment. II. Developmental flexibility and niche selection. Amer. Nat. 97, 75 - 90.

Lewontin, R.C. (1957). The adaptations of populations to varying environments. Cold Springs Harbor Symp. Quant. Biol. 22, 395 - 408.

Lindeman, W. (1958). Observations on the behaviour of phosphate compounds on Chlorella at the transition from dark to light. Proc. IInd Intern. Conf. of the UN on the peaceful Uses of Atomic Energy 24, 8 -15.

Lowry, O. H., Rosebrough, N. J., Farr, A. L. & Randall, R. J. 1951. Protein measurement with the Folin phenol reagent. J. Biol. Chem. 193, 265 - 275.

Marshall, D.R. & Jain, S.K. (1968). Phenotypic plasticity of Avena fatua and Avena barbata. Amer. Nat. 102, 457 - 467.

Martinez, J.A. & Armesto, J.J. (1983). Ecophysiological plasticity and habitat distribution in three evergreen sclerophyllous shrubs of the Chilean matorral.Acta Oecologia 4,211 - 219.

Mook, J.H., Haeck, J. & Van der Toorn, J. (1981). Survival of Plantago lanceolata in vegetations of varying structure. Verh. Kon. Ned. Akad. Wetensch. afd. Natuurk. II 77, 32 - 35.

Nevo, E. (1978). Genetic variation in natural populations: patterns and theory. Theor. Pop. Biol. 13, 121 - 177.

Noe, R. & Blom, C.W.P.M. (1980). Occurrence of three Plantago species in coastal dune grasslands in relation to pore volume and organic matter content of the soil. J. Appl. Ecol. 19, 177 - 182.

Pollard, A.J. (1980). Diversity of metal tolerances in Plantago lanceolata L. from the Southeastern United States. New Phytol. 86, 109 - 117.

Primack, R.B. (1978). Evolutionary aspects of wind pollination in the genus Plantago. New Phytol. 81, 449 - 458.

Primack, R.B. (1980). Phenotypic variations of rare and wide spread species of Plantago. Rhodora 82, 87 - 95.

Price, H.J., Bachman, K., Chambers, K.L. & Riggs, J. (1980). Detection of intraspecific variation in nuclear DNA content in Microseris douglasii. Bot. Gaz. 141, 195 - 198.

Price, H.J., Chambers, K.L. & Bachman, K. (1981). Genome size variation in diploid Microseris bigelovii (Asteraceae). Bot.

Gaz. 142, 156 - 159.

Quarrie, S.A. (1980). Genotypic differences in leaf water potential, abscisic acid and proline concentrations in spring wheat, during drought stress. Ann. Bot. 46, 383 - 394.

Quarrie, S.A. (1981). Genetic variability and inheritability of drought-induced abscisic acid accunulation in spring wheat. Plant Cell Environ. 4, 147 - 151.

Sagar, G.R. & Harper, J.L. (1969). Biological Flora of the British Isles: *Plantago major*, *Plantago lanceolata* and *Plantago media*. J. Ecol. 52, 189 - 221.

Smakman, G. & Hofstra, J.J. (1982). Energy metabolism of *Plantago lanceolata* as affected by change in root temperature. Physiol. Plant. 56, 33 - 37.

Somero, G.N. (1978). Temperature adaptation of enzymes. Annu. Rev. Ecol. Syst. 9, 1 - 29.

Stokes, P.M. & Dreier, S.I. (1981). Copper requirement of a copper-tolerant isolate of *Scenedesmus* and the effect of copper depletion on tolerance. Can. J. Bot. 59, 1817 - 1823.

Stulen, I., Lanting, L.,Lambers, H., Posthumus, F., Van de Dijk, S. & Hofstra, R. (1981). Nitrogen metabolism of *Plantago lanceolata* as dependent on the supply of mineral nutrients. Physiol. Plant. 51, 93 - 98.

Teramura, A.H. (1983). Experimental ecological genetics in *Plantago* IX: Differences in growth and vegetative reproduction in *Plantago lanceolata* (Plantaginaceae) from adjacent habitats. Amer. J. Bot. 70, 53 - 58.

Teramura, A.H. & Strain, B.R. (1979). Localized populational differences in the photosynthetic response to temperature and irradiance in *Plantago lanceolata*. Can. J. Bot. 57, 2559 - 2563.

Teramura, A.H., Antonovics, J. & Strain, B.R. (1981). Experimental ecological genetics in *Plantago* IV: Effects of temperature on growth rates and reproduction in three populations of *Plantago lanceolata*. Amer. J. Bot. 68, 425 - 434.

Van der Aart, P.J.M. (1984). Demographic, genetic and ecophysiological variation in *Plantago major* and *Plantago lanceolata* in relation to habitat type. Handbook of Vegetation Science: Vol. 3 " Population structure of vegetation " (in press).

Warwick, S.I. & Briggs, D. (1979). The genecology of lawnweeds III: Cultivation experiments with *Achillea millefolium* L., *Bellis perennis* L., *Plantago lanceolata* L., *Plantago major* L. and *Prunella vulgaris* L. collected from lawns and contrasting grassland habitats. New Phytol. 83, 509 - 536.

Wild, A. (1979). Physiologie der Photosynthese hoeherer pflanzen. Die Anpassung an die Lichtbedingungen. Ber. der Deutsch. Bot. Ges. 92, 341 - 364.

Ion transport differentiation among plants from four contrasting soils in the mediterranean ruderal *Dittrichia* (ex-*Inula*) *viscosa* W. Greuter.

J.P. Wacquant and N. Bouab*

C.N.R.S, Centre Louis Emberger, Unité B2P, 34033 Montpellier, France

ABSTRACT

In the area of Montpellier (France) *Dittrichia viscosa* grows naturally on prevailing calcareous soils and on salty and acid soils which are scarce. The study was carried out with the progenies derived from seeds of one plant from each of four contrasting soils (alkaline-calcareous, calcareous-salty, moderately acid and very acid). Progenies from the 4 habitats were germinated on 3 natural soils (calcareous, salty and acid) and grown in 3 solutions (calcareous, salty and acid) simulating natural soils (4x3x3x7 replicates). Biomass measurements and xylem sap analysis of plant individuals were performed. The variance analysis showed that differences observed between plants were more marked when considering the cation content of the xylem sap than the biomass. Most differences appeared with the solution that simulated an acid soil indicating that plants originated from the two calcareous habitats and those from the two acidic habitats differed genetically in their ability to select cations, mainly K and Ca, like calcicolous and calcifugous species respectively. For any of the 3 growth solutions there was less functional variability among plant individuals from calcareous sites than from the acidic ones. The study showed that the germination soil can have a screening effect (genetic) and a long term depressive or stimulating effect (phenotypic) on *Dittrichia viscosa*. The acidic habitat plants appear physiologically polymorphic in

---

* Present address : Faculté des Sciences, Rabat, Maroc.

NATO ASI Series, Vol. G5
Genetic Differentiation and Dispersal in Plants
Edited by P. Jacquard et al.

contrast to those of the calcareous habitat. They possibly contain calcicole and calcifuge genotypes. Results suggest that calcifuge populations may be in the process of differentiating from the prevailing calcicole type.

INTRODUCTION

Plant population differentiation in response to natural mineral stresses has been intensively studied on serpentine and heavy metal containing soils and to a lesser extent on salty, acidic and calcareous soils using demographic and genetic approaches (Rorison, 1969; Antonovics et al., 1971; Bradshaw, 1971).

Nutritional investigations are comparatively scarce (Rorison, 1969; Epstein, 1972) and are often descriptive rather than analytical and mechanistic (Epstein, 1972; Wright, 1976; Levitt, 1980). Nutritional differentiation has been studied from a functional view-point in *Anagallis arvensis* (Wacquant et al., 1981). Plants from two contrasting soil habitats, calcareous and acidic, were genetically different in the ability of their roots to select major cations. The difference suggested that the former selected cations like calcicole and the latter like calcifuge type species. The differentiation was demonstrated by growing the plants under the same hydroponic conditions and comparing in vivo the mineral content of their root xylem sap.

Our current research goal is to utilize the same experimental approach to examine *Dittrichia viscosa*, another ubiquitous species having plants distributed in contrasting mineral habitats. The species is a circumeditteranean ruderal bush that rapidly colonizes disturbed areas. In southern France near Montpellier *D. viscosa* is a very common plant on road embankments, waste-lands, pits and fallows, regardless of the chemical nature of the soil. Calcareous soils (marl, calcareous colluviums) are dominant in the area while acidic or salty soils are of minor importance. In the recent past *D. viscosa* was described only in plant communities of calcareous soils (Braun-Blanquet et al., 1952). During the last 20 years, however, a greater area and more diversified soil types have apparently opened up for colonization by *D. viscosa*. The plant is now found on acid soils and sometimes on salty soils. Snaydon (1970) showed in a naturally acid area of England that population differentiation of *Anthoxanthum odoratum* could occur on limed plots in less than 40 years. We assume that *D. viscosa* has produced distinct calcifugous populations on acidic soils and halophytic populations on salty soils. Our study is to determine if plants from calcareous, acidic and salty soils differ genetically in the ability of roots to select cations and if the differences are ecologically significant.

MATERIAL AND METHODS

Studies were conducted with progenies from seed obtained from plants growing in four of the most contrasting soils near (10 to 60 km) Montpellier, France. The seed was collected in November 1980 from the habitats, named, located and characterized as follow:

M (Mauguio), a sloping wall of an abandoned gravel pit (10 m in vertical height) with very acid (pH 4 to 5) clay soil on which *D. viscosa* plants were well developed ;

F (Faugères), an olive-tree fallow with a well developed stand of *D. viscosa* on a moderately acid schistous soil (pH 5 to 6) ;

G (Saint-Gély), a fallow with a dense stand of *D. viscosa* on a highly calcareous soil (marl) with pH 8-8.2 ;

P (Palavas), a marl bank (pH 8) of a salty lagoon with *D. viscosa* bushes scattered amongst a halophytic vegetation.

In each site the seeds of one plant (4 to 6 years old) were collected for the experiment. The breeding system of *D. viscosa* has not yet been precisely studied. It can be presumed that at least partial inbreeding occurs since the pollen is sticky and pollinators (bees) make several consecutive visit to flowers on the same plant.

Bouab (1982) found no differences between D. viscosa plants from 4 contrasting soils when the experiment was carried out using quartz sand and deionized water for germination and a non stressful nutrient solution for growth (conditions similar to the experiment with Anagallis arvensis, Wacquant et al., 1981). However differences appeared when plants were grown in a salty nutrient solution. This suggested the possible use of harsh germination and growth media for the current study.

Germination. The seeds from each plant M, F, G and P were germinated in trays on 3 kinds of soils, acid (a), calcareous (c) and salty (s). The soils were collected from the Mauguio, Saint-Gély and Palavas habitats respectively. The proportion of germinating seeds on a, c and s was respectively 48, 49 and 29% for M; 33, 35 and 24 for F; 66, 73 and 52% for G and 34, 36 and 20 for P.

Growth. The seedlings were grown in a greenhouse on sand culture with automatic subirrigation (Wacquant, 1974). Fourteen plants of each type of seedling Ma*, Mc, Ms; Fa, Fc, Fs; Ga, Gc, Gs and Pa, Pc, Ps, were transplanted, one per pot, on quartz sand in each of three independent hydroponic units (14 individuals X 4 sites X 3 germination soils X 3 units).

In the first unit the irrigation solution was acid (Table 1) and was composed to simulate acid soil. In the second and third units the solutions (Table 1) both with alkaline pH were respectively calcareous and salty, each being composed to simulate calcareous and salty soils. Because soluble phosphate and iron would not be available in the high pH of the two latter solutions, a capsule of 1g of insoluble phosphate powder (tricalcium orthophosphate) was

* Ma indicates plant from Mauguio (M) germinated on acid soil (a)

included in the sand of each pot for root contact and iron was supplied for both solutions as Fe-EDDHA chelate. Solutions were replaced once a week.

Table 1. pH and major ion concentration (mN) of the Acid, Calcareous and Salty growing solutions. In each solution micronutrient concentration (mM. 10-3) was as follows: Mn 0.7, Zn 0.6, B 1.8, Cu 0.2, Mo 0.01

	pH	NO_3^-	NH_4^+	$H_2PO_4^-$	$SO_4^{=}$	K^+	Ca^{++}	Mg^{++}	Al^{+++}	Fe^{+++}	$CO_3^{=}$	Na^+	Cl^-
Acid	4.2	1.4	0.35	0.55	1.5	1.0	1.0	0.4	1.1	0.1**	0	0.3	0
Calcareous	7 to 9	3.4	0	0.2+	2.4	0.6	5.5	0.4	0	0.009*	HCO_3^-	0.006	0
Salty	9	1.4	0	0.2+	1.4	0.6	2.0	0.4	0	0.009*	10.0	20.0	10.5

+ Plus Ca-3 orthosphosphate powder (1g per pot)

** as citrate, * as Fe-EDDHA chelate

Plants were grown during spring and early summer in a shaded and mist cooled greenhouse. All plant material examined was non-flowering.

Xylem sap collection and analysis. When plants were 72 days old, 7 plants (replicates) of each of the 36 plant types (4 sites X 3 germination soils X 3 growing solutions) were chosen for sap collection. Each plant was detopped in the morning and the xylem sap collected during the following 7 hours (see Fig 1). Collected sap was immediately frozen (- 18°C). Ion content was determined after the sap was thawed and diluted (50 to 200 fold, up to 400 in the case of Mg). Cations were assayed by flame emission (K, Ca and Na) and atomic absorption (Mg) spectrophotometry. For the statistical interpretation of sap analysis data, 5 missing values have been estimated using the mean of 6 replicates instead of 7.

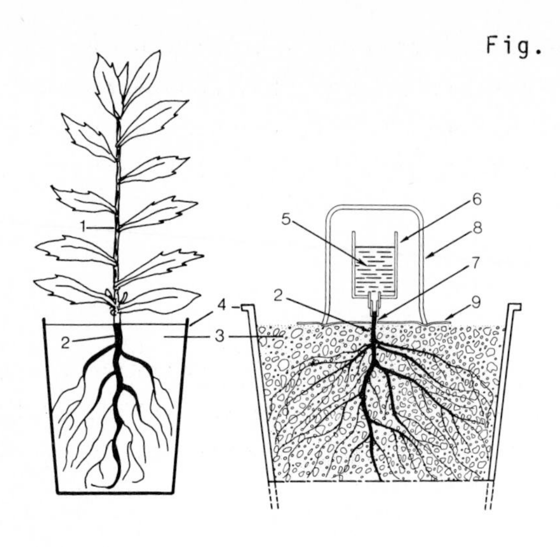

Fig. 1. A plant before (left) and during xylem sap collection (right). 1 shoot, 2 root, 3 silicious sand, 4 plastic pot (1L), 5 xylem sap, 6 sap container, 7 sap collecting tube, 8 beaker to prevent evaporation, 9 wet filter paper.

Biomass and mineral content of plant shoots and roots. Among the 7 plants of each type used for sap determination, 5 were analyzed for biomass and mineral content. Shoots of these plants were collected before sap collection and roots, after. As soon as collected they were rinsed with deionized water, weighed (Fresh weight, FW), dried at 80°C for 24 hours, weighed (Dry weight, DW) and stored.

The content of potassium, calcium, magnesium and sodium in dried shoots and roots was determined in a 0.1 N HNO_3 extract (28°C) for 36 hours (60 ml per g. of fresh weight). Cations were assayed as described for sap, and mean values for three replicates of each extract were calculated.

Cation content units and selectivity. The content y in each cation (K, Ca, Mg or Na) was determined as $\mu eq.ml^{-1}$ for xylem sap and $\mu eq.g^{-1}$ DW for shoots and roots. The proportion of each cation relative to the others was calculated as follows. For K, Ca or Mg as a percent of S the total amount of the three cations: $\%S=(y/S)100$. For Na as a percent of S', the total amount of K, Ca, Mg and Na: $\%S'=(y/S')100$. Since y and S or S' are expressed with the same unit ($\mu eq.ml^{-1}$ or $\mu eq.g^{-1}$), volume (ml^{-1}) or weight (g^{-1}) terms cancel and %S or %S' represent a charge percentage of a given cation relative to the total charge of the whole concerned cation species. This charge percentage is a convenient unit for comparing ion selective ability of plants since the uptake of a given ion is dependent of that of the other ions (Wacquant,1974; Wacquant et al., 1981).

RESULTS

All individual data for biomass (4 sites X 3 germination soils X 3 growth solutions X 5 replicates) and xylem sap (4X3X3X7) have been submitted to an analysis of variance according to a factorial model (Table 2).

Table 2 - Statistical significance (F test) of the analysis of variance for biomass and cation content of xylem sap.

		Sap			
			% S		
	Biomass	Na %S'	K	Ca	Mg
Sites	N.S.	9.69 ***	21.25 ***	11.16 ***	18.23 ***
Germination soils	N.S.	3.48 *	5.36 ***	6.68 ***	N.S.
Growing solutions	69.72 **	1476.41 ***	924.67 ***	1739.71 ***	70.74 ***
Sites X Germ. soils	3.62 ***	N.S.	3.78 ***	2.79 *	2.41 *
Sites X solutions	N.S.	5.10 ***	4.69 ***	3.77 ***	3.50 ***
Germ. soils X sol.	N.S.	N.S.	N.S.	N.S.	N.S.

*** P < 0.5%, ** P < 1%, * P < 5%

VARIABILITY AMONG PLANTS FROM CONTRASTING HABITATS

Comparisons are for plants germinated and grown under similar chemical conditions: plants germinated on the acid soil (Ma, Fa, Ga and Pa) and grown in Acid solution, plants germinated on the calcareous soil (Mc, Fc, Gc and Pc) and grown in Calcareous solution, and plants germinated on the salty soil (Ms, Fs, Gs and Ps) and grown in Salty solution.

Biomass performance in the various growth solutions. In general (Table 3) the best growth was made in the Acid solution (normal green leaves), the poorest in the Calcareous solution (green lighter leaves) and those in the salty solution (green-yellowish leaves) were intermediate. No differential deficiency or toxic symptoms appeared in leaves of the M, F, G and P plants, although differences were observed in their biomass (Table 3).

Table 3 - Biomass of the M, F, G and P plants in the Acid, Calcareous and Salty solutions. Means ± SE, n=5, in g FW (whole plant).

	Acid	Calcareous	Salty
M	11.37 ± 0.62	7.19 ± 0.51	9.05 ± 0.69
F	9.54 ± 1.13	6.87 ± 0.46	9.69 ± 0.77
G	8.45 ± 0.37	7.59 ± 0.37	8.59 ± 0.49
P	9.71 ± 0.61	7.95 ± 0.65	7.61 ± 0.48

In the Acid solution the greatest growth was by M plants originating from the most acidic habitat and the least by G plants from the highly calcareous habitat, while the F plants from moderately acid soil and the P plants from the calcareous-salty soil had an intermediate performance. In the Calcareous solution there was a tendency for the plants originating from the calcareous habitats (G and P) to grow better than those from the acidic ones (M and F). In the Salty solution the worst performance was observed for the P plants, the progenitor of which was from the salty habitat.

Sodium content of xylem sap in the three solutions. In general the sap of plants in the Salty solution contained much more sodium (52.3%) than that of plants in the Acid (14.2%) or Calcareous (1.55%) solutions (table 4). This is related to the sodium concentration of the solutions (Table 1). There was a tendency for the P plants from the calcareous-salty habitat and the G plants of the calcareous site to contain proportionately less sodium than the acidic habitat plants M and F.

Table 4. Mean proportion of Na (%S') in the xylem sap of plants M, F, G and P when growing in the Acid, Calcareous or Salty solution. S' is the total amount of K, Ca, Mg and Na in µeq. ml-1 and ± the Standard Error.

	Acid		Calcareous		Salty	
	Na	S'	Na	S'	Na	S'
M	15.6 ± 0.8	15.9 ± 0.7	1.7 ± 0.2	12.9 ± 0.6	53.5 ± 1.5	24.7 ± 0.7
F	13.3 ± 0.6	17.2 ± 0.9	1.7 ± 0.1	12.6 ± 0.3	53.8 ± 0.9	28.1 ± 1.2
G	13.6 ± 0.3	16.2 ± 1.0	1.4 ± 0.1	11.9 ± 0.3	50.3 ± 1.2	25.7 ± 0.7
P	14.3 ± 0.6	14.7 ± 0.6	1.4 ± 0.1	11.9 ± 0.5	51.4 ± 1.0	24.4 ± 0.8

Major cations of xylem sap in the alkaline solutions. In the Calcareous solution (Fig 2) the seven G individuals are close in their selection of cations K, Ca and Mg as well as the seven P individuals and most of the M and F individuals. The G and P form superposed clusters. There is less individuals variability among the calcareous plants G and P than among the acidic plants M and F. In the Salty solution (Fig 2) there is a tendency for the P plants to differ from the G plants (shift of the clusters) but there is still less variability within the P and G plants than within the M and F ones.

Major cations in xylem sap, shoot and root in the Acid solution. In the Acid solution (Fig 3) the M, F, G and P plants differ in their xylem sap content. The largest difference is observed between the G plants originated from the calcareous habitat and the M plants from the more acid habitat.

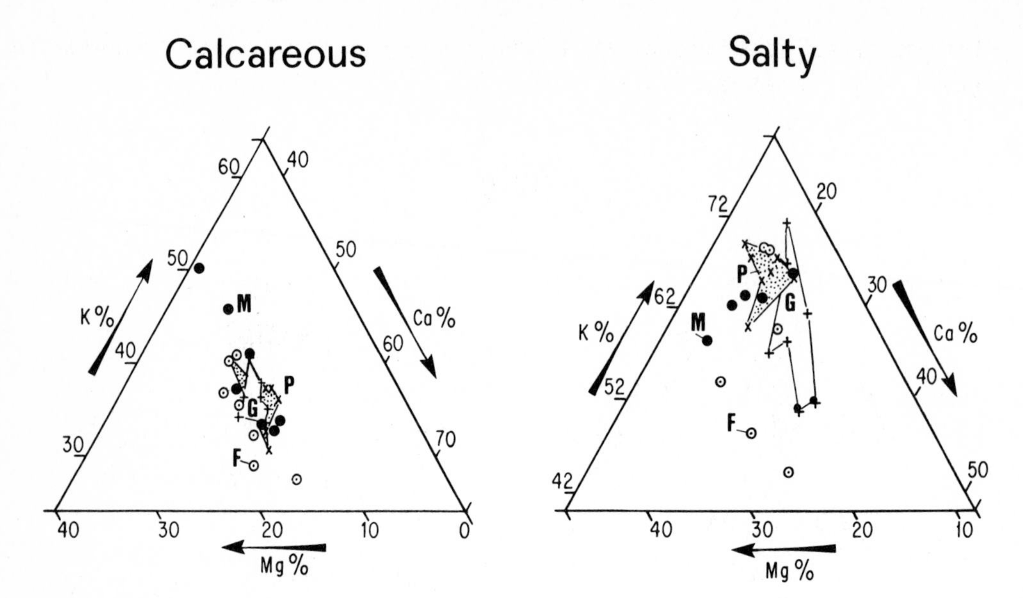

Fig 2 - Proportion of K, Ca and Mg (%S) in xylem sap of individual plants M, F, G and P in the Calcareous and Salty solutions. S, total amount of K, Ca and Mg in µeq. ml-1. Each dot in the triangle correspond to an individual value (6 to 7 per plant type). Dots enclosed are for plant types (here G and P) that form superposed or separated clusters. For a given dot, the K percentage is determined by drawing a line to the K% axis, paralelle to the Mg% axis, the Ca percentage by drawing a line to the Ca% axis, paralelle to the K% axis, and the Mg percentage by drawing a line to the Mg% axis, paralelle to the Ca% axis.

They differ mainly in their ability to select the monovalent K and divalent cations Ca and Mg. The xylem sap of the G plants contains proportionately more potassium and less calcium and magnesium than that of the M plants. The P plants, also originating from calcareous (and salty) soil, respond like the G plants and the F plants (the soil pH of the F plants was intermediate to those of G and M) show an intermediate response to the G and M plants.

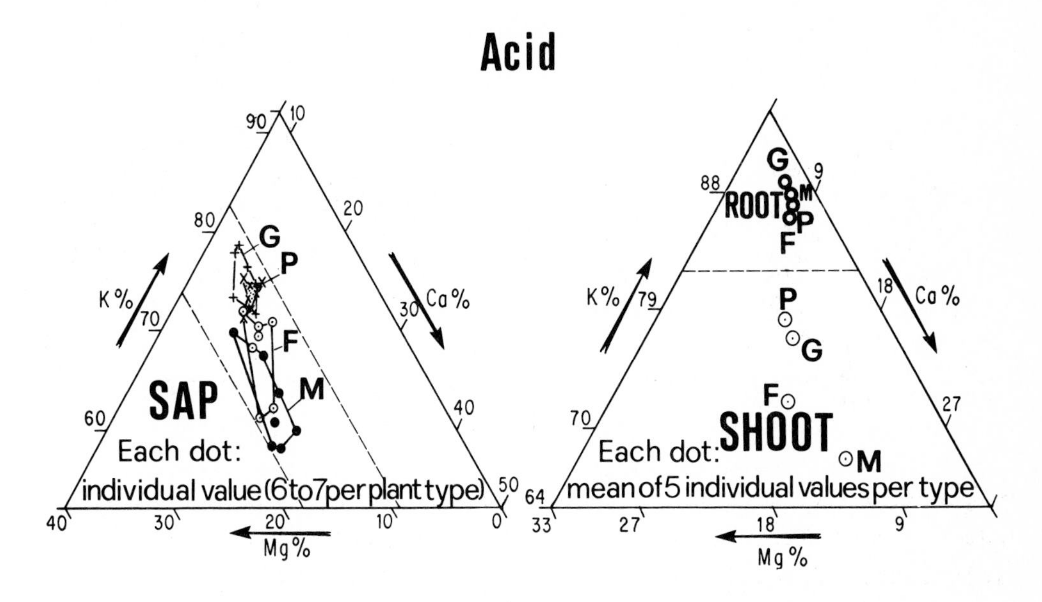

Fig 3 - Proportion of K, Ca and Mg (%S) in root xylem sap and in shoot and root tissues of the M, F, G and P plants grown in Acid solution.

The comparison of proportion of cations in xylem sap and tissues (Fig 3) show that the relationships among M, F, G and P plants are similar for root sap ion content and shoot tissue ion accumulation and suggest a genetic differentiation in the capacity of the roots to transport ions.

VARIABILITY BETWEEN PLANTS FROM THE SAME HABITAT

Are the progenies of the adult plants selected in each of the four habitats physiologically similar or polymorphic i.e. composed of various specialized individuals, some functioning like calcicolous plants, the others like calcifuge or halophytes? To answer this question we compared progeny from parent plants from sites M, F, G and P, germinated on acid (a), calcareous (c) and salty (s) soils and grown in a similar culture solution. The reason for utilizing various germination media was to examine the possibility of germination media screening specific genotypes. We assumed that each soil would select its own specialized individuals. If the hypothesis is supported, the progenies a, c and s of a given plant (M, F, G and P) would differ in their biomass yield and sap cation content.

Biomass performance according to the germination soil. According to the assumption, biomass yield would be best for plants a in the Acid solution (A), best for c in the Calcareous solution (C) and best for s on the Salty solution (S). This is partially confirmed (Fig 4) by the results with Ma plants on A, Fs on S, Pc on C and to a lesser extent with Gc on C and Gs on S. Another assumption is suggested (Fig 4) when comparing, for a given progenitor, the performance of each of the plants a, c or s in the three solutions. For M the a, compared to the c or s, grows well on A but also on C or S. It is as if M plants were stimulated on the acid germination soil contrary to G plants that are depressed and which in turn appear equally stimulated by the calcareous (c) and calcareous-salty (s) soils. For the P plants it would be the calcareous soil (c) which stimulate particularly on C and S and not the salty (s). For the F plants the stimulating soil would be the salty (s), especially on C and S. It is concluded that the chemical nature of the germination soil can induce a long term stimulating or depressing effect on further growth.

Cations in xylem sap according to the germination soil. Figure 5 shows that some differences appear among progenies a, c and s of the parent M in the Acid solution as well as in Calcareous and Salty solutions and, of the parent F in Acid solution i.e. for the parents of the acid habitats. Conversely the progenies a, c and s the parent of which were in the calcareous (G) and calcareous-salty (P) habitats show (Fig 5) less differences in any of the three solutions except for G progenies in the Salty solution.

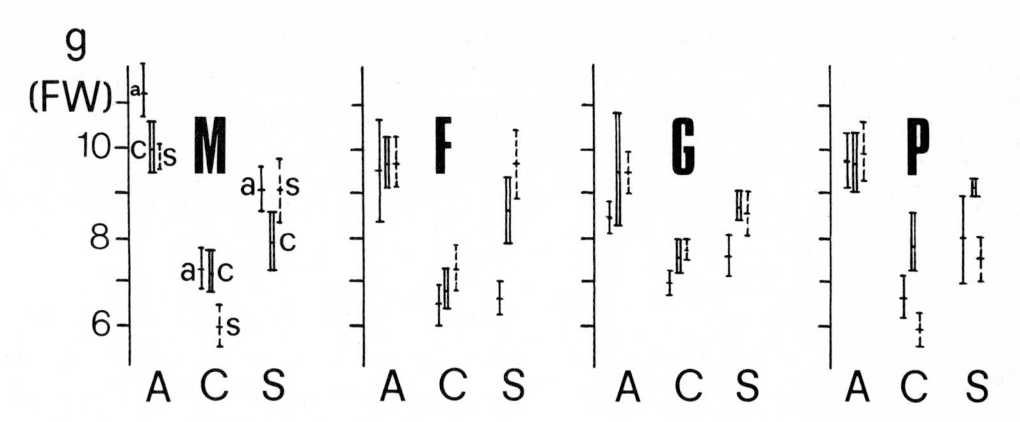

Fig 4 - Mean biomass (g FW ± SE, n=5) of the M, F, G and P progeny. Variation according to the germination soil: acidic (a), calcareous (c) or salty (s) and the growth solution: Acid (A), Calcareous (C) or Salty (S).

When the progenies a and c differ (M/Acid, Calcareous and Salty solutions; F/Acid solution) there is a tendency for the a to select more calcium and less potassium than the c. It is noticeable in Acid solution that plants Ma and Mc (Fig 5) differ in the same manner as plants Ma and Ga (Fig 3). In the same conditions Fa and Fc differ like Fa and Ga. There is a tendency for the M and F plants to behave like the G plants if germinated on the calcareous soil. The progenies c and s behave similarly, except in three cases (M/Calcareous, F/Acid, G/Salty). It is noticeable that in the Acid solution the Mc and Ms behave likely (Fig 5) as observed for the G and P plants (Fig 3). There is a tendency for the M plants to behave like the P plants if germinated on the salty soil.

DISCUSSION AND CONCLUSION

The variance analysis showed that differences observed between plants, according to their origin habitat, germination soil and growth medium, were more marked by considering the content of the xylem sap than the biomass.

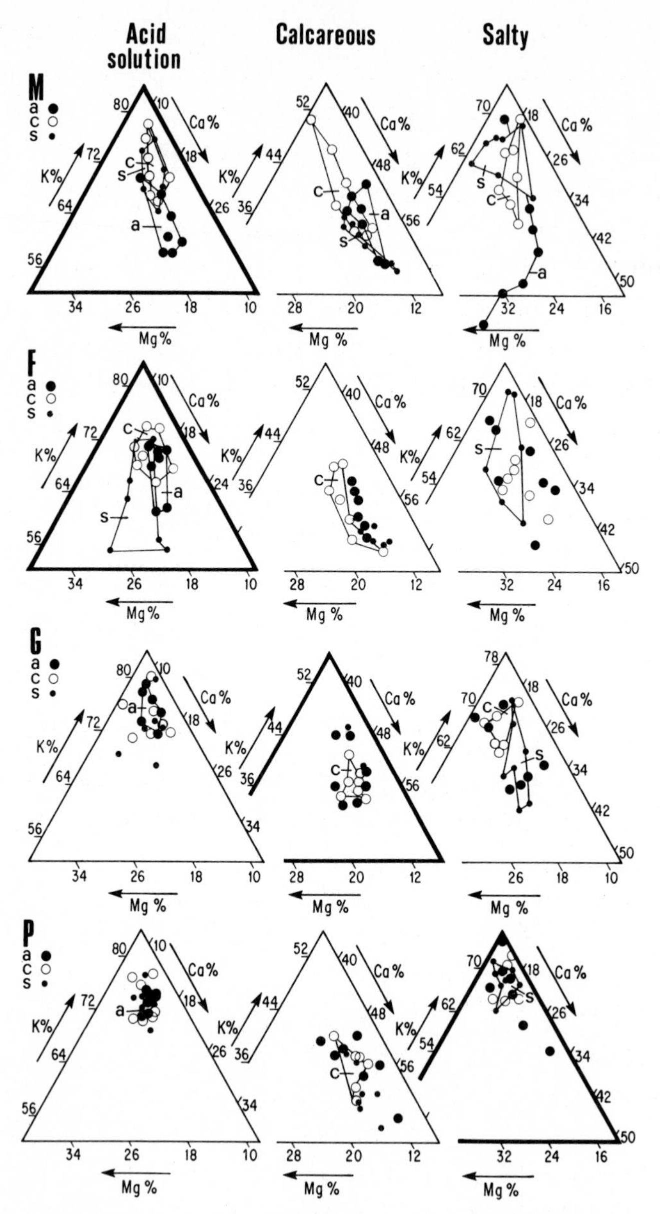

Fig 5 - Proportions of K, Ca and Mg (%S) in the root xylem sap of the a, c and s progenies of the plant M, F, G and P in the three growth solutions (Acid, Calcareous and Salty). The progenies a were germinated on acidic soil, the c and s respectively on calcareous and salty soils.

For interpretation of this figure the important concept is to observe in each triangle the separation (or lack of) of clusters of the three kind of progenies

PLASTICITY OF THE SPECIES. The growth of Dittrichia viscosa was better in the Acid solution than in the Calcareous one. The reversed was expected since the species normally occurs and grow well on calcareous soil in the study areas. We think the Calcareous solution did not adequately simulate the calcareous soil. This was because the solution, containing chelated iron, could not induce iron-deficiency and the pH 9 proved difficult to maintain. Daily hydroxide additions to control pH caused precipitations. The performance in the Acid solution (pH 4.2) showed that the plant tolerated a relatively high level of NH4+ (0.35 mN) and a toxic Al+++ level (1.1 mN). The intermediate performance in the Salty solution showed that D. viscosa tolerated salt (20 mN) and high alkalinity (pH 9) even though the plants had green yellowish leaves. It can be concluded that D. viscosa tolerates a wide range of artificial chemical conditions even when plants respond differently according to the habitats of origin of their parents.

DIFFERENTIATION AMONG PLANTS FROM CONTRASTING HABITATS. Some differentiation of adaptative significance appeared among the plants M, F, G and P mainly when grown in the Acid solution.

The largest difference occured between the M plants from the more acidic habitat and the G plants from the highly calcareous site. They differed in their biomass as shown in reciprocal transplant studies for populations of Trifolium repens (Snaydon, 1962) and Teucrium scorodonia (Hutchinson, 1967). The acidic plants M, like typical calcifuge species grew better in the Acid medium than the calcareous plants G. However it must be pointed out that in the Calcareous solution the M plants did not show any iron-deficiency chlorosis as observed in the extreme case of calcifuge ecotype differentiation (Snaydon, 1962; Hutchinson, 1967).

Reciprocal transplants of the M and G plants of D. viscosa on calcareous and acid soils showed (Wacquant et al., 1983) that the plants differed essentially in their ability to accumulate calcium and potassium as described by other workers for calcicolous and calcifugous species (Duvigneaud et al., 1964; Passama, 1970; Hamzé et al., 1980; Salsac 1980; Hamzé, 1983) and populations (Wacquant et al., 1981). It was suggested that the M plants, like calcifuges, have a high affinity for potassium and the G plants, like calcicoles, for calcium since on acidic soil the M accumulated more potassium and less calcium than the G.

In Acid hydroponics (Fig 3), the M and G plants also differed markedly in their ability to accumulate K and Ca in xylem sap as well as in shoots but results were reversed to those observed in soils. The M selected more calcium and less potassium than the G. This change suggests that in the Acid solution, possibly richer in NH4+ (20% of the N as NH4+ and 80% as NO3-) than the acid soil, the M plants absorbed more NH4+ as observed for calcifuges (Haynes et al., 1978), than the G plants. Because of competitive interaction between NH4+ and K+ (Haynes et al., 1978), the M plants could absorb more NH4+ and then less K+ than the G plants. Even though the assumption has to be experimentally confirmed the differences observed in sap and shoot cation content between the M and G plants indicate a clear differentiation of physiotypes . On the other hand the differences in biomass in the Acid solution and in cation content on acidic and calcareous soils (loc. cit.) strongly suggest that the plants M behave physiologically as calcifuges and that plants G as calcicoles.

It is noteworthy that the plants F from the moderately acid habitat had in the Acid solution an intermediate response between the more acid plants M and the alkaline calcareous plants G. The intermediate response was observed for biomass performance and ion selectivity as well, suggesting a calcifuge type behavior.

The plants P although originated from a salty habitat did not perform better in the Salty solution than the other plants. They had the least biomass yield in the Salty solution and sodium or major cation transport did not suggest a halophytic behavior.

However the P plants originated from the salty soil which was also calcareous behaved in many respects rather like the calcareous plants G.

VARIABILITY WITHIN THE PROGENY. Progenies of a given plant, germinated on acid, calcareous and salty soils and grown in the same solution differed in their biomass suggesting that germination soils could select different genotypes. We also observed that a given germination soil could induce in some plants a long term stimulating or depressing effect on further growth and this almost independently of the future growth medium. As far as we know those effects are not mentioned in the literature.

There is little doubt that such carryover effects of the soil can render less evident a possible genetic selection of ecotypes by screening plants on soils at the germination level. As a consequence differences observed between plants germinated on acid, calcareous and salty soils can be genotypic and phenotypic and at present there is no way to distinguish the two effects. However it appears that the germination soil carryover effect could have an ecological significance since it was the acid germination soil that stimulated growth of the acid habitat plants (M) while it depressed that of the calcareous habitat plants (G) whatever the growth solution. It is also the calcareous germination soil that seems to influence more favorably the plants originated from the calcareous habitats (G and P) than those from the acid habitats (M and F).

Progenies germinated on acid, calcareous and salty soils also differed in their xylem sap composition (Fig 5). Again they differed more for acid habitat progenitors (M and F) than for calcareous progenitors (G and P). The largest difference was between progenies germinated on the acid and calcareous soils and particularly for the more acid plants (M). The acid progenies (Ma) tended to select more calcium and less potassium than the calcareous one (Mc). The same differences were observed (Fig 3) between the acid habitat plants (M) and the calcareous habitat plants (G) suggesting that the acid habitat plants (M and F) if germinated on calcareous soil function like the calcareous habitat plants G or P i.e. that they could contain calcicolous type plants in their progeny.

It could be argued that functional differences observed between Ma and Mc plants (Fig 3) in Acid solution and maintained in both Calcareous and Salty solutions could be phenotypic if resulting from a germination soil carryover effect. Physiological arguments suggest that similarity in the three solutions could be coincidental. If Ma plants select less potassium that Mc in the Acid solution it would be as previously discussed because Mc if a calcicole would absorb less $NH4^+$ and more K^+ than Ma. However on both alkaline solutions it would be because Ma like a calcifuge, contrary to Mc like a calcicole, would not limit Ca^{++} entrance into the plant at the expenses of K^+ (Salsac, 1980). So the observed physiological differences between progenies germinated on acid and calcareous soils would be genotypic rather than phenotypic. It can be concluded that the progenies of the acid habitat plants M and F possibly contain calcicole and calcifuge genotypes. They are physiologically polymorphic contrary to that of the calcareous habitat plants G and P.

This functional variability only observed among the acid habitat plants seems consistent with the fact that the only clear functional differentiation among the plants of the four habitats was observed in the Acid solution. Why such a functional diversity among the plants of the acid habitat? Physiological polymorphism often suggest a better capacity for colonizing new environments (Bradshaw, 1971; Wacquant et al., 1981). Possibly D. viscosa, a common plant on calcareous soil in the area of Montpellier, is now differentiating calcifuge genotypes on the acid soils.

Our work achieved with progenies of only one plant from each of four contrasting habitats shows that D. viscosa plants can differ genetically in the ability of their root systems to select cations in response to mineral constraints. Interpretation of the results in the context of populations is limited but the experiment is a first step toward population physiology.

REFERENCES

Antonovics J, Bradshaw A D, Turner R G (1971) Heavy metal tolerance in plants Adv in Ecol Res 7: 1-85

Bouab N (1982) Nutrition minérale d'Inules visqueuses (Dittrichia viscosa W Greuter) provenant de différents biotopes Thèse Spé Univ II, Montpellier

Bradshaw A D (1971) Plant evolution in extreme environments In: Creed (ed) Ecological genetics and evolution Blackwell Scientific Pub: 20-50

Braun-Blanquet J, Roussine N, Negre R (1952) Les groupements végétaux de la France Méditerranéenne CNRS, Paris, p 297

Duvigneaud P, Denaeyer-de Smet S (1964) Le cycle des éléments biogènes dans l'écosystème forêt Lejeunia 28: 1-148

Epstein E (1972) Mineral nutrition of plants: principles and perspectives J Wiley and Sons, New York, p 412

Hamzé M, Salsac L, Wacquant J-P (1980) Recherche de tests pour déceler précocement l'aptitude des agrumes à résister à la chlorose calcaire 1 capacité d'échange cationique et degré d'estérification des racines Agrochimica 24: 432-442

Hamzé M (1983) Recherches sur la nutrition et la chlorose des agrumes en sols calcaires Thèse Doc ès-Sci Univ II, Montpellier

Haynes R J, Goh K M (1978) Ammonium and nitrate nutrition of plants Biol Rev 53: 465-510

Hutchinson T C (1967) Ecotype differentiation in Teucrium scorodonia with respect to susceptibility to lime-induced chlorosis and to shade factors New Phytol 66: 439-453

Levitt J (1980) Responses of plants to environmental stresses, vol 2, Academic Press, London, p 607

Passama L (1970) Composition minérale de diverses espèces calcicoles et calcifuges de la Région méditterranéenne française Oecol Plant 5: 225-246

Rorison I H (1969) Ecological aspects of the mineral nutrition of plants Blackwell Scientific Pub, p 484

Salsac L (1980) L'absorption du calcium par les racines des plantes calcicoles et calcifuges Bull Ass Fr Et Sol 1: 45-77

Snaydon R W (1962) The growth and competitive ability of contrasting natural populations of Trifolium repens L. on calcareous and acid soils J Ecol 50: 439-447

Snaydon R W (1970) Rapid population differentiation in a mosaic environment 1 The response of Anthoxanthum odoratum populations to soils Evolution 24: 257-269

Wacquant J-P (1974) Recherches sur les propriétés d'adsorption cationique des racines (Rôle physiologique et importance écologique) Thèse Doc ès-Sci Univ II, Montpellier

Wacquant J-P, Hochepot M, Valdeyron G (1981) Variation dans la composition cationique des sèves xylémiques d'Anagallis arvensis L. provenant de deux sols, acide et calcaire, cultivés dans les mêmes conditions C R Acad Sc, Paris, 293: 813-816

Wacquant J-P, Bouab N (1983) Nutritional differentiation within the species Dittrichia viscosa W Greuter, between a population from a calcareous habitat and another from an acidic habitat Plant and Soil 72: 297-303

Wright M J (1976) Plant adaptation to mineral stress in problem soils Cornell Univ Pub, New York, p 420

Dispersal: Gene Flow

GENE-FLOW IN TRIFOLIUM REPENS
- AN EXPANDING GENETIC NEIGHBOURHOOD

C. GLIDDON & M. SALEEM
School of Plant Biology,
University College of North Wales,
Bangor, Gwynedd LL57 2UW, U.K.

ABSTRACT

Effective levels of outcrossing, pollen from flower and seed from pod dispersal distances are measured in two natural populations of *Trifolium repens*. A computer model of plant growth (Bell *et al*., 1979) is used to estimate the dispersal of inflorescences from the parental germination site for white clover of different ages. A model is presented which allows calculation of the parent-offspring dispersal variances from a combination of the above components of gene flow and neighbourhood area is calculated according to a modification of the formula of Crawford (1984a).

It is shown that the vegetative growth component of gene dispersal causes the genetic neighbourhood area to increase with age of plant. The possible range of neighbourhood area is calculated and compared with the observed distribution of genetic variation in the natural populations studied. Some of the problems associated with the neighbourhood approach are discussed.

INTRODUCTION

For almost two decades, since the application of gel-electrophoresis to natural populations of animals and plants, it has been apparent that an extremely large amount of genetic variability exists within species. This observation has been of some embarrassment to theoretical population geneticists of both

NATO ASI Series, Vol. G5
Genetic Differentiation and Dispersal in Plants
Edited by P. Jacquard et al.

selectionist and neutralist schools and has been termed the paradox of variation (Lewontin, 1974). However, these estimates of genetic variation have often been made on assemblages of individuals in nature which do not necessarily conform to the requirements of a population in the strict genetical sense and thus the interpretation of such data in terms of theoretical population genetic models is frequently invalid. Crawford (1984a) directly asked the question "What is a population?" and attempted to reconcile the different operational definitions used by ecologists and geneticists. The structuring and cohesiveness of species has been examined by Olivieri _et al_.(1984) who pointed out that there is a structural hierarchy within a species. In evolutionary terms, these hierarchical levels are related both spatially and temporally by gene-flow. Restrictions imposed on gene-flow mediated by both gametes and zygotes (migration) can lead to non-random patterns of mating (Schaal, 1980), which may result in both spatial and temporal structuring within a species. There is clearly a need to define precisely the nature of these spatial and temporal structures within species if meaningful evolutionary hypotheses are to be tested, using data on genetic variability.

In general, population geneticists define a population as a set of individuals connected through parenthood or mating and a Mendelian population as a set of individuals that mate effectively at random with each other (Crawford, 1984a). Natural populations, that is assemblages of individuals which appear more or less isolated from other such groups of the same species, may well share a common gene-pool, but often deviate from the random mating requirement of the Mendelian population. That is, populations in nature are frequently structured.

It is apparent, therefore, that there is a need to define the biological features connecting natural populations if evolutionary questions concerning between-population differentiation are to be addressed and to identify the constraints acting within these populations if intra-population variation is to be examined.

Inter-population variation has been examined theoretically using island and stepping-stone models (Wright, 1940; Kimura, 1953; Malecot, 1955, 1959; Moran, 1959; Weiss & Kimura, 1965). These models make a general prediction that for genetic drift to cause between-island differentiation $4\ Nm < 1$, where N is the effective population size of the islands and m the migration rate per generation. More recently, Slatkin (1981) has described a method using rare alleles to estimate Nm and thereby predict the degree of genetic connection between islands.

Structuring within continuous populations was examined theoretically by Wright (1943,1946,1969,1978) in his isolation by distance model and by Van Dijk (this issue), both of whom attempt to define a genetic neighbourhood. The neighbourhood may be defined operationally as that area within which individuals behave as a panmictic unit (i.e. within which there is no sub-structuring).

Many workers have attempted recently to estimate genetic neighbourhoods in plant species (Levin & Kerster, 1968,1971,1974; Schaal, 1975, 1980). Despite the fact that the use of Wright's formula was incorrect in these studies (Crawford, 1984b), even the corrected estimates of neighbourhood size indicate that, at least in the species studied to date, there is considerable scope for within-population structuring. A study of *Primula vulgaris* (Cahalan & Gliddon, 1984) indicated that neighbourhood sizes in this species are frequently lower than 30 individuals using Wright's method of estimation, and that these never rise above two individuals using the method of Van Dijk (this issue).

It is the purpose of this paper to attempt to extend the isolation by distance model of Wright (1946) to allow its application to plants which have vegetative growth as one of their means of dispersal and to apply data on gene-flow and mating systems in natural populations of *Trifolium repens*, white clover, to this model to allow an estimate of the scale of structuring which would be expected in natural populations of this species. These expectations are then tested against the observed distribution of genetic variation in the populations.

THE MODEL

Wright (1943, 1946, 1969, 1978) defined a genetic neighbourhood as being that area within which the parents of central individuals may be treated as if drawn at random. The neighbourhood area N_A, is then given by:

$$N_A = 4 \pi \sigma^2$$

where σ^2 is the parent offspring dispersal variance. The meaning, and many of the problems attendant in measuring σ^2 in plant populations, are given in Crawford (1984a, 1984b).

In plants, it is convenient to consider parent-offspring dispersal as involving two distinct phases: gametic dispersal; and zygotic (post-fertilisation) dispersal. The means whereby these two components may be combined to give the single parent-offspring dispersal variance of Wright (1946, 1969) is given in Crawford (1984b).

Crawford (1984a) extended the model of Wright (1946, 1969) to include the effects of partial outcrossing and his amended formula is:

$$Na = 4\pi \left[\frac{t \sigma p^2}{2} + \sigma_0^2 + \sigma_S^2 \right]$$

where t = frequency of outcrossing,

σ_p^2 = pollen dispersal variance,

σ_o^2 = ovule dispersal variance,

and σ_s^2 = seed dispersal variance.

A comparison of the components of parent-offspring dispersal, with and without vegetative growth, is given in Table 1.

The assumption made in Crawford (1984b) that ovule dispersal variance is zero, however, is not valid for plant species which show extensive vegetative growth prior to flowering. In such cases, vegetative growth is a component of dispersal of both male and female gametes. Assuming that growth follows a similar pattern in the male and female parents, the components of dispersal may be seen in Figure 1.

Table 1: Components of parent-offspring dispersal with and without vegetative growth.

Component	Without vegetative growth	With vegetative growth
Gametic:		
Male	σ_p^2	$\sigma_p^2 + \sigma_o^2$
Female	0	σ_o^2
Average gametic	$\frac{1}{2}\sigma_p^2$	$\frac{1}{2}\sigma_p^2 + \sigma_o^2$
Zygotic:	σ_s^2	σ_s^2
Total	$\frac{1}{2}\sigma_p^2 + \sigma_s^2$	$\frac{1}{2}\sigma_p^2 + \sigma_o^2 + \sigma_s^2$

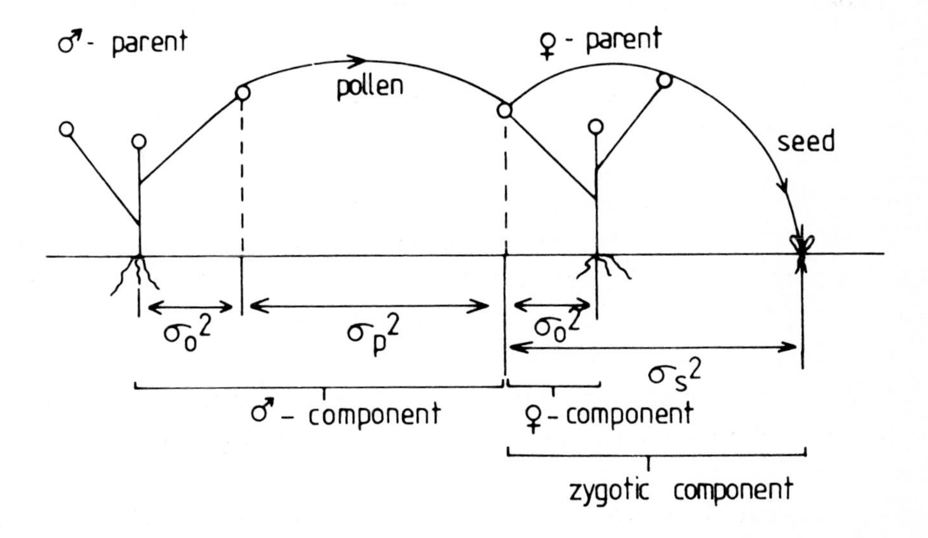

Figure 1: Schematic diagram of gene-flow in plants with vegetative dispersal.

MATERIALS AND METHODS

Two sites containing natural populations of T. repens were chosen for study, close to the Botanic Gardens of the University College of North Wales, Bangor. Neither site had been used for agricultural purposes in recent times and both were on land with controlled access, allowing marking of plants without fear of removal of labels. Both sites were relatively isolated from other areas containing T. repens and contained over 100 individuals within an area of 4 m^2. In each population, fifty adults were chosen at random, labelled, and leaf tissue collected from them in early spring and stored for future electrophoresis. In late summer, the sites were revisited and mature seed heads collected from the labelled plants. These seeds were scarified with sandpaper and sown, in family units, in a heated greenhouse. All families, which contained at least 8 progeny, were then electrophoresed. Electrophoresis was carried out on 7.5% polyacrylamide slab-gels using LKB multiphor equipment, the details of which are given in Saleem (1984).

The electrophoresis data for mothers and progeny was used to estimate the per locus effective level of outcrossing using the maximum likelihood method of Clegg et al. (1978) and the goodness of fit of the model was tested using a heterogeneity chi-square (Brown et al., 1975).

Pollen flow was estimated by following pollinator flights within each population, making the assumption that pollinator-flight distances are equivalent to pollen transfer distances.

Seed dispersal proved impossible to study directly in natural populations and an artificial plot of dimensions 5 m x 5 m was set up at the U.C.N.W. Experimental Field Station. The area was covered with a 5-8 cm thick layer of sterilised soil to suppress weeds and provide a good substrate for seedling establishment. In the autumn several plants, bearing many mature seed heads, were placed in the centre of the plot and left to allow the seeds to disperse naturally. The following spring, the positions of all T.

repens seedlings were marked and the dispersal distribution was calculated. This single distribution of seed dispersal was then used as the estimate of gene-flow through seed in both populations.

Estimation of the growth parameter of dispersal is again not possible to achieve in natural populations. This is due primarily to the rotting of stolon connections in *T. repens* between one year class and another. A single genet, in nature, therefore appears as a collection of groups of ramets which are not physically connected with each other. To overcome this problem, a computer plant growth simulator was used (Bell *et al.*, 1979). The parameters for specifying the growth of *T. repens* were obtained by Sackville-Hamilton (unpublished data) in a permanent pasture in North Wales. The model was modified to give the coordinates of the year classes of inflorescences. Thirty replicate runs allowed calculation of the distribution of ovules for each year class of growth and, as with seed-dispersal, this single parameter was used as the estimate of the vegetative growth component of dispersal in both populations.

RESULTS AND DISCUSSION

A summary of genetic variation in the two maternal populations of *T. repens* is given in Table 2.

Table 2: Summary of different measures of genetic variation.

Population	Mean prop$^{\underline{n}}$ loci polymorphic (PLP)	Average No. alleles per locus (k)	Mean prop$^{\underline{n}}$ loci heterozygous (H_o)	Average outcrossing rate (t)	Number of genotypes (n = 50)
Pop 1	0.25	1.25	0.118	0.90	43
Pop 2	0.27	1.27	0.124	0.89	46
Means	0.26	1.26	0.121	0.895	45

Table 3 allows a comparison of allele frequencies, at polymorphic loci, between maternal populations and progeny sets derived from them. It is apparent that the raw estimates of genetic variation in these populations are in accord with previously published data on natural plant populations (Hamrick _et al_., 1979). In addition, the striking similarity between parental and progeny gene frequencies suggests the absence of any significant selective force acting prior to the maturation of seed.

Table 3: Maternal and progeny gene frequencies for polymorphic loci.

	Frequency							
	Pop 1				Pop 2			
Locus	Parents (n=50)		Progeny (n=300)		Parents (n=50)		Progeny (n=400)	
	a	b	a	b	a	b	a	b
APH_1	0.60	0.40	0.58	0.42	0.56	0.44	0.54	0.46
$ATPase_2$	0.48	0.52	0.45	0.55	0.55	0.45	0.53	0.47
EST_1	0.61	0.39	0.60	0.40	0.52	0.48	0.49	0.51
LAP_1	0.47	0.53	0.46	0.54	0.46	0.54	0.46	0.54
MDH_4	0.56	0.44	0.57	0.43	0.53	0.47	0.51	0.49
PER_3	Not interpretable				0.57	0.43	0.56	0.44

The estimates of effective outcrossing rate for all polymorphic loci are given in Table 4, together with their associated heterogeneity chi-squares. It is clear that the mixed-mating system model fits the data extremely well and, as would be expected in a self-imcompatible species, the effective outcrossing rates are close to unity, indicating that very little inbreeding, including mating between relatives, takes place in these populations.

A summary of the distribution of pollinator flights, seed dispersal distances and density measurements is given in Table 5.

Table 4: Outcrossing rate estimates (t) and heterogeneity chi-square in two populations of Trifolium repens.

Locus	Pop 1 t ± S.E	Pop 1 chi-sq. (d.f)	Pop 2 t ± S.E	Pop 2 chi-sq. (d.f)
APH_1	0.891 ± 0.071	7.29 (12)	0.900 ± 0.069	6.18 (20)
$ATPase_2$	0.872 ± 0.092	8.50 (11)	0.858 ± 0.072	8.24 (18)
EST_1	0.941 ± 0.099	10.31 (13)	0.887 ± 0.074	5.15 (17)
LAP_1	0.900 ± 0.088	11.83 (12)	0.920 ± 0.072	7.31 (19)
MDH_4	0.926 ± 0.085	10.95 (14)	0.896 ± 0.064	9.70 (23)
PER_3	- -	-	0.892 ± 0.069	6.24 (21)

Table 5: Means and one-way variances of pollen and seed dispersal distances and density of flowering plants (d) in populations of Trifolium species.

Species	Population	Pollen Mean (m)	Pollen Variance (m^2)	Seed Mean (m)	Seed Variance (m^2)	d (plants/m^2)
	Pop 1	0.38	0.12	-	-	*46-+150.16
T. repens		-	-	0.36	0.1	-
	Pop 2	0.39	0.149	-	-	*46-+138.26

* based on number of unique genotypes
\+ density of inflorescences m^{-2}

The pollinators which were observed foraging on T. repens were two long-tongued bumble-bees, Bombus pascorum and B. lapidaris, and one short-tongued bee, B. lucorum. In order to be able to derive pollen dispersal distributions from pollinator flight distances, it is necessary to assume that all pollen grains collected from one flower are deposited on the next flower visited (Levin, 1981). This assumption ignores possible pollen carry-over and thus tends to underestimate pollen dispersal. However, the floral structure of T. repens is such that many individual flowers are mature and open, on a single inflorescence, at the same time. The behaviour of pollinators on these inflorescences is to visit at least four flowers on each inflorescence, which will result in any pollen carry-over being much reduced. Therefore, although pollen carry-over was not directly measured, it is likely that using pollinator flight distance as a direct measure of pollen flow, in these populations, will not substantially underestimate the true value.

The inflorescence/seedling dispersal data was found to have no directional bias. The maximum distance any seedling was dispersed was 1.2 m which was substantially smaller than the size of the experimental plots. It is likely, therefore, that no seeds were dispersed outside the range of sampling. The seedling dispersal data may well be a substantial underestimate of that found in nature. In particular, if there are herbivores eating mature capsules, hard-coated seed of T. repens can pass through the mammalian gut and remain viable, increasing the dispersal variance by a large amount. However, in the natural populations, there was no evidence of eating of mature seed-heads but the possibility of this occurring cannot be excluded.

The computer simulation of growth, allowing estimation of ovule dispersal variance (σ_o^2), showed an approximately linear increase over time. The neighbourhood areas and sizes using the estimates of all components of gene flow, measured in the two populations, together with σ_o^2 are given in Table 6.

It is apparent that vegetative growth becomes an important factor in determining gene-flow in T. repens after approximately 5 years of growth, with neighbourhood area being doubled in 10 years.

Table 6: One-way ovule variances, estimates of neighbourhood area (A), and neighbourhood size (N_e) in T. repens populations at different ages.

Generation	ovule variance (m^2)	A (m^2) Pop 1	Pop 2	*N_e Pop 1	Pop 2	+N_e Pop 1	Pop 2
1	0.046	2.77	2.92	121	127	395	382
2	0.057	2.97	3.12	130	136	424	408
3	0.0795	3.38	3.53	148	154	482	461
4	0.0805	3.4	3.55	149	155	485	464
5	0.0915	3.6	3.75	157	163	513	490
6	0.1185	4.09	4.24	178	184	583	554
7	0.126	4.23	4.38	185	190	603	572
8	0.160	4.85	4.99	212	217	692	652
9	0.164	4.92	5.07	215	220	702	662
10	0.176	5.14	5.29	225	230	733	691
11	0.1895	5.38	5.53	235	241	767	723

* Using minimum estimate of d (Table 5).
+ Using maximum estimate of d (Table 5).

The simulation of growth is likely to give a conservative estimate of the dispersal variance of inflorescences. This is because there is no interference between stolons in the model, resulting in a massive biomass around the centre of origin. Any reduction in this central biomass, for example due to interference, would result in a substantial increase in the dispersal variance. Therefore, the importance of vegetative growth as a means of accomplishing gene-flow has been underestimated in the results of this study.

The neighbourhood area estimates given in Table 6, even for young plants, would suggest that the two populations studied here should not show any structuring within them, as their total area is approximately equal to the neighbourhood area. To test this prediction, the genetic variation estimates of adult plants, given in Table 2, were analysed using F-statistics (Wright, 1965,1969), where F, the fixation index, is given by:

$$F = 1 - \frac{\text{Observed Heterozygosity}}{\text{Heterozygosity predicted under inbreeding equilibrium}}.$$

A positive value of F indicates a deficiency of heterozygotes, a negative value an excess, due to population sub-division. The results for both populations are given in Table 7.

Table 7: Estimates of Wright's fixation index for each polymorphic locus.

Locus	Pop 1		Pop 2	
	F value	$\pm$ S.E	F value	$\pm$ S.E
APH_1	0.083	$\pm$ 0.05	0.026	$\pm$ 0.04
$ATPase_2$	-0.042	$\pm$ 0.06	0.151	$\pm$ 0.06
EST_1	0.033	$\pm$ 0.04	-0.042	$\pm$ 0.05
LAP_1	0.077	$\pm$ 0.06	0.114	$\pm$ 0.05
MDH_4	0.026	$\pm$ 0.05	0.157	$\pm$ 0.04
PER_3	-	-	0.062	$\pm$ 0.05
Means	0.035		0.078	

There is no evidence of any sub-structuring of population 1 as all F-values are not significantly different from 0. However, in population 2, 3 out of 6 loci show significant positive values of F. This heterozygote deficiency may have been caused by sub-structuring of the population, the Wahlund effect (Wahlund, 1928), indicating that the neighbourhood area for population 2 was an overestimate.

The increase in neighbourhood area, as a result of growth, has been presented assuming that populations consist of plants of the same age. In most natural populations, the age structure wil never be as simple as this, with plants of many different ages and sizes coexisting at a particular point in time. However, if different populations of the same species are categorised with respect to turnover, certain generalisations about neighbourhood area and, therefore, local gene-flow can be made. Populations with a high rate of turnover will contain, on average, younger and smaller individuals than those with a low rate of turnover. Therefore, the average neighbourhood area will be smaller in populations with a high rate of turnover when compared with those with a low rate of turnover, with obvious implications for local gene-flow.

Consideration of a growth component to gene dispersal highlights two major problems attendant upon the application of Wright's neighbourhood model to natural populations of plants. Firstly, it is apparent that plants of different sizes will tend to disperse their genes, through pollen, ovules and seed, by different amounts. Since neighbourhood area is a population attribute, it can only give some average measure of gene-flow and will not allow precise predictions of population structuring in natural populations. In particular, the patches in a natural population that form panmictic units will have a variable size within that population. Secondly, it is clear that the single parameter, σ^2, for the parent-offspring dispersal distribution is essentially generated by averaging the male and female components. Providing there is a non-zero pollen dispersal, male parent/offspring dispersal will always be greater than female parent/offspring dispersal. Using an average of these two distributions to define a neighbourhood area will therefore result in an under-representation of the male component of the panmictic unit and an over-representation of the female component in the neighbourhood. Both of these problems associated with neighbourhood estimation may be more generally considered as a criticism of using point estimators rather than interval estimators in describing complex biological situations (see, for example, Connolly & Gliddon, 1984).

An alternative approach to describing the structure of

populations is clearly desirable, given the limitations of Wright's neighbourhood model. Such a possible alternative would be to examine the expected gene-flow per generation in males and females separately, as has been done by Van Dijk (this issue). It is clear, however, that some measure of local gene-flow needs to be made, before any evolutionary inferences may be made concerning the genetic structuring of natural populations.

ACKNOWLEDGEMENTS

The authors wish to thank Dr. Mark MacNair and Dr. Terry Crawford for helpful discussions concerning the calculations of gene-flow. M. Saleem was supported by Roberts Cotton Associates, Khanewal, Pakistan.

REFERENCES

Bell, A.D., Roberts, D. & Smith, A., 1979 - Branching patterns: the simulation of plant architecture. J. Theor. Biol., 81, 351-375.

Brown, A.H.D., Matheson, A.C. & Eldridge, K.G., 1975 - Estimation of the mating system of *Eucalyptus obliqua* L'Herit. by using allozyme polymorphisms. Aust. J. Bot., 23, 931-949.

Cahalan, C.M. & Gliddon, C., 1984 - Neighbourhood sizes in *Primula vulgaris*. Heredity, in press.

Clegg, M.T., Kahler, A.L. & Allard, R.W., 1978 - Estimation of life cycle components in an experimental plant population. Genetics, 89, 765-792.

Connolly, J. & Gliddon, C., 1984 - On the estimation of viabilities in competition experiments. Heredity, in press.

Crawford, T.J., 1984a - What is a population? In B. Shorrocks (ed.) Evolutionary Ecology, pp.135-173. Blackwells, London.

Crawford, T.J., 1984b - The estimation of neighbourhood parameters for plant populations. Heredity, 52(2), 273-283.

Hamrick, J.L., Linhart, Y.L. & Mitton, J.B., 1979 - Relationships between life-history characteristics and electrophoretically detectable genetic variation in plants. Ann. Rev. Ecol. Syst., 10, 173-200.

Kimura, M., 1953 - "Stepping stone" model of population. Ann. Rep. Nat. Inst. Gent., Japan, 3, 62-63.

Levin, D.A., 1981 - Dispersal versus gene flow in seed plants. Ann. Rev. Miss. Bot. Gard., 68, 233-253.

Levin, D.A. & Kerster, H.W., 1968 - Local gene dispersal in _Phlox_. Evolution, 22, 130-139.

Levin, D.A. & Kerster, H.W., 1971 - Neighborhood structure in plants under diverse reproductive methods. Amer. Natur., 105, 345-354.

Levin, D.A. & Kerster, H.W., 1974 - Gene flow in seed plants. Evol. Biol., 7, 139-220.

Lewontin, R.C., 1974 - "The genetic basis of evolutionary change". Columbia University Press, New York.

Malecot, G., 1955 - Decrease of relationship with distance. Cold Spring Harbor Symp. Quant. Biol., 20, 52-53.

Malecot, G., 1959 - Les modeles stochastiques en genetique de population. Pub. Inst. Stat. Univ. de Paris, 8(3), 173-210.

Moran, P.A.P., 1959 - The theory of some genetical effects of population subdivision. Aust. J. Biol. Sci., 12, 109-116.

Olivieri, I. & Gouyon, P.H., 1984 - A theoretical approach to dispersal polymorphism. Proc. 2nd. Int. Symp. on the Structure and Functioning of Plant Populations, in press.

Saleem, M., 1984 - Gene-flow & breeding system in _Trifolium_ species. Ph. D. Thesis, Univ. Wales.

Schaal, B., 1975 - Population structure and local differentiation in _Liatris cylindracea_. Amer. Natur. 110, 511-528.

Schaal, B., 1980 - Measurement of gene flow in _Lupinus texenis_. Nature, 284, 450-451.

Slatkin, M., 1981 - Estimating levels of gene flow in natural populations. Genetics, 99, 323-335.

Van Dijk, H., 1985 - The estimation of gene flow parameters from a continuous population structure. This issue.

Wahlund, S., 1928 - Zusammensetzung von populationen und korrelationserscheinnung von standpunkt der verebungslehre aus betrachtet. Hereditas, 2, 65-106.

Weiss, G.H. & Kimura, M., 1965 - A mathematical analysis of the stepping stone model of genetic correlation. J. Appl. Prob., 2, 129-149.

Wright, S., 1940 - Breeding structure of populations in relation to speciation. Amer. Natur., 74, 232-248.

Wright, S., 1943 - Isolation by distance. Genetics, 28, 114-138.

Wright, S., 1946 - Isolation by distance under diverse systems of mating. Genetics, 31, 39-59.

Wright, S., 1965 - The interpretation of population structure by F-statistics with special regard to systems of mating. Evolution, 19, 395-420.

Wright, S., 1969 - Evolution and the genetics of populations. Vol. 2. The theory of gene frequencies. Univ. of Chicago Press.

Wright, S., 1978 - Evolution and the genetics of populations. Vol. 4. Variability within and among natural populations. Univ. of Chicago Press.

The estimation of gene flow parameters from a continuous population structure

H. van Dijk

Department of Genetics, University of Groningen
Kerklaan 30, 9751 NN Haren, The Netherlands

ABSTRACT

Restricted gene flow, allowing genetic drift to generate local differences in allele frequencies, will cause a type of population structure which is characterized by increasing genetic similarity with decreasing spatial distance. From such a population structure, assuming that selective effects are absent and that an equilibrium situation has been reached, the plant gene flow parameters σ_p (pollen dispersal), σ_s (seed dispersal) and t (outcrossing rate) can be estimated by a new method. This method is an extension of Wright's analysis of discontinuously structured populations by the use of F-coefficients. A new set of F-coefficients has been derived for the continuous situation.

Although elaborated for wind-pollinated plant species, the method can be modified for insect pollination and for animal species. Allozyme variation is the most suitable form of genetic variation for the population structure analysis. A few hundreds of individuals have to be examined.

The terms "local differentiation", "neighbourhood size and area" and "isolation by distance" are discussed for the special situation of seed plants.

INTRODUCTION

In a path population of *Plantago major* a strongly non-uniform distribution of allozyme alleles over the population area was found. Three different phenomena were met which indicated this

NATO ASI Series, Vol. G5
Genetic Differentiation and Dispersal in Plants
Edited by P. Jacquard et al.

deviation from a uniform situation:

1). Homozygosity was far beyond the level expected under Hardy-Weinberg conditions for all loci examined: Phosphoglucomutase-1 (*Pgm*-1), Esterase-4 (*Est*-4) and Shikimate dehydrogenase (*Shdh*). The F-values appeared to be 0.85 approximately.

2). The population exhibited a certain degree of patchiness in the distribution of alleles. This is best described by saying that neighbouring individuals had a far higher probability to be of the same genotype than individuals from different parts of the population.

3). Alleles of the three different loci showed a strong apparent local linkage disequilibrium i.e. a particular allele of the one locus was almost always combined with a particular allele of a second locus. This linkage disequilibrium is called apparent because it is known that the three loci are not linked chromosomally (Van Dijk 1981), and local because other combinations of alleles were found in other parts of the population so that there was almost no net linkage disequilibrium in the population as a whole.

A first question is whether such a situation can be the result of random processes only, or must be brought about at least partially by selectional differences between parts of the population. To answer this question a series of computer simulations has been set up to imitate the random processes occurring in nature. Results very similar to the situation described for *Plantago major* could be reached by introducing low levels of gene flow and a high selfing rate into the simulation model. Therefore, local selectional differences seem to be not necessary to explain the situation observed. This conclusion is in agreement with the observed homogeneity of the population with respect to genetically based morphological characters: all plants belonged to the same trampling resistant ecotype. The allozyme loci, although (in the case of *Pgm*-1 and *Est*-4) linked with loci coding for quantitative characters which may influence fitness (Van Dijk 1984), are considered to represent selectively neutral variation in this morphologically homogeneous population.

A second question is whether it would be possible to estimate the gene flow parameters from the population structure observed.

Apparently the introduction of a particular set of gene flow parameters brought about a particular population structure in the computer simulations, but to be able to go the other way around a necessary condition is that different gene flow regimes never can lead to the same population structure. In this paper an analysis of the underlying forces which determine local differentiation: gene flow and genetic drift, is made possible by a new method. Existing methods were only suitable for populations subdivided into distinct subpopulations. In addition to the estimation method of the gene flow parameters σ_p (pollen dispersal), σ_s (seed dispersal) and t (outcrossing rate), the neighbourhood model of Wright (1943) will be discussed for the special situation of seed plants because the conceptions of neighbourhood size and area are very often used in describing these types of population structure. Mathematical details will not be presented in this paper, but will be published elsewhere (Van Dijk submitted).

LOCAL DIFFERENTIATION

The situation of distinct subpopulations with different allele frequencies due to restricted gene flow and genetic drift has its parallel in a continuous population with different local allele frequencies. Because no subpopulations can be defined, however, these local allele frequencies cannot be determined as such. Of course it is possible to try artificial subdivisions of variable size, in the way it was done by Schaal (1975), who chose a subdivision into squares having an area equal to the putative neighbourhood area. On condition that there are sufficient numbers of individuals in each square, allele frequencies can be determined and a statistical test will provide information about the heterogeneity between squares. Also deviations from Hardy-Weinberg ratios within squares may be established, indicating that a too coarse grid had been chosen. Direct information about the processes of gene flow and genetic drift cannot be obtained in this way. A severe problem arises when the number of individuals per square becomes too low for an accurate determination of allele frequencies while there is still an excess of homozygotes within the squares. A different approach is, therefore, necessary to analyse local differentiation in a continuous population.

As mentioned, local differentiation due to restricted gene flow and genetic drift can be described by the relationship between genetic similarity and spatial distance. The genetic similarity, I, of two individuals at a given locus is defined as the probability of producing the same gamete with respect to that locus. This is illustrated by the following table:

genotype individual 1	genotype individual 2	genetic similarity $I_{1,2}$
aa	aa	1
aa	ab	$\frac{1}{2}$
ab	ab	$\frac{1}{2}$
ab	ac	$\frac{1}{4}$
aa	bb	0

When all individuals of the population under examination are scored for position coordinates and genotype, the wanted relationship can be established. The most convenient way is to take all possible pairs of plants, and group them into distance classes, e.g. 0-5 cm, 5-10 cm, etc. For each distance class a mean value of I can be calculated, after which the function, denoted as I_r, can be plotted. The number of plants in each distance class is also scored by means of a function N_r, and gives information about the mean number of individuals standing within a certain distance transect around any of the plants. In a 2-dimensional uniform distribution N_r will increase linearly with r, whereas in a 1-dimensional uniform distribution (all individuals standing on one line) N_r is a constant. In a path situation like the *Plantago major* population N_r will increase linearly with r for the lower distances, but will reach a maximum value at distances higher than the path width.

The $I_r(r)$ function depends on the allele frequencies in addition to the degree of local differentiation. This is obvious in a population without any local differentiation: the probability of any pair of individuals of producing the same gamete is then equal to the sum of the squares of the allele frequencies (Σp_i^2 when the allele frequencies are p_1, p_2, ..., p_i, ...). Thus, the $I_r(r)$ curve appears to be a horizontal line at the level Σp_i^2. In considering this level as the zero level of local differentiation and the maximum I value, 1, as the maximum level of local differentiation, the following conversion of scale will make genetic similarity independent of allele frequencies:

$$F_r = \frac{I_r - \Sigma p_i^2}{1 - \Sigma p_i^2}$$

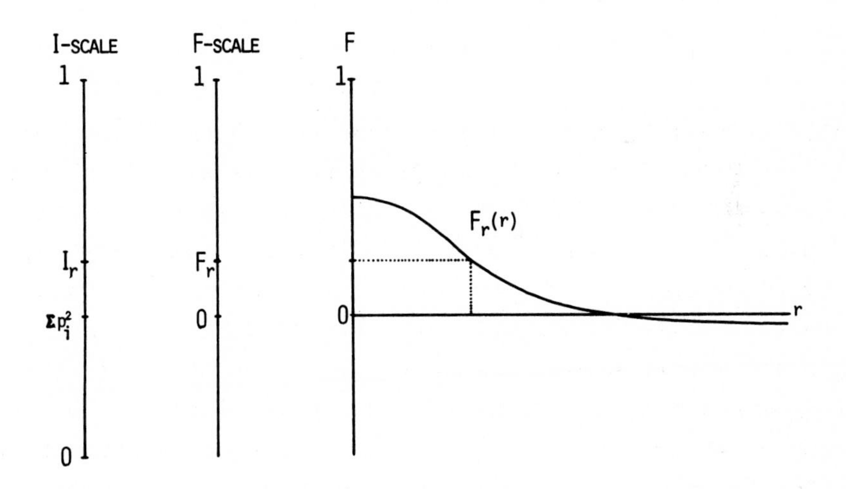

Fig. 1 - The change-over from I-scale to F-scale; the shape of the $F_r(r)$ curve.

Figure 1 illustrates the relationship between I-scale and F-scale together with the shape of the $F_r(r)$-curve. The change-over of scale has two advantages:

1). The degree of local differentiation can now be defined quantitatively as the intersection point of the $F_r(r)$-curve with the F-axis, this is the point $F_r(0)$.

2). The F_r values of different loci have the same expectation, and by this reason the more accurate average F_r values of a series of loci can be used to determine a more precise $F_r(r)$-curve and a more reliable value for the degree of local differentiation.

POLLEN DISPERSAL

Pollen dispersal influences both local differentiation and homozygosity. Homozygosity is the result of combining two identical gametes from both parents, and, by that reason, the expected homozygosity of the descendants is equal to the I value of the parents. When matings only occur between individuals standing in each others' neighbourhood, the relatively high I values will lead to high levels of homozygosity. Changing-over to the F-scale, the higher than zero F values will lead to a higher than zero F_{IT} (Wright's fixation index for the whole population), because

$$F_{IT} = \frac{H_{exp} - H_{obs}}{H_{exp}} = \frac{Hom_{obs} - \Sigma p_i^2}{1 - \Sigma p_i^2}$$

(H_{exp}, H_{obs} and Hom_{obs} being the heterozygosity expected under Hardy-Weinberg conditions, the heterozygosity observed and the homozygosity observed in the population, respectively). Obviously F_{IT} is directly related to F_r, and this is the reason why the symbol F was chosen for F_r.

The excess of homozygosity due to local differentiation appears to be equal to the average F_r value of the parental pairs of all individuals in the population. The expectation of that average F_r value, F_x, considered for a particular female individual, depends on both the number of (male) individuals in the various distance classes, and on the probability of a male individual at distance r becoming the male parent. This probability is described by a function $P_r(r)$ and depends on how far pollen is dispersed, because $P_r(r)$ is proportional to the pollen density at distance r. Now

$$F_x = \sum_r F_r N_r P_r, \text{ with } \sum_r N_r P_r = 1.$$

Pollen dispersal in two-dimensional space is often considered to follow an axial normal distribution with standard deviation σ_p. This distribution will also be used here for the calculation of the P_r values, but any other distribution can be used instead on condition that it contains only one unknown parameter like σ_p in the normal distribution. For a particular value of σ_p, all terms of F_x can now be calculated, and the value of σ_p leading to an F_x which is equal to F_{IT} is the σ_p value representing the level of pollen dispersal in the population.

Selfing has been neglected so far, but has of course a strong influence on homozygosity. The F caused by selfing alone has been established as $F_s = (1-t)/(1+t)$. The two different influences on F are combined according to the relationship

$$1 - F_{IT} = (1 - F_s)(1 - F_x)$$

When F_s is not zero, but has a known value, σ_p can still be calculated. When both t and σ_p are unknown, however, F_{IT} depends on two unknown variables and a second, different, relationship between σ_p and t is necessary to be able to solve them both. The third phenomenon observed in the *Plantago major* population, the apparent local linkage disequilibrium, is used to give additional information. At least two chromosomally unlinked loci are required.

The calculation of a local linkage disequilibrium as such has the same objection as the calculation of local allele frequencies: this needs the subdivision of the population into distinct parts. A similar approach as for the separate loci allows a solution for combinations of loci: the level of homozygosity on all loci simultaneously is explained by the influences of selfing and local differentiation. The relationship found will be a different one than the one for separate loci. This is because the influence of local differentiation on its own on multi-locus homozygosity is nothing more than the multiplied probabilities of being homozygous on the separate loci. When selfing occurs instead of outcrossing, all loci become more homozygous simultaneously. The multi-locus homozygosity will, therefore, exceed the level predicted by the multiplication of the separate homozygosities when selfing occurs.

The two different relationships between σ_p and t can be plotted graphically, and, as shown in the example in Fig. 2, the intersection point of the two curves indicates the best estimates of both σ_p and t (note that t cannot exceed 1 in reality!). High selfing rates cause high levels of homozygosity. F_{IT} is then almost completely determined by F_S, and local differentiation, irrespective of its strength, has a neglectible influence.

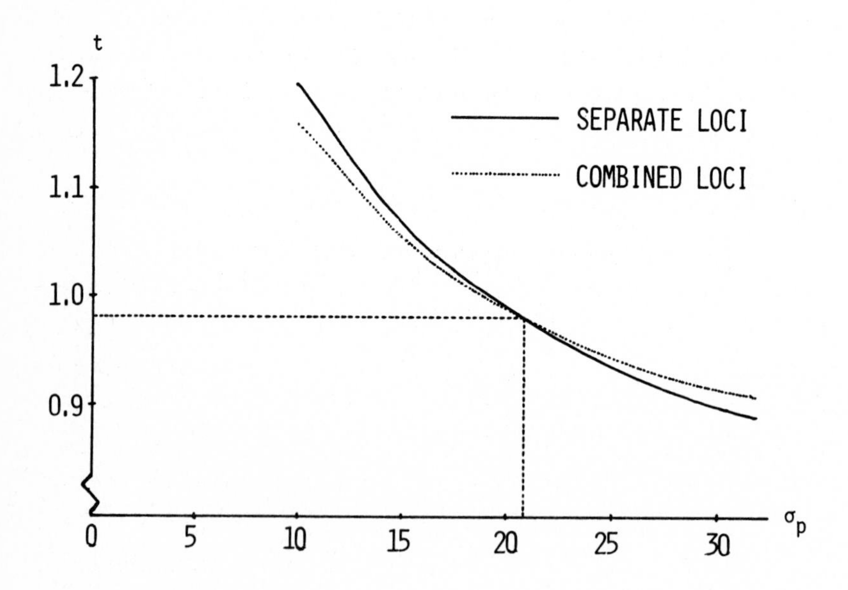

Fig. 2 - An example of the determination of σ_p and t simultaneously.

High selfing rates make reliable estimation of σ_p very difficult by this reason, and much larger numbers of individuals have to be examined in such situations. Experimental measurements of σ_p may then be preferable.

LOCAL EFFECTIVE POPULATION SIZE AND SEED DISPERSAL

Pollen dispersal is a form of gene dispersal by gametes, which explains its influence on homozygosity. Gene dispersal in general is described by the distances travelled by genes from parents to their descendants having reached the same life stage. In the diploid phase, seed dispersal is the only mechanism by which (sessile) plants will migrate. The genes obtained from the female parent are only dispersed by seed, whereas the genes from the male parent are transported by both pollen and seed. Gene flow in general has a negative effect on local differentiation. If no counteracting force existed, local differentiation should disappear finally. In an equilibrium situation gene flow is counterbalanced by (local) genetic drift, causing local loss of genetic variation in a random direction and consequently generating local differences.

What is the mechanism by which genetic drift enhances local differentiation? To understand the underlying processes two individuals standing at the same place will be considered with respect to their probability of being genetically similar. Their genetic similarity is on the average equal to $F_r(0)$ when measured on the F-scale. If the two individuals do not have any parents in common, their expected F_r will be lower than $F_r(0)$ because the respective sets of parents are situated at a larger than zero distance and will have a mean F_r belonging to a lower part of the curve (see fig. 1). On the other hand, if one or both parents are the same for both individuals, the F_r value of the individuals will on the average exceed $F_r(0)$. This probability of sharing parents forms the mechanism of genetic drift.

Genetic drift is directly related to effective population size. The probability of sharing parents in a population of N individuals in which each individual has the same chance to become the parent, equals 1/N. This probability increases when the chances of becoming the parent are less equal for the various individuals, and reaches the value $1/N_e$, N_e being lower than N. N_e is called the effective population size. In considering the situation of two individuals at the same place and a limited number of possible parents due to restricted gene flow, the probability of having parents in common can also be quantified as $1/N_e$; that N_e is the "local effective population size", and is synonymous to Wright's "neighbourhood size". Wright (1943) calculated N_e as $4\pi d\sigma^2$, d being the population density of reproducing individuals and σ^2 being the variance of gene dispersal in both sexes. Apparently local genetic drift depends on density and the gene flow parameters, exclusively. This means that all forces acting on the maintenance of the point $F_r(0)$ can be described by d, σ_p, σ_s and t. Knowing d, σ_p and t, the only unknown variable is σ_s which can be calculated now.

The way in which N_e depends on the three gene flow parameters is given by the average of the probabilities that two plants at the same place have the same female parent, the same male parent or that the female parent of the one plant is the male parent of the other one or vice versa. Derived in analogy to Wright's $N_e = 4\pi d\sigma^2$,

$$\frac{1}{N_e} = \frac{1}{4\pi d}\left\{\frac{(1-\frac{1}{2}t)^2}{\sigma_s^2} + \frac{t(1-\frac{1}{2}t)}{\sigma_s^2 + \frac{1}{2}\sigma_p^2} + \frac{\frac{1}{4}t^2}{\sigma_s^2+\sigma_p^2}\right\}$$

For both t = 0 and $\sigma_p = 0$, N_e becomes $4\pi d\sigma_s^2$ because then, like in Wright's situation, both sexes have the same gene dispersal parameter namely σ_s.

When σ_p becomes very large with respect to σ_s, the second and third term become very much smaller than the first term, so that $N_e \approx 4\pi d\sigma_s^2/(1-\frac{1}{2}t)^2$. This means that N_e is almost exclusively determined by σ_s when σ_p is relatively large. Two neighbouring plants will then often be related by sharing the female parent, but almost never by having the same male parent, except in the case of selfing.

ISOLATION BY DISTANCE

Next to the neighbourhood size N_e, "isolation by distance" is a consequence of restricted gene flow. Wright (1943) described this isolation by distance by showing that the parents come from inside a circle with radius 2σ with a probability of 86.5%. This circle has an area of $4\pi\sigma^2$, and the number of individuals within it is equal to N_e; it is called the "neighbourhood area". In considering all three gene flow parameters, the area of $4\pi\sigma^2$ has to be replaced by $4\pi(\sigma_s^2+\frac{1}{2}t\sigma_p^2)$ as shown by Crawford (1984), who also pointed out that the probability of finding a parent within that circle is no longer 86.5% and may be completely different for male and female parents. The number of individuals within the circle is also different from N_e, because the circle area is mainly determined by the largest gene flow parameter σ_p, whereas N_e, as mentioned earlier, is governed mainly by σ_s.

Instead of the neighbourhood area, isolation by distance is better quantified by the mean distance covered by genes each generation. Using a standard deviation σ, the mean gene transport per generation M equals $\sqrt{\frac{1}{2}\pi\sigma^2} = 1.2533\sigma$. After g generations, the mean gene transport is $M\sqrt{g}$ as illustrated in fig. 3(a). For seed plants M appears to be $\sqrt{\frac{1}{2}\pi(\sigma_s^2+\frac{1}{2}t\sigma_p^2)}$. The area within the various circles

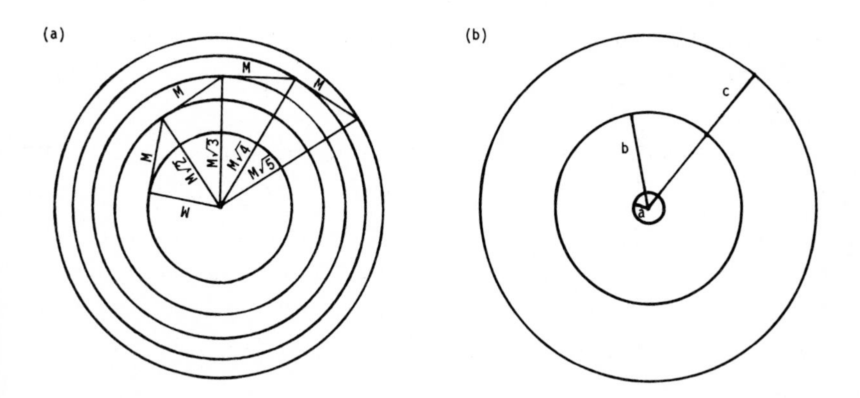

Fig. 3-(a) The mean gene transport in 1-5 generations. (b) The mean distance from the grandparents (two generations): female line = a; mixed line = b; male line = c.

of fig. 3(a) is $\pi g M^2$, increasing by πM^2 each generation. This is equal to $\frac{1}{2}\pi^2(\sigma_s^2+\frac{1}{2}t\sigma_p^2)$ and is, therefore, proportional to Crawford's neighbourhood area.

The mean gene transport per generation M is a long-term measure. It assumes an equal number of male and female dispersion phases. In reality, and of particular interest for the explanation of short-term dispersal effects, the numbers of male and female phases follow a binomial distribution. This is illustrated by fig. 3(b) which shows an example of the mean distances from individuals to the three categories of their grandparents: in female line, in male line and in mixed line.

COMPUTER SIMULATIONS

To check the method for the estimation of σ_p, t and σ_s, a series of computer simulations has been carried out in the way already mentioned in the introduction, using various gene flow regimes. The artificial populations resulting from those simulations, which were continued until the populations were obviously in equilibrium (after 25-50 "years" using a turnover value of 25%), were analysed by a computer program based on the ideas presented in this paper. Three loci were used with two alleles of frequency 0.5 each. The number of individuals was 200 for each population and 5 simulations were carried out with each regime. Only the results of the combined 5 populations (1000 individuals) will be presented for nine gene flow regimes.

The next table shows that the recoveries of σ_s and N_e, and also of σ_p when the value of t is known, are satisfying:

Input parameters and derived N_e			Estimated parameters and N_e (t is known)		
σ_p	σ_s	N_e	σ_p	σ_s	N_e
20	5	4.2	29.3	3.8	4.3
20	5	4.3	18.7	4.6	4.1
20	5	3.7	21.4	6.8	5.0
20	5	2.6	17.4	8.6	3.9
20	5	1.8	20.1	5.6	2.1
5	15	8.5	4.0	13.3	7.9
100	20	33.8	70.5	24.9	38.2
100	20	19.6	88	14.9	12.3
∞	1	4.0	∞	1.0	4.0

When t is unknown, the estimates of σ_p are less reliable for lower t values, as has been discussed. The estimated value of M appears to be very sensitive to deviations in σ_p as can be seen in the next table ("-" means that no value could be obtained):

Input parameters and derived M			Estimated parameters and M		
σ_p	t	M	σ_p	t	M
20	1	18.8	36.5	0.964	32.1
20	1	18.8	22.2	0.955	20.1
20	0.856	17.6	26.0	0.811	22.4
20	0.564	14.7	8.4	0.750	12.6
20	0.125	8.9	8.5	0.260	8.0
5	0.263	18.9	90	0.160	36.0
100	1	92.1	-	-	69.8
100	0.437	63.7	∞	0.408	∞
∞	0.997	∞	-	-	∞

Of course the reliability of the estimates can be improved by using larger numbers of individuals or by using more - chromosomally unlinked - loci with intermediate allele frequencies.

DISCUSSION

The ideas about the estimation of gene flow presented in this paper, when applied in practice using the formulae of Van Dijk (submitted), will lead to reliable values of the various gene flow parameters only if several conditions have been fulfilled. Most important is that selection is absent and that the population investigated is more or less in equilibrium. Tests should be carried out to check these conditions. The method has been elaborated for wind-pollinated plant species, making modifications necessary when other species are investigated. Refinements should also be introduced when population characters occur which interfere with population structure, like, for instance, vegetative propagation, long distance seed transport or the existence of a seed bank. The computer simulations made clear that the presence of a self-incompatibility system did not influence the validity of the method, which also appeared to be rather robust with respect to the dimensionality of the population and the uniformity of the distribution of its individuals.

Since the introduction of Wright's neighbourhood model a lot of confusing ideas have been presented in literature about this subject. When Wright's model is applied to seed plants, the idea that the neighbourhood represents a panmictic unit, already not very useful in the classic model, appears to be completely inappropriate. The here presented separate treatment of neighbourhood size (representing local genetic drift) and neighbourhood area (describing isolation by distance), may probably contribute to a better understanding of underlying mechanisms of this kind of population structure.

REFERENCES

Crawford T.J., 1984. - The estimation of neighbourhood parameters for plant populations. *Heredity*, 52, 273-283.

Schaal B.A., 1975. - Population structure and local differentiation in *Liatris cylindracea*. *Amer.Natur.*, 109, 511-528.

Van Dijk H., 1981. - Genetic variability in *Plantago* species in relation to their ecology. 1. Genetic analysis of the allozyme variation in *P. major* subspecies. *Theor.Appl.Genet.*,60,285-290

Van Dijk H., 1984. - Genetic variability in *Plantago* species in relation to their ecology. 2. Quantitative characters and allozyme loci in *P. major*. *Theor.Appl.Genet.*, 68, 43-52.
Van Dijk H., (submitted) - A method to estimate gene flow parameters from a population structure caused by restricted gene flow and genetic drift.
Wright S., 1943. - Isolation by distance. *Genetics*, 28, 114-138.

Protogyny in *Plantago lanceolata* populations: an adaptation to pollination by wind?*

M. Bos, R. Steen and H. Harmens

Department of Genetics,
University of Groningen,
Kerklaan 30, 9751 NN Haren (Gn)
The Netherlands

ABSTRACT

Windpollination seems a wasteful process, providing non-directional transfer of male gametes and consequently a low proportion of effective fertilization. Properties that can enhance this process have evolved and are called the "syndrome of anemophily" (Faegri and van der Pijl 1979). *Plantago lanceolata* displays the syndrome perfectly, as will be shown.

Protogyny (the stigma is receptive before the pollen is liberated) is considered as a property that increases cross-pollination. In a species like *P. lanceolata*, with self-incompatibility, protogyny seems a superfluous character. We suggest that in this wind-pollinated species, protogyny optimizes fertilization. This effect is obtained by the combination of protogyny and elongation of the scape during the time the spike changes from female to male flowering.

We will present arguments for this hypothesis:

1. Differences between eight *P. lanceolata* populations.
2. Differences between *Plantago* species.
3. Indications of inadequate seed set in *P. lanceolata* field populations.
4. Influence of pollen release and -capture on dispersal.

*Grassland Species Research Group Publication no. 85.

NATO ASI Series, Vol. G5
Genetic Differentiation and Dispersal in Plants
Edited by P. Jacquard et al.

INTRODUCTION

The biology of *Plantago lanceolata* is relatively well-known and reported in various studies e.g. Sagar and Harper (1964), Cavers et al. (1980). The species is a widespread, self-incompatible, perennial, hermaphrodite herb, found in contrasting habitats: wet hay-fields, dry dune grasslands, road sides etc. Aspects of reproduction are described in the above, but also in Ross (1970), Hammer (1978) and Primack (1978). Besides gynodioecy (Ross 1970; van Damme and van Delden 1982; van Damme 1983 and 1984) protogyny is observed frequently in the genus *Plantago*.

In *P. lanceolata* (Fig. 1) the stigma appears first from the

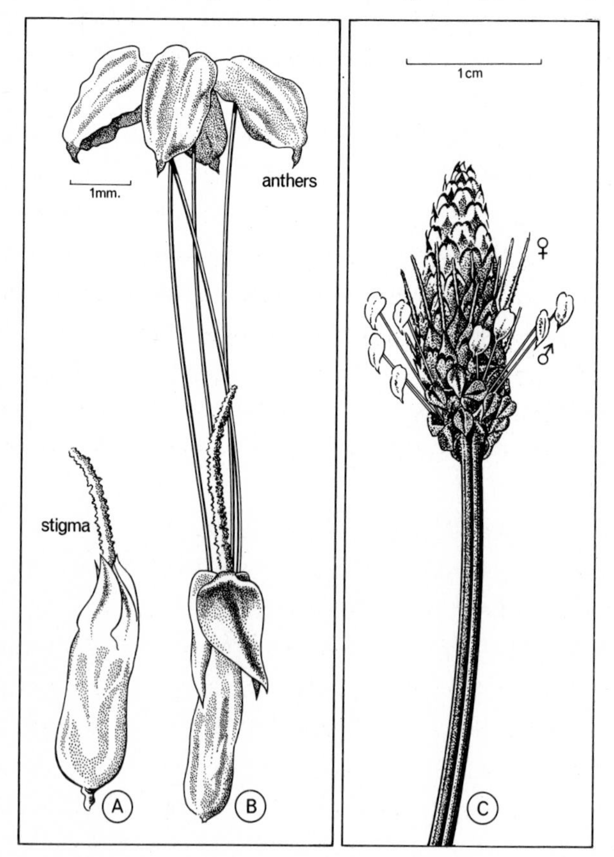

Figure 1 - Flower and spike in *Plantago lanceolata*: (A) flower in female stage (B) flower in hermaphrodite stage (C) spike with flowers in different stages (design E. Leeuwinga).

individual hermaphrodite flower through the opening between the four closed corolla lobes.

After several days the corolla opens, the four anthers appear and release pollen within twenty four hours. This protogyny shows female and hermaphrodite functioning separated in time within a flower, while gynodioecy shows the two morphs separated in space in different individuals. Female flowering starts at the base of the spikes and progresses upwards. Male flowering of the flowers at the base starts when female flowering has extended to roughly 70% of the spike.

The primitive Angiosperm (Crowe 1964) is considered to be a self-incompatible, hermaphrodite, insect pollinated species. *P. lanceolata* shows most of the primitive characters, but the species is known as wind-pollinated with incidental pollination by insects (Clifford 1962; Stelleman 1982).

Protogyny is often considered as a property that increases outcrossing. The presence of protogyny is then somewhat unexpected in *P. lanceolata* since the species is self-incompatible, resulting in all individuals being obligately out-crossed. From a genetical point of view protogyny seems therefore superfluous. The same holds for the explanation of gynodioecy as an outbreeding mechanism in populations of *P. lanceolata* (Primack 1978). Van Damme (1984) has shown that the polymorphism is maintained by the enhanced fecundity and survival of female plants in gynodioecious populations.

The question arises whether protogyny in *P. lanceolata* also is maintained in the species because it confers a higher fitness and what is the nature of the selective process?

Several hypotheses about the selective process are possible:

1. *"Pollen-competition"-hypothesis*. Protogyny prevents, at least for some days, pollination within a flower and even within a spike. If a foreign pollen grain then arrives the incompatibility mechanism within the style has not yet been stimulated and fertilization by the foreign pollen grain and subsequent seed production will be improved.
2. *"Allocation of energy"-hypothesis*. Protogyny provides an individual with the possibility to invest first in female functioning exclusively. Resources are then available for increased

seed-set and/or larger seeds.

3. *"Specialisation to wind pollination"-hypothesis.* The species is extremely well fitted for wind-pollination: 1. Flowering occurs above the vegetative parts. 2. Anthers are exposed outside the reduced flower. 3. Stigmas are sticky and provided with increased effective stigmatic surface (feathered). 4. Exposed anthers open only in good weather conditions. 5. Pollen grains are small (20-30 μ), smooth and non-sticky and the number per anther is high. 6. The cylindric form of the spikes is highly efficient to impact pollen.

The combination of protogyny in the spike with continued elongation of the scape (protogyny-growth syndrome) results in the stigmas being presented at a lower level in the vegetation than that at which the pollen is released. This provides a more directional and effective flow of male gametes to the stigmas and the possibility of higher seed production.

In this paper we present some arguments in favour of the third hypothesis, but we would emphasize that the arguments are partly speculative and not decisive and it is impossible to decide which (combination of) mechanism(s) maintains protogyny in the species.

ARGUMENTS

1. *Differences between eight P. lanceolata populations.* In order to determine the genetic variation in protogyny, the floral development of 2 or more scapes in each of 20 plants from seed collected from each of eight populations in the Netherlands was examined in a glasshouse environment (60% R.H.; 16 h. light and 21°C, 8 h. dark and 19°C).

To show the extent of protogyny along a spike floral development in two of the eight populations (Heteren, a hay-field; Westduinen, a dune pasture) will be considered in some detail here and is shown in Fig. 2.

The lower part gives scape-height in time; the upper part the development of female- and male-flowering along a spike in two populations in time. Individual scapes grew to maximal

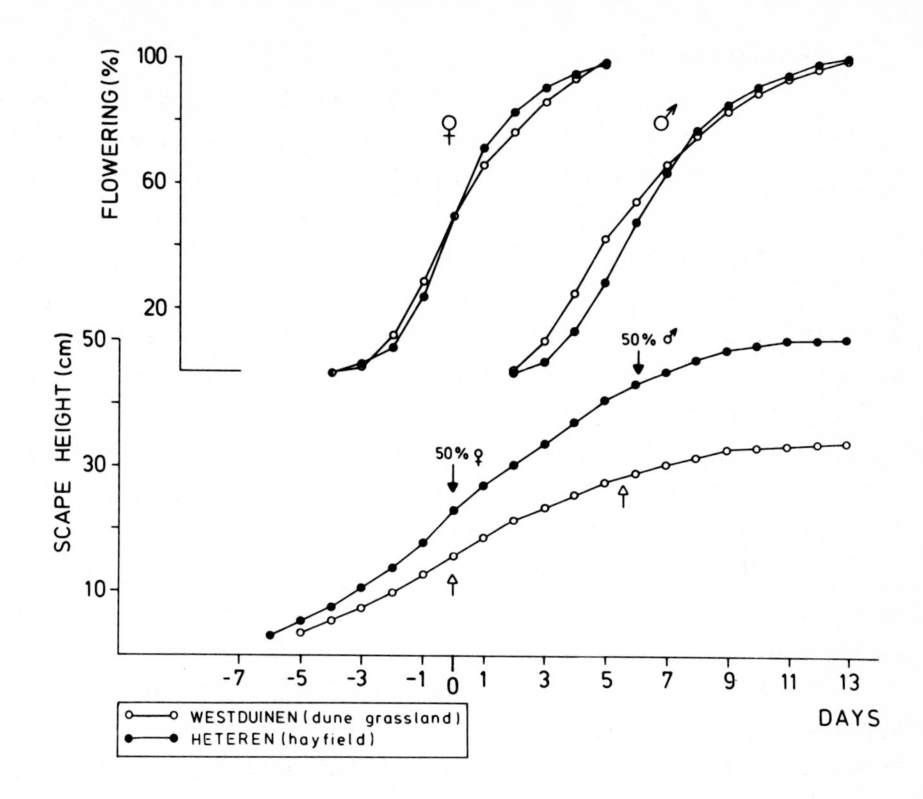

Figure 2 - Scape height and floral development along the spike upwards in two populations of *P. lanceolata*. Explanation in text.

height in about 18 days. When they were 10 cm high spikes started female-flowering of the basal flowers. Scapes were about 20-30 cm (depending on the population) when male-flowering started. Arrows in Fig. 2 indicate the scape-height at which 50% of the flowers within the spikes had become female-flowering or male-flowering respectively. The time at which this occurs is called zero. The flower development upwards along the spikes in the two populations was rather similar, although the populations differed significantly in height. The time between first female- and first male-flowering of the spikes was about 5 days in the pasture and 6 days in the hayfield population. Also in the other 6 populations this period was 5-6 days.

Figure 3A shows scape-height in the eight populations at the

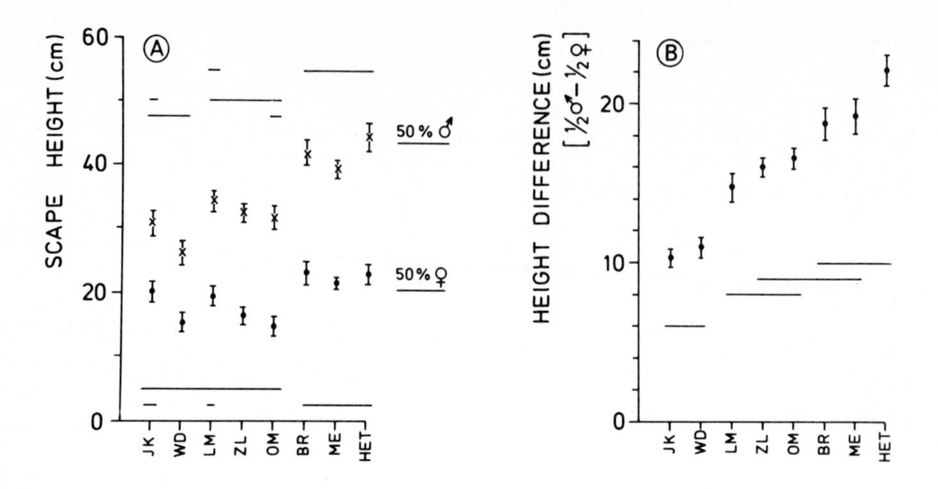

Figure 3 - A. Scape-height in 8 *P. lanceolata* populations. Explanation in text. B. Differences in scape-height between the moment ♀ flowering had advanced to half way up the spike and the moment ♂ flowering was half way. Vertical lines indicate 95%-confidence intervals. Horizontal: one level indicates populations which are not significantly different ($P < 0.01$).

moment spikes were 50% ♀ flowering (lower points) and at the moment of 50% ♂ flowering (upper points). Management of the populations is given in Table 1. Scape-height in the three old hayfields was the same, but differed significantly from the other fields at 50% ♀ as well as 50% ♂ flowering (for significance of differences: see horizontal bars in lower and upper part of the graph respectively. Student-Newman-Keuls' t-test). Fig. 3B detects differences in scape-height between the moment of 50% ♀ and 50% ♂ flowering in the eight populations and the significance of differences between populations (horizontal bars). It is clear from these figures that, in all populations, scapes were short when spikes were ♀ flowering and long when releasing their pollen as males. These differences differed significantly depending on the management: high in hayfields, low in pastures.

Table 1 shows that the ratio of the height difference ($\frac{1}{2}$♂-$\frac{1}{2}$♀) and final height of the scapes was rather constant in the eight populations: ranging from 0.31 to 0.44.

Table 1 - Ratio of scape-height differences and final height in the eight populations (from Fig. 3B). The management of the populations is indicated.

JUNER KOELAND	(JK)	0.31	Pasture
WESTDUINEN	(WD)	0.35	Dune pasture
LOENERMARK	(LM)	0.44	Pasture
ZEEGSER LOOPJE	(ZL)	0.41	Pasture/Hayfield
OUDEMOLEN	(OM)	0.40	Wet Hayfield
BRUUK	(BR)	0.42	Old Hayfield
MERREVLIET	(ME)	0.40	Old Wet Hayfield
HETEREN	(HE)	0.43	Old Hayfield

These observations show that, in all the populations of *P. lanceolata*, spikes flower about six days as females, far below the height level of spikes which flower as hermaphrodites. In a downward flow of pollen from source to earth this seems a functional position.

2. *Differences between Plantago species* (Table 2). Protogyny is a wide-spread phenomenon in the genus *Plantago*. The five species in the Netherlands are all protogynous, although in different degree. The degree of protogyny as indicated by stigma length is positively associated with the degree of anemophily as indicated by the number of pollen per anther.

Table 2 - Some reproductive characteristics in five *Plantago* species (after Ross 1970; Hammer 1978; Bos, this paper). SI = selfincompatibility; * "pollenreich" (Hammer); Time [♀-♂] = time between first ♀ and first ♂ flowering of the spike; W = by wind, I = by insects; # = number of pollen per anther.

	Protogyny	Gynodioecy	SI	Stigma-length(mm)	#Pollen/anther	Time[♀-♂] (days)	Pollination
lanceolata	+	+	+	4.57	20.400	5-6	W(I)
maritima	+	+	+	3.66	10.640	2.0	W(I)
media	+	-	+	4.13*	14.100	2.8	W/I
coronopus	+	+	-	3.53	5.170	4.8	W
major	+	-	-	2.10	6.350	2.7	W

The "time" data in Table 2 come from a preliminary experiment in which scape-growth and spike development were followed in *P. media*, *P. coronopus*, *P. lanceolata* and *P. major spp. major* (Fig. 4). Time zero gives scape-height at the time spikes started female-flowering, the asterisk * indicates the time spikes started

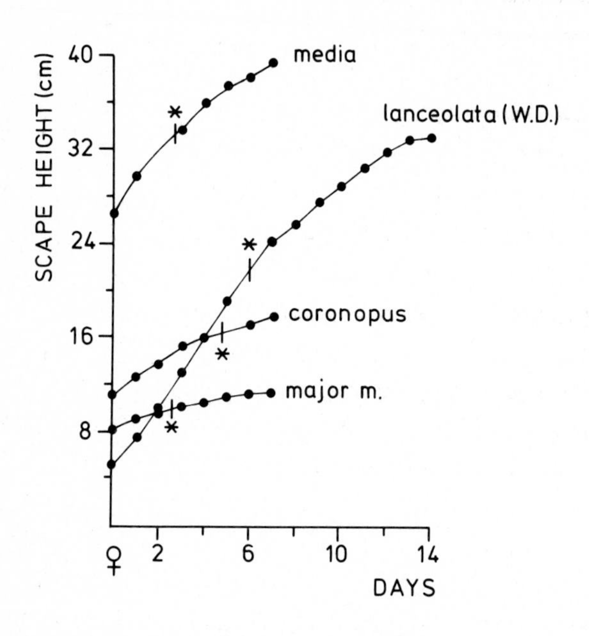

Figure 4 - Scape height development in relation with ♀ and ♂ flowering in four *Plantago* species. Explanation in text.

male flowering. Of the four species *P. lanceolata* (WD-population) started ♀ flowering at the lowest scape height and lasted longest: 6 days. Growth curves in the predominantly selfing species *major* and *coronopus* were rather flat (linear regression coefficient b = resp. 0.47 and 0.89) during flowering, compared to the two selfincompatible species *media* and *lanceolata* (b = resp. 2.22 and 2.22). Of the species compared in this way *P. lanceolata* shows, in addition to the most extreme characteristics related to wind-pollination, also the most extreme combination of strong growth of the scape (b = 2.22) with long duration of the protogyny-period (6 days).

3. *Indications of inadequate seed set in P. lanceolata field populations*. Selection for higher efficiency of wind pollination

by protogyny (as supposed in our hypothesis) can only be effective if less than maximal seed production/capsule in the field (2 seeds/capsule) is found. Table 3 presents relevant data. It is shown that in all field populations the number of seeds/capsule was < 2.

Table 3 - Number of seeds/capsule in *P. lanceolata*, as derived from three publications. † = range of 8 populations in North-Carolina; F = field, L = low nutrition, H = high nutrition. He, Gr, Me, Ha = four populations in the Netherlands.

author	van Damme (1984)			Bos (this paper)	Primack and Antonovics (1981)†			Dowling (1935)
pop.	He	Gr	Me	Ha	F	L	H	-
	1.59	0.87	0.95	0.71	0.22-1.05	1.32-1.48	2.00	1.44

The experiments of Primack and Antonovics (1981) surely suggest that malnutrition is the main cause of poor seed set and also that seed production can be improved in controlled conditions even with low nutrition (L). So, ovule sterility seems not to be the restrictive character. Stelleman (1982) noticed in some *P. lanceolata* populations that seed set percentages decreased, if insect visits were prevented. This shows that at least in some populations pollen concentrations in the air are limiting. We intend to test this with extra pollen supply in field populations.

4. *Influence of height of pollen release and -capture on dispersal.* In a series of experiments (Bos et al.; in preparation) we studied gene-dispersal in *P. lanceolata*. Windtunnel experiments showed that the distances over which pollen was dispersed were dependent on conditions like presence or absence of vegetation, pollen-source height and wind-speed. Fig. 5 shows relative pollen dispersal (proportion of total number of pollen-grains caught on the given receptor height) for two different scape heights. By raising the source height the peaks of the graphs shifted to the right and the tails became longer. Pollen dispersal distances are influenced not only by changes in the height of pollen release but also of pollen collection.

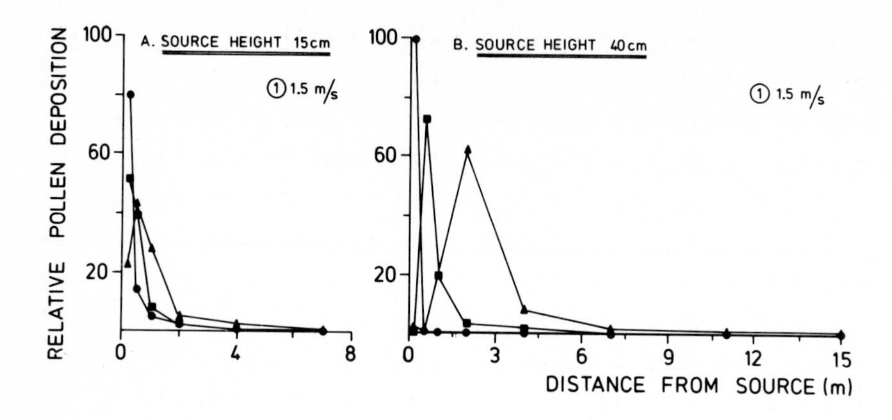

Figure 5 - Pollen dispersal in a windtunnel after release of pollen grains from 15 cm (A) or 40 cm (B) height. Receptor heights were (in A or B resp.): ●---● 15 or 40 cm; ■---■ 10 or 25 cm; ▲---▲ 5 or 10 cm.

DISCUSSION

In this paper we describe protogyny in the individual flower and development of sexuality along the spike during the growth of the scape in *Plantago lanceolata* (Fig. 1 and 2). We conclude: 1. An individual hermaphrodite flower of the brush-like inflorescence starts with the extension of the stigma and after 5-6 days the four anthers appear. 2. This is also reflected in the flowering pattern along the spike: flowering starts at the base of the spike and moves upwards along it, first ♀, then ♂. 3. In eight populations scape-height was significantly less when the spike started female flowering compared with the moment male flowering started. The populations showed absolute differences (Fig. 2 and 3) but no great relative differences (Table 1).

The result of these characteristics is that spikes function for several days as females at a lower height than as hermaphrodites releasing pollen. Since the species is wind-pollinated and pollen on the average flows downwards in the vegetation this appears functional for pollen collection by the stigma during its life-time. With the "protogyny-growth-syndrome" pollen is dispersed over a longer distance than if there was synchrony between male and female flowering. This may be a device to reduce the level of fertilization between relatives, if close neighbours tend to be related.(Bos et al., in prep.).

Compared to some other *Plantago* species in the Netherlands *P. lanceolata* shows the sexual developmental characters in the most extreme form (Fig. 4 and Table 2). *P. major ssp. major* ranks last. In the families of *Cyperaceae* and *Graminae* also, wind-pollinated species with brush-like inflorescences are found, in which ♀ flowering takes place at a lower height level than ♂ flowering; often combined with some kind of protogyny or unisexuality. In the open spring vegetation we found flowering simultaneously with *P. lanceolata* e.g. *Carex acutiformis* (monoecious) with 1-2 male spikes above 3-5 female spikes per scape; *Alopecurus vulgaris* and *geniculatus* (both protogynous) with scapes growing during development from ♀ to ♂ flowering. A genus in which the two syndromes of wind pollination and unisexuality developed hand-in-hand is *Thalictrum* (Kaplan and Mulcahy 1971). Improvement of the efficiency of pollen utilization is seen as an effect of such evolution. So, the positive association between characters of the "protogyny-growth-syndrome" and the "anemophily-syndrome" is not restricted to the *Plantago* genus.

We think that the above facts and that in the field there seems to be room for selection for a higher number of seeds per capsule are arguments in favour of the hypothesis that in the self-incompatible species *P. lanceolata* the above "protogyny-growth-syndrome" could be a specialisation to wind-pollination.

The present description of the syndrome is part of our study of gene-flow in the species.

REFERENCES

Bos M., Harmens H. & Vrieling K., in prep. - Gene flow in *Plantago* 1. Dispersal and neighbourhood in *P. lanceolata.*

Cavers P.B., Bassett I.J. & Crompton C.W., 1980. - The biology of Canadian weeds 47. *Plantago lanceolata* L. *Can.J.Plant.Sci.*, 60, 1269-1282.

Clifford H.T., 1962. - Insect pollinators of *Plantago lanceolata. Nature*, 193, 196.

Crowe L.K., 1964. - The evolution of outbreeding in plants. I. The angiosperms. *Heredity*, 19, 435-457.

Dowling R.E., 1935. - The structure of the ovary in the genus *Plantago* L. - I. The British species. *J.Linn.Soc.Lond.Bot.*, 50, 323-336.

Faegri K. & van der Pijl L., 1979. - *The principles of pollination ecology*. Perg.Press, Oxford.

Hammer K., 1978. - Entwicklungstendenzen blütenökologischer Merkmale bei *Plantago. Flora*, Bd. 167, 41-56.

Kaplan S.M. & Mulcahy, 1971. - Mode of pollination and floral sexuality in *Thalictrum. Evolution*, 25, 659-668.

Primack R.B., 1978. - Evolutionary aspects of wind pollination in the genus *Plantago* (Plantaginaceae). *New Phytol.*, 81, 449-458.

Primack R.B. & Antonovics J., 1981. - Experimental ecological genetics in *Plantago*. V. Components of seed yield in the ribwort plantain *Plantago lanceolata L. Evolution*, 35, 1069-1079.

Ross M.D., 1970. - Breeding systems in *Plantago. Heredity*, 25, 129-133.

Sagar G.R. & Harper J.L., 1964. - Biological flora of the British Isles. *Plantago major L., P. media L.* and *P. lanceolata L. J. of Ecology*, 52, 189-221.

Stelleman P., 1982. - De betekenis van biotische bestuiving bij *Plantago lanceolata. Thesis*, University of Amsterdam.

Van Damme J.M.M. & Van Delden W., 1982. - Gynodioecy in *Plantago lanceolata L.* I. Polymorphism for plasmon type. *Heredity*, 49, 303-318.

Van Damme J.M.M., 1983. - Gynodioecy in *Plantago lanceolata L.* II. Inheritance of three male sterility types. *Heredity*, 50, 253-273.

Van Damme J.M.M., 1984. - Gynodioecy in *Plantago lanceolata L.* III. Sexual reproduction and the maintenance of male sterility. *Heredity*, 52, 77-93.

Van Dijk H., in prep. - The estimation of gene flow parameters from a continuous population structure.

Adjacent Populations of Cocksfoot (*Dactylis Glomerata* L.): A Detailed Study of Allozyme Variation across Contrasting Habitats.

M. Valero* and I. Olivieri

Biologie des Populations et des Peuplements, Centre L. Emberger, CNRS, F-34033 Montpellier Cedex

ABSTRACT

This study of gene flow between adjacent populations of cocksfoots (*Dactylis glomerata* L.) takes place in the plain of "la Crau" (Bouches-du-Rhône, south of France). This xeric stony land is in some places irrigated and exhibits two very different environments separated from each other only by a few meters. Using allozymes, genetic differences between cocksfoots of the two environmental types (xeric and mesic) were shown in previous works.

The aim of this study was to determine the extent to which dry and moist Crau individuals were genetically differentiated, and to examine if there was any evidence that the direction and amount of gene flow was influencing the pattern of differenciation. It is firstly shown that no introgression occurs from the moist Crau towards the dry Crau populations of cocksfoot. Secondly, allelic frequencies and allelic association show that direct gene flow occurs from the dry Crau towards the moist Crau populations, but that they are very limited and restricted to places where the trees separating the two habitats are well spaced.

---

* Present address: Laboratoire de Génétique écologique et de Biologie des Populations Végétales, U.S.T.L., F-59655 Villeneuve d'Asq Cedex.

NATO ASI Series, Vol. G5
Genetic Differentiation and Dispersal in Plants
Edited by P. Jacquard et al.

INTRODUCTION

In a series of papers (Antonovics 1966; Jain and Bradshaw 1966; Mc Neilly 1967; Mc Neilly and Bradshaw 1968) on the evolution of closely adjacent grass populations at metal mine boundaries, it has been shown that genetic divergence could occur by disruptive selection over short distances even if gene flow was considerable. Evidence of the development of breeding barriers between these adjacent plant populations has been reported: differences in flowering time (Mc Neilly and Antonovics 1968) and increasing selfing rate (Antonovics 1968), for both *Anthoxanthum odoratum* and *Agrostis tenuis*. According to the authors, reproductive isolation evolves because gene flow is inherently disadvantageous (since it leads to less well adapted offspring) and thus any gene which reduces gene flow has a selective advantage. Using a computer simulation, Caisse and Antonovics (1978) have shown that both genetic divergence and reproductive isolation may occur between adjacent populations connected by gene flow. However, in this case, reproductive isolation as a response to divergent selection requires very high selective pressures. The authors concluded that "conditions leading to isolation are far more stringent than those permitting divergence". This raises the question of how much genetic divergence is associated with reproductive isolation.

Recently, Stam (1983) proposed a theoretical model, and pointed out that "reproductive isolation in plants through differences in flowering time needs not necessarily be a response to divergent selection but may essentially be non Darwinian if it is triggered by non selective environmental differences". Notice that Mc Neilly and Antonovics (1968) had already approached this point, but they always assumed the existence of a local adaptation of flowering time, prior to the evolution of reproductive isolation. The originality of Stam's model is the assumption that there is the same genetic polymorphism for flowering time in both adjacent habitats of a population, meaning that the differences in flowering time between these habitats are only environmental. These differences do not even need to be important. Gene flow between the two habitats will preferentially occur between earlier flowering plants of the habitat determining a late flowering, and later flowering plants of the habitat determining an early flowering; in such a way that "any shift in the allelic frequency for flowering time in one part of the population is exactly counter-balanced by an opposite shift in the other. This means that as an average over the two habitats, the allelic frequencies will not change. "The alleles are merely sorted out by the mechanism (...). Although migration is automatically preferential, the mechanism involves no selection when the (sub)populations are considered as a single unit."

On the other hand, a case of fairly little genetic divergence between two reproductively isolated, sympatric populations of brown trout (*Salmo trutta*) has been reported by Ryman et al. (1979).

Situations of closely adjacent plant populations found in contrasting habitats are ideal for studying the interplay of genetic divergence, selection and gene flow at the population level. This paper reports on genetic studies on adjacent populations of cocksfoot (*Dactylis glomerata* L.) carried out in the Crau (southern France, near Salon-de-Provence), where two contrasting habitats (dry and moist) are separated by only a few meters.

The alluvial plain of the Crau is very dry, stony and windy; the prevailing wind during the flowering season is the strong northerly "mistral". This plain is partially irrigated so that natural meadows have established after stone removal and silt accumulation. The irrigated ("moist") Crau then appears as an island inside the "dry" Crau. Lumaret and Valdeyron (1980) and Valero (1983) have shown that the dry Crau cocksfoot belongs to the narrow leaved mediterranean species *Dactylis glomerata* ssp. *hispanica*, while the moist Crau cocksfoot is intermediate between the broad leaved northern species *Dactylis glomerata* ssp. glomerata and the mediterranean subspecies hispanica, but is closer to the former with regard to several enzymatic, morphological and physiological characteristics. Experimentally, there is no reproductive barrier, between either the individuals of the two subspecies glomerata and hispanica, or dry and moist Crau populations (Valero 1983).

The object of this paper is to determine the extent to which dry and moist Crau individuals are genetically differentiated, and to examine if there is any evidence that the direction and amount of gene flow is influencing the pattern of differenciation.

MATERIAL AND METHODS

Location of Study Sites and Transects

Moist Crau meadows are surrounded by wind-breaking tree-rows and by irrigation ditches. Four transects were studied (Fig. 1), located at the boundary of dry and moist Crau. Two of these transects (MC1a and MC1b) were inside an old meadow (at least 100 years old) previously described (Valero, 1983) as consisting of individuals intermediate between subspecies glomerata and hispanica, but closer to the former one. Both transects were parallele and oriented from north to south. They were located behind a tree-row that formed a barrier between the dry and moist environments. This tree-row was dense beyond transect MC1b while it showed spaced trees beyond transect MC1a. One individual was sampled every meter on each transect (91 plants in transect MC1a and 71 in MC1b). The two other transects (DCa and DCb) were located inside the dry Crau, on the same boundary (Fig.1), 21 or 20 plants were sampled on each transect.

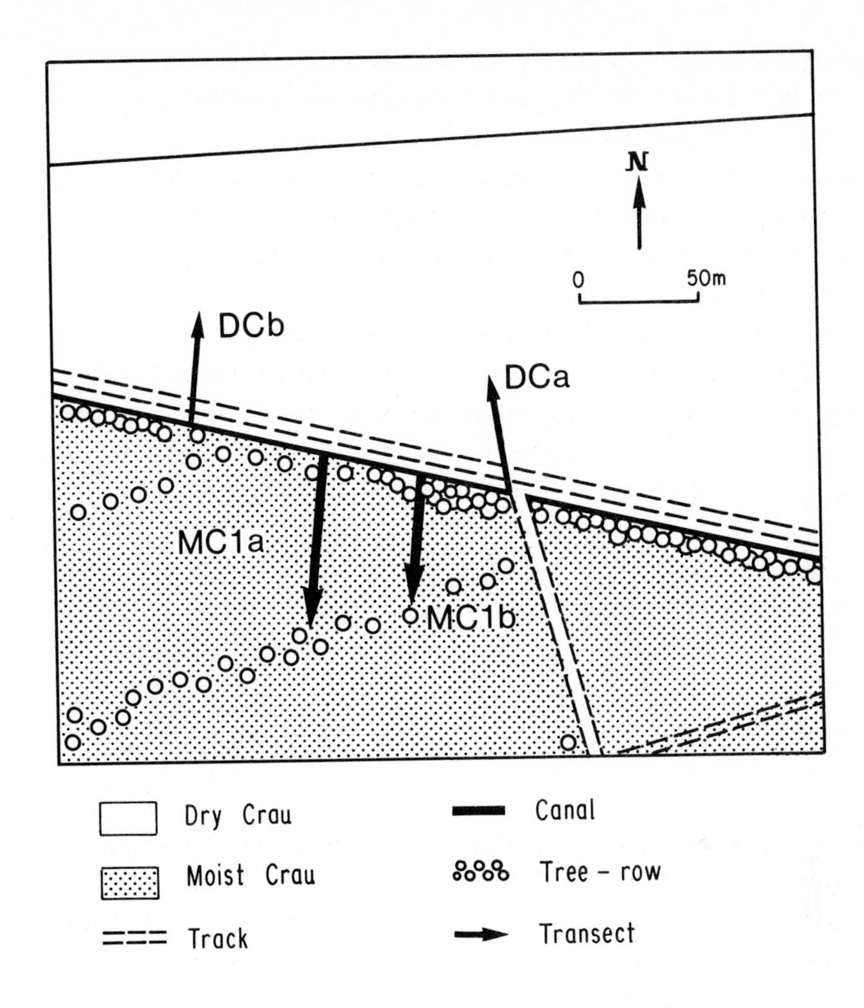

Figure 1: LOCATION OF THE STUDIED TRANSECTS

Other independant samples, taken from dry and moist Crau (Valero 1983), will be used as "moist Crau" (MC) and "dry Crau" (DC) references. "Typical" allelic frequencies for D. glomerata ssp. glomerata (DGG) and ssp. hispanica (DGH) are those obtained by Lumaret (1981a) by averaging allelic frequencies of several populations close to the region where the present study was conducted.

Electrophoretic Analyses

Analyses were performed upon leaves collected from the individuals sampled from each transect and grown inside a greenhouse. The extraction method has been described by Lumaret and Valdeyron (1978). Six loci were studied for enzymatic polymorphism. Genetic determinism and technics used for cocksfoot have been described in other papers: glutamate-oxaloacetate transaminases (GOT1) (Lumaret and Valdeyron, 1978); acid phosphatases (ACPH1) (Lumaret, 1981b); peroxidases (PX1) (Lumaret, 1982); tetrazolium oxidases (TO1) (Valero, 1980); phosphogluco-isomerases (PGI1 and 2) (Lumaret, 1981a). In one system (PGI), the two loci result from the duplication of an ancestral locus. Therefore, they have common isozymes and it was not possible to distinguish between alleles of the two loci, meaning that only the allelic compositon (and not the genotypic composition) of an individual could be determined. These loci are therefore not considered in some calculations. In this paper, alleles are identified by a number which indicates allelic mobility relative to the most common allele (assigned mobility=1.00).

RESULTS

Transects in the dry Crau

The allelic frequencies obtained in transects DCa and DCb, and in other dry Crau populations (Table 1) and Nei's genetic distances (Nei 1972) (Table 2) calculated for the six loci show that the dry Crau individuals belong to the mediterranean subspecies hispanica. No influence of the moist Crau populations can be detected. The few small differences observed between transects do not reflect any geographical influence.

TABLE 1. Allelic frequencies found in the two subspecies of cocksfoot (*glomerata* (DGG) and *hispanica* (DGH)), in the populations of the Crau (dry and moist), and in the four transects
studied in the Crau.

		SUBSPECIES		POPULATIONS		TRANSECTS			
LOCI	ALLELES	DGG	DGH	MOIST CRAU	DRY CRAU	MC1a	MC1b	DCa	DCb
	1.00	.77	.13	.60	.07	.59	.59	.08	.05
	.72	.23	.33	.26	.39	.25	.28	.44	.33
GOT1	.38	-	.40	.12	.46	.12	.10	.42	.55
	.10	-	.12	.01	.07	.03	.02	.06	.07
	1.26	-	-	.01	.01	.01	.01	-	-
	N	x	x	x	x	368	276	84	80
	1.00	.59	.64	.74	.73	.76	.79	.73	.72
	.88	.40	.08	.13	.08	.15	.15	.01	.08
TO1	1.12	-	.28	.10	.14	.07	.06	.24	.14
	others	.01	-	.03	.05	.02	-	.02	.06
	N	x	x	x	x	368	284	80	76
	1.00	.70	.43	.63	.30	.64	.56	.34	.45
PX1	.93	.18	.46	.24	.46	.26	.31	.49	.38
	others	.12	.11	.13	.24	.10	.13	.17	.17
	N	x	x	x	x	368	280	84	80
	1.00	.67	.85	.65	.70	.62	.56	.87	.80
ACPH1	.88	.27	.13	.24	.13	.31	.36	.10	.14
	others	.06	.02	.11	.17	.07	.08	.03	.06
	N	x	x	x	x	368	276	68	44
	1.00	.60	.67	.60	.66	.59	.63	.70	.65
	1.25	.14	.10	.17	.12	.14	.17	.13	.04
PGI1	.75	.13	.17	.17	.17	.20	.15	.15	.23
and	.50	-	.02	.05	.01	.01	.04	.01	.02
PGI2	others	.13	.04	.01	.04	.06	.01	.01	.06
	N	x	x	x	x	712	450	168	120

N = Number of alleles ; x Allelic frequencies averaged over 10 populations for each subspecies (Lumaret, 1981), and over 2 and

3 independant populations of the moist and dry Crau, respectively (Valero 1983).

Transects in the moist Crau

The allelic frequencies observed for transects MC1a and MC1b (Table 1) and Nei's genetic distances (Table 2) confirm that individuals of these transects are intermediate between the two subspecies *glomerata* and *hispanica*, closer to the northern subspecies *glomerata*. Indeed, they possess some typical mediterranean alleles of the subspecies *hispanica* (alleles 0.38 and 0.10 of locus GOT1 for instance, Table 1).

TABLE 2. Nei's genetic distances (x1000) averaged over 5 loci.

	SUBSPECIES		POPULATIONS		TRANSECTS			
	DGG	DGH	MC	DC	MC1a	MC1b	DCa	DCb
DGG		266	33	349	27	36	342	369
DGH			144	21	148	150	14	18
MC				191	4	8	184	194
DC					198	190	8	16
MC1a						5	153	198
MC1b							187	204
DCa								20

DGG: *Dactylis glomerata* ssp. *glomerata*
DGH: *Dactylis glomerata* ssp. *hispanica*
MC: moist Crau
DC: dry Crau

The distribution of these allelic frequencies is similar for both transects at all loci studied (Table 3), except for loci PGI1 and PGI2 where differences concern essentially rare alleles (0.5 and "others", Table 1). When the allelic frequency distribution of the moist Crau is compared to that of MC1a, MC1b and MC1a + MC1b pooled together (Table 3), some significant differences appear. However, these also concern rare alleles, or alleles for which the frequencies do not discriminate between the two ssp. *glomerata* and *hispanica*.

TABLE 3. Significance of differences between moist Crau(MC), MC1a+MC1b, MC1a and MC1b (Chi-square test).

LOCUS	MC1a /MC1b	MC1a/MC	MC1b/MC	MC1a+MC1b/MC
GOT1	N.S.	N.S.	N.S.	N.S.
TO1	N.S.	N.S.	**	***
PX1	N.S.	N.S.	N.S.	N.S.
ACPH1	N.S.	*	***	***
PGI1-2	**	***	N.S.	***

N.S. = non significant(p_10%), * = p_5%, ** = p_1%, *** = p_0.1%.

The high similarity between the allelic frequency distribution of these two transects may indicate the similarity of ecological conditions inside this meadow. However, it might simply show that genetic mixage is very important between the individuals of meadows.

Since the position of the individuals was known along both transects, it was possible to test the randomness of the allelic distribution in the meadow. Two methods were used.

Analysis of the allelic frequencies distribution along the two transects: Each transect was divided into two equal parts. The first half consisted of individuals that were the closest to the dry Crau, and the second half, the farthest. Allelic frequencies in these two parts (Table 4) were compared by Chi-square tests.

For most loci (PX1, AcPH1, TO1, PGI1 and PGI2), no significant difference was observed between the two halves of each transect (Table 4).

TABLE 4. Distribution of allelic frequencies in each of the two parts of transects MC1a and MC1b.

LOCI	ALLELES	MC1a 1st h.n* DC (1)	MC1a 2nd h.f from DC (2)	X2 (1)/(2)	MC1b 1st h.n DC (3)	MC1b 2nd h.f from DC (4)	X2 (3)/(4)
	1.00	.60	.59		.56	.63	
	0.72	.21	.29		.28	.28	
GOT1	0.38	.15	.08	(*)	.12	.07	NS
	0.10	.03	.03		.02	.01	
	1.26	.01	.01		.02	.01	
	N	184	184		136	136	
	1.00	.77	.76		.83	.74	
	0.88	.16	.15		.13	.18	
TO1	1.12	.06	.08	NS	.04	.08	NS
	others	.01	.01		-	-	
	N	184	184		140	144	
	1.00	.66	.61		.52	.60	
	0.93	.24	.29		.32	.29	
PX1	others	.10	.10	NS	.16	.11	NS
	N	184	184		136	144	
	1.00	.65	.59		.52	.60	
	0.88	.28	.34		.39	.33	
ACPH1	others	.07	.07	NS	.09	.07	NS
	N	184	184		132	144	
	1.00	.60	.58		.62	.64	
PGI1	1.25	.14	.15		.18	.17	
and	0.75	.20	.20		.16	.14	
PGI2	0.50	.01	.02	NS	.03	.04	NS
	others	.05	.05		.01	.01	
	N	368	344		120	280	

(*) = p 6% NS = non significant (p 10%)
N = number of alleles
DC: dry Crau.
* : h.n (half near); h.f (half far)

At the GOT1 locus, frequency variations were observed essentially for the mediterranean allele 0.38, it being slightly more frequent in the first half, for both transects MC1a and MC1b. These variations are almost significant in MC1a (a=6%) and not at all in MC1b. This result may reveal gene flow from the dry Crau population; the difference observed between the two transects could result from their different locations: MC1a being behind a spaced tree-row, gene flow might be more important there than behind the dense tree-row where MC1b was located. In order to test this hypothesis, a second method of analysis was used.

Test of allelic association between loci (Table 5): If the slight increase of the allele 0.38 frequency at locus GOT1 was a consequence of gene flow from dry Crau plants, an association between the loci studied should be observed. The analysis used indirectly here reflects linkage disequilibrium. Since Lumaret and Valdeyron (1978) and Lumaret (1981a) have shown that alleles 0.38 and 0.10 of the locus GOT1 were only found in the subspecies *hispanica*, the presence or the absence of one of these alleles was chosen as a criterion to separate the cocksfoot of the moist Crau transects into two groups, "*glomerata*" and "*hispanica*". Differences in the allelic composition between these two groups of plants at other loci were tested, using contingency tables (the duplicated system, PGI1 & 2, was not taken into account in this test).

TABLE 5. Test of allelic association in the two transects MC1a and MC1b.

		TRANSECT MC1a INDIVIDUALS MORE			TRANSECT MC1b INDIVIDUALS MORE				
		DGG AT LOCUS GOT1	DGH AT LOCUS GOT1	X2	DGG AT LOCUS GOT1	DGH AT LOCUS GOT1	X2		
LOCI	ALLELES	1	2	1 / 2	3	4	3 / 4	DGG	DGH
	1.00	.71	.83		.79	.77		.59	.64
	0.88	.19	.10		.15	.19		-	.08
TO1	1.12	.08	.06	*	.07	.04	NS	.40	.28
	others	.02	.01		-	-		.01	-
	N	208	160		180	92		x	x
	1.00	.60	.62		.59	.50		.70	.43
	0.93	.29	.23		.31	.31		.18	.46
PX1	others	.11	.15	NS	.10	.19	NS	12	.11
	N	208	160		180	92		x	x
	1.00	.57	.69		.55	.57		.67	.85
ACPH1	0.88	.34	.27	*	.37	.37	NS	27	.13
	others	.09	.04		.08	.06		.06	.02
	N	208	160		180	92		x	x

* = p 5% NS = non significant (p 10%)
DGG: D. glomerata ssp. glomerata, DGH: D. glomerata ssp. hispanica.
N = number of alleles, x = cf. Table 1.

No significant difference was observed, for any of the studied loci, in transect MC1b located behind the dense tree-row (Table 5). In transect MC1a were observed significant differences towards an association between *glomerata* alleles at locus GOT1 and overall *glomerata* tendency at loci AcPH1 and TO1, and between *hispanica* alleles at locus GOT1 and overall *hispanica* tendency at loci AcPH1 and TO1.

DISCUSSION

Lumaret and Valdeyron (1978) and Lumaret (1981a) have shown that the genetic structure of locus GOT1 had very likely been under a selective pressure related to the hydric status of the habitat. Whether the effect is directly or not related to the function of the enzyme specified by the locus is unknown. First, alleles 0.38 and 0.10 are only found in the mediterranean subspecies *hispanica*. Second, in Europe, clines oriented from north to south were shown for alleles 1.00 and 0.72. These frequency changes were found again on the small scale of the site (50 m): the frequency of allele 1.00 is very high in moist habitats and regularly decreases with increasing dryness, while the frequency of the allele 0.72 increases. Since the loci other than GOT1 seem to be neutral (Lumaret 1981a and Valero 1983), a consequence of the occurence of gene flow from the dry Crau in transects MC1a and MC1b will be that the individuals classified as "*hispanica*" for locus GOT1 overall resemble the subspecies *hispanica* more than those from the "*glomerata*" class, and reciprocally. No formal genetic work has been carried out so far on an eventual genetic linkage between locus TO1 and the other studied loci. But Lumaret (1981a) has shown that loci GOT1 and ACPH1 were genetically independant. An allelic association is therefore observed between the two independant loci GOT1 and ACPH1 in transect MC1a (Table 5). The different result obtained for this transect and transect MC1b is probably a consequence of their different location: an allelic association is observed between the three loci GOT1, ACPH1 and TO1 in the transect located behind a spaced tree-row, when no association could be detected in the transect behind a dense tree-row. This indicates that migration alone, and not selection operating on coadapted multilocus units (such as those described on Avena barbata by Clegg et al. 1972) is responsible for the allelic association. This is emphasized by the fact that the reproductive system of cocksfoot, i.e. cross-pollination, is not likely to encourage the maintenance of such coadapted units.

CONCLUSION

Two important points have been shown. Firstly, no introgression occurs from the moist Crau towards the dry Crau populations of cocksfoot (Table 1). Secondly, allelic frequencies (Tables 1 and 4) and allelic association (Table 5) show that direct gene flow occurs from the dry Crau towards the moist Crau populations, but that they are very limited and restricted to places where the trees separating the two habitats are well spaced.

The difference in the direction of introgression may result from the geographical location of the study sites: the prevailing, northerly wind more likely brings pollen from the dry Crau towards the moist Crau. However, other studies (Kozumplick & Christie 1972, Naghedi-Ahmadi 1977) have shown that migration by pollen was very low in the cocksfoot, and this was verified inside the dry Crau populations (Valero 1983). Also, interpollination between adjacent populations of the dry and moist Crau is possible during the last two weeks of June only. The lack of gene flow from the moist Crau towards the dry Crau is probably a consequence of this low migration of pollen and of the selection against any allele from moist Crau. The occurence of a restricted gene flow from the dry Crau towards the moist Crau shows that migration is not selected against in that case, since migration by pollen is still likely to be very low. It might even be selected for, given that the ecological conditions of the moist Crau present some habitat characteristics of subspecies glomerata and some of subspecies hispanica, so that the introgression of certain genes of the hispanica type could have been favorable for the glomerata populations established in the moist Crau. This hypothesis could explain why this moist Crau population does not present any tendency towards isolation in comparison with the heavy metal tolerant and non-tolerant adjacent grasses populations studied by Antonovics (1966), Jain and Bradshaw (1966) and Mc Neilly (1968). Physiological studies are strongly needed here, for a better understanding of ecotypic differenciation.

ACKNOWLEDGMENTS

The authors are indebted to J. Antonovics for reviewing a later draft of this paper. P. Boursot, D. Couvet, P.H. Gouyon, J.M. Labrosse, R. Lumaret, M. Raymond, and G. Valdeyron are thanked for their helpful comments and field assistance.

REFERENCES

Antonovics J (1966) The genetics and evolution of differences between closely adjacent plant populations, with special references to heavy metal tolerance. PhD thesis, University of Wales

Antonovics J (1968) Evolution in closely adjacent plant populations. V.Evolution of self-fertility. Heredity 23:219-238

Caisse M, Antonovics J (1978) Evolution in closely adjacent plant populations. IX.Barriers to gene flow. Heredity 40, 3:371-384

Clegg MT, Allard RW, Kahler AL (1972) Is the gene the unit of selection? Evidence from two experimental plant populations. Proc Nat Acad Sci, Vol 69:2474-2478

Jain SK, Bradshaw AD (1966) Evolutionary divergence among adjacent plant populations. I.The evidence and its theoretical analysis. Heredity 21:406-411

Kozumplink V, Christie BR (1972) Dissemination of orchard grass pollen. Can J Plant Sci 52:977-1002

Lumaret R (1981a) Structure génétique d'un complexe polyploide Dactylis glomerata L.. Thèse de Doctorat d'Etat, mention Sciences, USTL Montpellier

Lumaret R (1981b) Etude de l'hérédité des phosphatases acides chez le Dactyle (Dactylis glomerata L.): triploide et tétraploide. Can Genet Cytol 23:513-523

Lumaret R (1982) Proteins variations in diploid and tetraploid orchard grass (Dactylis glomerata L.): formal genetics and population polymorphism of peroxidases and malate dehydrogenases. Genetica 57:207-215

Lumaret R, Valdeyron G (1978) Les glutamate-oxaloacétate transaminases du Dactyle (Dactylis glomerata L.); génétique formelle d'un locus. C R Acad Sci:705-708

Lumaret R, Valdeyron G (1980) Les Dactyles de la Crau: mise en évidence de relations entre les différents écotypes par le polymorphisme enzymatique. C R Acad Sci:229-238

Lumaret R, Valero M (in prep) Evidence for duplication of a gene coding for PGI in both diploid and tetraploid plants of Dactylis glomerata L.

Naghedi-Ahmadi I (1977) Zur Frage pollenflug wate bei Dactylis glomerata L.. Z Pflanzenzuchtg 78:163-169

Nei M (1972) Genetic distances between populations. Amer Nat 106: 283-292

McNeilly TS (1967) Evolution in closely adjacent plant populations III.Agrostis tenuis on a small copper mine. Heredity 23:99-108.

McNeilly TS, Bradshaw AD (1968) Evolution and processes in populations of copper tolerant Agrostis tenuis Sibth. Evolution 22, 1:108-118

McNeilly TS, Antonovics J (1968) Evolution in closely adjacent plant populations. IV.Barriers to gene flow. Heredity 23:205-218

Ryman N, Allendorf FW, Stahl G (1978) Reproductive isolation with little genetic divergence in sympatric populations of brown trout (Salmo trutta). Genetics 92:247-262

Stam P (1983) The evolution of reproductive isolation in closely adjacent plant populations through differential flowering time. Heredity 50, 2:105-118

Valero M (1980) Contribution à l'étude du polymorphisme enzymatique chez *Dactylis glomerata* L. . Recherche de nouveaux marqueurs et leurs utilisation dans les populations naturelles. Mémoire de DEA, USTL Montpellier

Valero M (1983) Etude du flux génique entre populations adjacentes de *Dactylis glomerata* L.. Thèse de Doctorat de 3è cycle, USTL Montpellier.

Gene flow and population structure in *Coffea canephora* coffee populations in Africa.

Julien BERTHAUD

O.R.S.T.O.M., 24, rue Bayard F-75008 PARIS
C.N.R.S/G.P.D.P. F-91190 GIF-sur-YVETTE

ABSTRACT

General characteristics of wild populations of African coffee trees are presented. A population of *C. canephora* situated in the Ivory Coast is taken as an example. Dynamic population flow within and between populations is estimated using genetic markers (incompatibility and enzyme alleles). Two main groups were defined in this geographical region ; a western (Ivory Coast) group, and central African forms. However a third form found in Ivory Coast seemed to be closer to the central African forms.

Population dynamics data and information on other coffee species led us to interpret the structure of *Coffea canephora* as resulting from bottlenecks provoked by the retreat of the rain forest over recent geological time.

INTRODUCTION

The Plant Breeding Department of the O.R.S.T.O.M. has long been interested in the genetic structure of certain plants, in relation to breeding programs. We have studied coffee trees, using material available in established collections and material collected by us in various African countries during survey missions. Part of our work, bearing on wild populations in Ivory Coast, is presented here. Characteristics of coffee populations were analysed in the wild. Transferring part of this material from the forest to a collection allowed us to undertake a genetic analysis, using S-incompatibility and isozyme alleles. Thorough reports of our investigations are in

NATO ASI Series, Vol. G5
Genetic Differentiation and Dispersal in Plants
Edited by P. Jacquard et al.

preparation. In this short paper, we wish to explain our general approach and show some significant results.

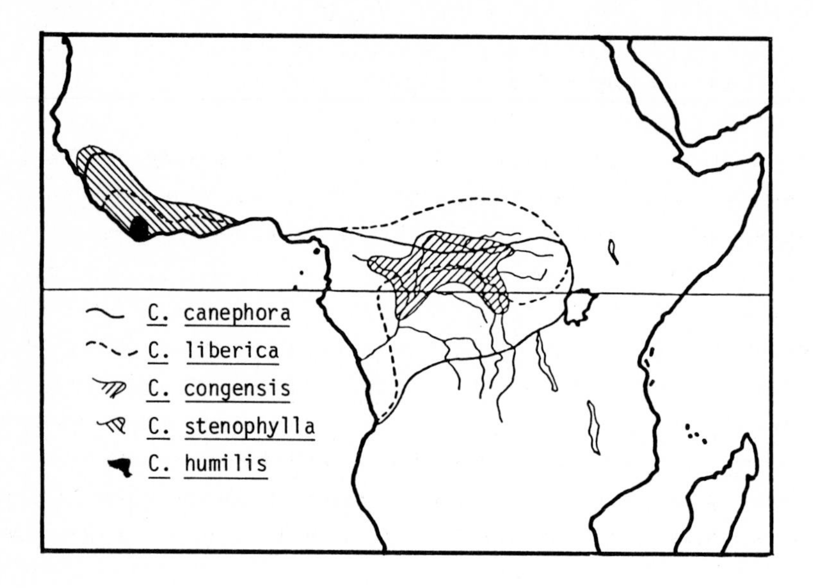

Fig. 1 - Distribution of wild coffee species in Africa.

PRESENTATION OF COFFEE TREES AND METHODS OF STUDY

Coffee trees are natives of African forests. Their natural area of distribution is limited to this continent. Coffea canephora will mostly be treated here. This species is found from Guinea to Angola, throughout the entire African forest block (Fig. 1). It is the coffee species most cultivated in Africa, and commercially known as "robusta" coffee.

Coffee trees are clustered in populations, that is to say for us, sets of ten to a hundred individuals, spread out over one or a few hectares, in the understorey of the primary forest. They are climax, not colonizing, species. Populations contain trees up to ten metres high. Their life span is about thirty years. Flowering occurs every year during the dry season and is triggered by a rain shower. Every plant of a species in an area flowers the same day. The flowers are hermaphrodite. Flowering lasts one or two days and pollination is preferentially entomophilous.

Except for *C. arabica*, all of the species are self-incompatible. Incompatibility in *C. canephora* was shown to be gametophytic, with one locus and a series of S alleles (Berthaud 1980). The species of *Coffea* form a multispecific complex. All the species share a basic genome (Charrier 1978), but each has very specific ecological properties.

To evaluate population dynamics and demography, trees were observed for several years in the wild. To estimate genetic diversity and gene flow within and between populations, however, genetic markers were needed. Two different approaches were chosen, in order to take advantage of their complementarity.

Enzyme polymorphism studies were initiated by Berthou *et al.* (1980) and the author continued this work, verifying and establishing the genetic control of the enzymes involved. Seven isozyme systems were studied for which nine loci were observed.

The second approach consisted in stuying S-incompatibility alleles in a population or among progeny. The estimation of the number of S alleles in a population permits evaluation of the number of founder trees and migration flow in this population. The number of mating groups (or genotypes) present in a sample of 17 to 26 trees in a population was studied using the same method as was previously established to test the genetic control of incompatibility (Berthaud 1980). The genotypes were determined by the observation of pollen tube growth through living excised carpels. Pollen tubes were stained in decoloured aniline blue and observed under an ultra-violet light microscope as described by Martin (1959).

RESULTS

Only results from *C. canephora* will be presented in some detail, specially as concerns Ivory Coast populations. However, as our studies were conducted on various species, we will cite some complementary results when needed for discussion.

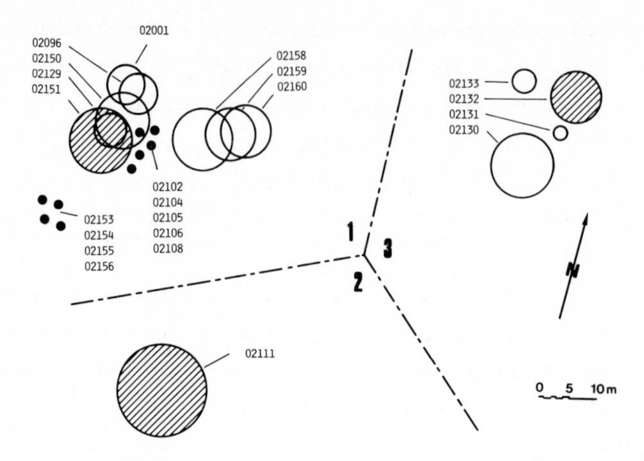

Fig. 2 - Map of the Ira forest C. canephora population (Ivory Coast). The centre of each circle represents tree position, the radius represents its height. The hatched circles denote the three trees that yielded the progenies analysed.

C. canephora Population in the Ira Forest.

This population was selected for its easy access, allowing regular visits. The trees were individually identified ; their positions are shown on Fig. 2. On the basis of tree distribution, we were able to distinguish three sectors, labelled 1, 2 and 3 on the map.

The period of observation was six years. The population size was 22 trees, of which 9 were young trees and 13 adults. We consider that every tree in this population was identified. During this period, two trees bore two thirds of the fruit production. Two thousand fruits were counted during these six years, i.e. a mean of 17 fruits per tree per year (on a basis of 20 trees in the population). Three trees died during this period. They had stood side by side, and their death might have been due to a soil fungus. Turn-over rate was estimated at one tree every two years, which gives a period of 44 years to renew the entire population.

In the population studied two loci, phosphogluconate dehydrogenase 2 (PGD2) and phosphogluco-isomerase 1 (PGI1), possessed respectively three and four alleles. These can be used to mark different trees. For other enzymes, Ivory Coast coffees were fixed for one allele, whereas central African coffees were fixed for another. Forms of this latter group were introduced into the Ivory Coast and constitute the basic material of commercial plantations there. Plantations one or two kilometres away gave us the means to detect spontaneous hybridization between these two groups and to measure the pollen gene flow from cultivated coffee to wild coffee.

Three progenies from different sectors in the Ira Forest population were analyzed. In the progeny of 02151 there was one Ivory Coast-central African hybrid among twenty seedlings ; for the progeny of 02111 it was ten in twenty-three seedlings, and for 02131, twelve in fifty-one. The tree receiving the highest proportion of alien pollen, 02111, was also the most "isolated". There was an important "position effect" for gene flow between trees of the two populations. This position effect was also present within the wild population, as confirmed by the analysis of the genotypes of progenies obtained by within-population pollination. More pollinators participed in fertilizing flowers of isolated trees. Even trees bearing very few flowers were involved in pollen flow. More trees furnish pollen than participate as females.

Table 1 - Mating groups and S-alleles in two progenies from the Ira Forest

parent tree		progeny size : n	Xo	group description*	N	Y
02151		17	9	(1x4)+(1x3) +(3x2)+(4x1)	11-13	6-7
02111	H	10	7	(1x4)+(6x1)	10-15	5-7
	W	10	10	(1x2)+(9x1)		5+

* (number of groups x number of plants in the group)
H : hybrids between cultivated central African and wild Ira Forest coffees
W : within-population hybrids
X : number of mating groups in sample (Xo : observed, Xe :estimated)
N : estimated number of mating groups in population
Y : estimated number of S-alleles of pollen

The number of mating groups per population N was estimated by iteration as Xe = N x (1-(1-(1-1/N)). Xe was supposed to be very close to Xo.
Y $= \frac{1}{2}$N in progenies

Two progenies obtained by picking seeds from two trees of the Ira Forest population were studied to estimate how many mating groups and S-alleles they presented. Results are shown on Table 1. It was confirmed by this approach as well, that the isolated tree in sector 2 received much more alien pollen than the tree located in a cluster in sector 1.

Species Structure of *C. canephora*.

From studies we made on fifteen populations, we found that between one and two thirds of the loci in each population were polymorphic, with a mean of 38% polymorphic loci per population. At the species level, polymorphism was found in all the enzyme loci studied. The mean number of alleles per locus in a population was 1.7 and per polymorphic locus 2.5. Mean heterozygosity was 0.14 (direct count) or 0.16 (expected, supposing panmictic fertilization). These values were similar to mean values calculated for 113 species by Hamrick *et al.* (1979). Nei's distance between fifteen populations was calculated from allelic frequencies for nine loci. The dendrogram shown on Fig. 3 was constructed from the distance matrix.

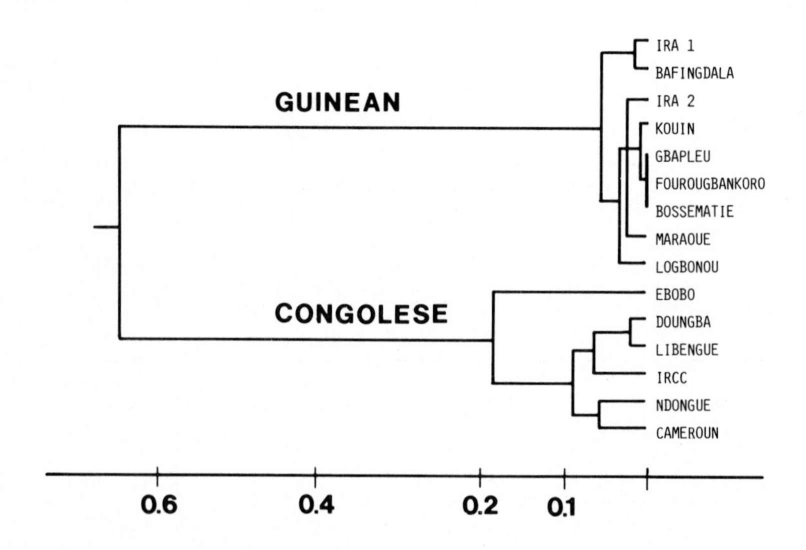

Fig. 3 - Genetic organization of *C. canephora*. The dendrogram is established using Nei's Distance between populations representing the entire distribution area.

The interpretation is very simple. Two groups were present in C. canephora, which we called "Guinean" and "Congolese", in analogy to the names found in botanical literature for the forest blocks of these two regions. Ivory Coast coffees are genetically separated from coffees of the Congolese forest block. This structure was based on data from loci with different fixed alleles.

Within the Guinean group the structure was not very meaningful, but in the Congolese group, the Ebobo population was isolated from the others. It would be very interesting to know more about this population but more material must be collect first in its region of origin (south-east Ivory Coast).

Table 2 - Distribution areas, genetic structure and fecundity of four Coffea species

Species	distribution	structure of genetic diversity	MNAP	MNAS	Fecundity (seeds per tree)
C.canephora	west and central African forest block	G — C	1.7	3.5	1000 : 1
C.liberica	west and central African forest block	G — G	1.8*	3.1*	500 : 1
C.stenophylla	west African forest block	WIC — EIC	1.1*	2.0*	10,000 : 1
C.humilis	Liberia Ivory Coast		2.4*	3.1*	20 : 1

MNAP(S) : mean number of alleles per population (species)
G : Guinean group
C : Congolese group
WIC : Western Ivory Coast form
EIC : eastern Ivory Coast form
* : calculated from data published by Berthou et al. (1980).

DISCUSSION

Genetic Diversity of *Coffea* species

The main characteristic of the organization of *C. canephora* was the existence of two groups, a Guinean group and a Congolese group. In the latter group some structuring was evident, indicating that more diversity was present in the central African forest block than in the west African forest block. In each of these groups, alleles were fixed at a few loci, but at the same time, various alleles were found at other loci, and these alleles were encountered within every population of a group. It would therefore be necessary to explain the allelic fixation at some loci, and the presence of a high number of alleles (up to 3 or 4) at other loci, in small sized populations (10 to 20 individuals). This same situation was also found in *C. liberica*, which has a similar distribution area (Fig. 1).

Two other species, found only in west Africa, presented different situations. Within *C. stenophylla*, two groups were distinguished, one grouping western Ivory Coast populations and the other eastern Ivory Coast populations. There was no polymorphism within each group, but fixed alleles were different from one group to the other (Berthou *et al*. 1980). The distribution of *C. humilis* was limited to western Ivory Coast and eastern Liberia. High genetic diversity was observed in this species (Berthou *et al*. 1980). The mean number of alleles per locus per population was 2.4 (compared to 1.7 for *C. canephora*). How can these various situations be explained ?

Population Structure and Dynamics

In *C. liberica* and *C. stenophylla* the same behavior was observed as in *C. canephora.* Few trees bore fruits ; only 5% of the population in the best of years. The fecundity rate was high, up to 500 seeds per tree in *C. liberica* and up to 10,000 in *C. stenophylla*. However, as with *C. canephora*, pollinating trees were more numerous than seed-bearing trees. The number of mating groups was established for the progeny of a wild isolated C. liberica tree, and it was shown that at least five S-alleles had been brought by pollen.

Differences were found for C. humilis. About one third of the 114 trees in the population studied bore fruits during six years of observation. Two thirds of the population flowered during that time. Each year, at least 1/10 to 1/6 of the population produced fruits. However, the fecundity was very low, with one to ten fruits per tree, to a maximum of twenty seeds per tree (see Table 2).

We do not know if different reproductive strategies are sufficient to explain the various and varied genetic structures observed in Coffea species. We think that a portion of these genetic structures could have originated from a more general trend observed in the Africain forest flora.

African Forest Flora Diversity and Palaeoclimates

In modern times, the African forest is formed by two blocks, the Congolese block in central Africa and the Guinean block in west Africa. In east Africa, the forest is split up into a mosaic of vegetation levels in the various mountainous regions. When the locations of endemic species or species richness are examined, the impact of the Congolese forest block is overwhelming. In this part of Africa there are more endemic species, and more species in general, than in any other part of the African forest. An example is described by Aubréville (1959) for Leguminosae. Out of a total of 454 species, the two blocks shared 62 species, 71 species were endemic to the Guinean block, while fully 321 species (71%) were found only in the Congolese forest block. This distribution can also be encountered in many different genera and families, as for example kola (Sterculiaceae) (Bodard 1962). Hamilton (1976) cited various other examples. The existence of species shared by the two blocks shows that the forest was once continuous, at a period to be defined. Connections are evident on the genus level, as shown in comparisons between the Guinean and Congolese blocks (Guillaumet 1967) or the Congolese forest block and east African forests (Hamilton 1974).

The structure of Coffea is very close to that of the

African forest flora taken as a whole. More species are present in the Congolese block (7) than in the Guinean block (4). Connections exist at the species level (Fig. 1), as *C. canephora* and *C. liberica* are shared by the two blocks, while on the genus level endemic *Coffea* species are present in every rain forest block in Africa. We can affirm that the main factor influencing flora and coffee structure was climatic variation.

Information on palaeoclimates draws its source from observations of geological deposits and analyses of pollen and diatomites. In Africa observations have mostly been made in the drier zones, from Senegal to Ethiopia, and very few in the wet zone (see however Maley and Livingstone 1983). Nevertheless, on the basis of the information avalaible, a history of forests in Africa can be tentatively approached.

A wet period extended from -3.3 to -2.6 million years, and a dry period began at -2.5 million years. These events were deduced from palynological studies and the observation of rodent fossils. The forest, which had been continous from the Atlantic to the Indian Ocean, was split during the dry period into smaller forest blocks and part of the forest was converted into savanna (Coppens *et al*. (eds.) 1976). Later, the climatic indicators showed a succession of wet and dry periods and temperature variations. At about -1.8 million years, *Podocarpus-Juniperus* forests were much more extended than they are now, indicating that the temperature at that time was cooler.

Thanks to the carbon-14 dating technique, the period starting forty thousand years ago (B.P.) is much better known. For twenty millenia the climate was relatively more humid than it is today, but from -20,000 to -15,000 it was very dry, approximately corresponding to the period of maximal glaciation. Dunes invaded the Senegal River bed and part of Zaïre. Wetter conditions reappeared between 15,000 and 10,000 years ago (Maley 1981), during the last world deglaciation. Forests became reestablished from -12,000 in East Africa (Hamilton 1974) and -9,000 in West Africa (Maley and Livingstone 1983). They spread out of refuge areas (Aubréville 1949) and colonized savanna areas very quickly. The advance of the forest front could have been as fast as 100m. per year, allowing a few specialized genotypes to colonize vast areas.

Floristic and coffee organization reflects this history.

The zone of richest species diversity are thought to have been refuge areas possessing favoured climatic conditions.

CONCLUSION

Climatic variations explain, on a large scale, the geography of African forest floristic richness, and the genetic diversity of coffee trees. The dispersion abilities of different species could have had a very great effect on the organization of the genetic structure during the periods of the forest expansion.

At the population level, we have pointed out the importance of migration in the creation and preservation of polymorphism. Gene flow is, or can be, important between populations. Migration does not equally affect each tree in a population. The position of an individual tree in a population could be the most critical factor determining gene flow between and within populations.

Forest species, as a whole, seem to have been influenced by the same constraints. If this is the case for different genera in other families, forest species found in several forest zones should exhibit a similar genetic organization.

The recognition of genetic groups can offer practical advantages. Worthy progenies with improved vigour and productivity often result from breeding between groups, due to the expression of heterosis or gene complementarity. This phenomenon has already been observed in one African tree species, the oil palm (Ghesquière 1983). A future paper will show that this pattern can also be exploited in <u>C. canephora</u> (Berthaud, thesis in preparation).

REFERENCES

AUBREVILLE A., 1949. - Climats, forêts et désertification de l'Afrique tropicale. Soc. Ed. géographiques, maritimes et coloniales, Paris, 351p

AUBREVILLE A., 1959. - Etude comparée de la famille des Légumineuses dans la forêt équatoriale africaine et dans la flore de la forêt amazonienne. C.R. Soc. Biogéo., 314-316, 43-57.

BERTHAUD J., 1980. - L'incompatibilité chez Coffea canephora. Méthode de test et déterminisme génétique. Café, Cacao, Thé, 24, 267-274.

BERTHAUD J., à paraître. Les ressources génétiques chez les caféiers africains : évaluation de la richesse génétique des populations sylvestres et de ses mécanismes organisateurs. Thèse, Paris XI.

BERTHOU F., TROUSLOT P., HAMON S., VEDEL F. & QUETIER F., 1980. - Analyse en électrophorèse du polymorphisme biochimique des caféiers : variations enzymatiques dans 18 populations sauvages : variation de l'ADN mitochondrial dans les espèces C. canephora, C. eugenioides et C. arabica. Café, Cacao, Thé, 24, 316-326.

BODARD M., 1962. - Contribution à l'étude systématique du genre Cola en Afrique Occidentale. Ann. Fac. Sci. Univ. Dakar, 7, 7-187.

CHARRIER A., 1978. - La structure génétique des caféiers spontanés de la région malgache (Mascarocoffea). Leurs relations avec les caféiers d'origine africaine (Eucoffea). Mémoire O.R.S.T.O.M. N°87 O.R.S.T.O.M. Paris, 223p.

COPPENS Y., HOWELL F.C., ISAAC G. Ll. & LEAKEY R.E.F., (eds.), 1976.- Earliest man and environments in the Lake Rudolf Basin. University of Chicago Press, Chicago.

GHESQUIERE M., 1983. - Contribution à l'étude de la variabilité génétique du palmier à huile (Elaeis guineensis) Jacq.). Le polymorphisme enzymatique. Thèse Docteur-Ingénieur, Paris XI, Orsay, 146p.

GUILLAUMET J.-L., 1974. - La végétation et la flore de la région du Bas Cavally. Mémoire O.R.S.T.O.M. N°20, O.R.S.T.O.M., Paris, 247p.

HAMILTON A.C., 1974. - The history of the vegetation. In : E.M. Lind & E.S. Morrison (eds.). East African Vegetation, London, Longman, 188-209.

HAMILTON A.C., 1976. - The significance of patterns of distribution shown by forest plants and animals in Tropical Africa for the reconstruction of Upper Pleistocene palaeo-environments : a review. Palaeooecologica of Africa, 9, 63-97.

HAMRICK J.L., MITTON J.B., LINHART Y.B., 1979. - Levels of genetic variation in trees : influence of life history characteristics. In : Proceedings of Symposium on Isozymes of North American Forest Trees and Forest Insects. July 27, Calif. Genet. Tech. Rep. PSW-48, Berkeley, 35-41.

MALEY J., 1981. - Etudes palynologiques dans le bassin du Tchad et Paleoclimatologie de l'Afrique nord tropicale de 30 000 ans à l'époque actuelle. Travaux et documents O.R.S.T.O.M. N°129, O.R.S.T.O.M., Paris, 586p.

MALEY J., LIVINGSTONE D.A., 1983. - Extension d'un élément montagnard dans le sud du Ghana (Afrique de l'Ouest), au Pléistocène supérieur et à l'Holocène inférieur : premières données polliniques. C.R. Acad. Sc., Paris, t. 296. série 2, 1287-1292.

MARTIN F.W., 1959. - Staining and observing pollen tubes in the style by means of fluorescence. Stain Technology, 39, 125-128.

Dispersal: Phenotype Dispersal

The Fitness of Dispersed Progeny: Experimental Studies with *Anthoxanthum odoratum*

Janis Antonovics* and Norman C. Ellstrand**

*Unite de Biologie des Populations et des Peuplements, Centre Louis Emberger, C.N.R.S., Montpellier, France, and Department of Botany, Duke University, Durham, North Carolina, U.S.A.

**Department of Botany and Plant Sciences, University of California, Riverside, California, U.S.A.

ABSTRACT

To examine how dispersal distance, density, and genetic variation influence progeny fitness, tillers of *Anthoxanthum odoratum* (Sweet Vernal Grass) were transplanted into their natural habitat (a mown field) at different distances from their parents. Tillers were either cloned from the parent and genetically identical, or, from seed-derived progeny of single parents and genetically variable. In a simulated seed shadow, competition among sibs reduced the fitness of the progeny in the dense regions of the gradient, but the reduction was small (two-fold) relative to the forty-fold change in density. Genetic variance did not reduce competition at high densities, but at lower densities genetically variable sibs had a greater fitness than asexual sibs. Long distance dispersal was generally unfavourable for asexually derived tillers. In contrast, some seed-derived tillers had a high fitness far from the parents. Seed dispersal measurements showed that most seeds fell close to the parent. The individual advantages of long distance dispersal of genetically variable progeny are therefore likely to be small, except in a colonising situation. The results show that the ecological consequences of dispersal depend on genetic variation among the progeny.

INTRODUCTION

The ubiquity and elaborate nature of dispersal mechanisms unequivocally argues that there are often large evolutionary advantages associated with dispersal. These advantages fall into two general classes (Howe and Smallwood, 1982): escape

NATO ASI Series, Vol. G5
Genetic Differentiation and Dispersal in Plants
Edited by P. Jacquard et al.

from the parent and colonisation of new areas. Advantages to escape from the parent include avoidance of competition with the parent or among sibs, avoidance of predators and pathogens concentrated around the parent, and avoidance of inbreeding. Advantages to colonisation include escape from deteriorating environments (as in a successional situation), and increased probability of finding new suitable habitats. In spite of a wealth of speculation, many of the hypothesised advantages to dispersal remain untested, particularly by direct studies of progeny experimentally dispersed to different distances around the parent plant (Howe and Smallwood, 1982; for exceptions see Janzen, 1972, Wilson and Janzen, 1972). In the present study we experimentally transplant tillers of Anthoxanthum odoratum (Sweet Vernal Grass) at different distances from the parent, to answer the following questions.

1. Does progeny fitness decrease as the density of progeny increases?
2. Does genetic variation among sib-progeny ameliorate the effects of density?
3. Does dispersal well away from the parent increase or decrease progeny fitness?
4. Does the outcome of long distance dispersal depend on the genetic variance of the progeny?

The experiments described here were part of a larger study of the evolutionary significance of sexual reproduction (Antonovics and Ellstrand, 1984; Ellstrand and Antonovics, 1985; Antonovics, 1984; Antonovics and Ellstrand, 1985). In these experiments, plantings were carried out with tillers rather than seeds so as to permit control of the genetic variance among the progeny: tillers cloned from the parent are genetically uniform, while tillers cloned from seed progeny of one parent provide a genetically variable family group. All experiments were carried out in a long-established mown field on the campus of Duke University, Durham, North Carolina. Tillers were planted without disturbing the surrounding vegetation. Initially, studies on seed dispersal are described as a context for the experimental work.

DISPERSAL IN ANTHOXANTHUM

Anthoxanthum posseses no obvious dispersal mechanisms. As in many other grasses, the caryopses ("seeds") plus associated glumes are shed when the inflorescence dries at maturity. The seeds are small (c. 1 x 1.5 mm) and the glumes are somewhat hairy with a hygroscopic awn. Dispersal occurs by wind and when seeds plus inflorescence fragments are scattered by mowing: such mowing occurs regularly in the spring when seeds are mature and has been a regular feature of the study site for at least thirty years.

To assess wind dispersal, the vegetation around eight large individuals of *Anthoxanthum* was mowed, and dispersed seeds were trapped on three plastic sheets (1m x 2m) radiating around each individual and covered with sticky 'tanglefoot'. To assess mower dispersal, the mature panicles of ten large individuals of *Anthoxanthum* were lightly spray-painted different colours prior to mowing. After mowing, the distances of inflorescence fragments from each parent were measured; all fragments, regardless of size, were counted equally. Mowing was carried out, as at other times, with a rotary power-mower with an unprotected side discharge.

Dispersal distances were relatively short, with 95 % of the seed dispersed less than 1.6 m by wind and less than 3.0 m by the mower (Fig. 1). The furthest dispersal recorded was an

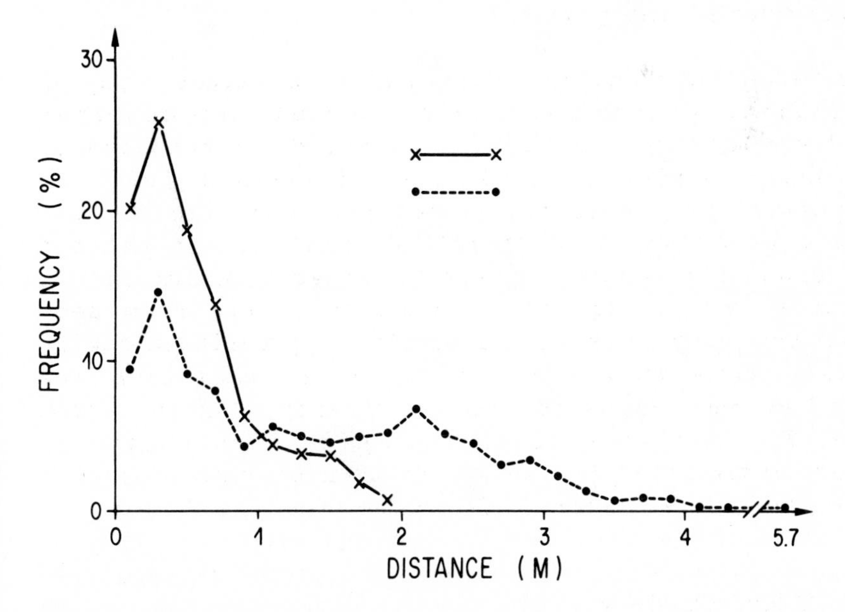

Fig. 1. Seed dispersal by wind (x---x) or mower (o---o). Graph shows frequency falling at a given distance from the parent, within a fixed arc angle. Totals: 572 for wind and 1313 for mower dispersal. Data from Kelley (unpublished results).

inflorescence fragment which was mower-dispersed 5.7 m from the parent.

SHORT DISTANCE DISPERSAL EXPERIMENTS

Methods. As part of a study to examine the effects of density on performance of variable and uniform progeny (Ellstrand and Antonovics, 1985), adult- and seed-derived tillers were grown in hexagonal fan arrays resembling seed dispersal patterns (Fig. 2). Arrays were planted into the natural community within three meters of the parents. The density within these arrays varied from 43 plants to 1 plant per 100 sq cm. There were eight replicates of variable (half-sib family) and non-variable (clonal) arrays within each of two sites. Tillers were planted in the autumn of 1979, and followed through two flowering seasons, by which time most individuals had died. Fitness was estimated as total inflorescence production over two years.

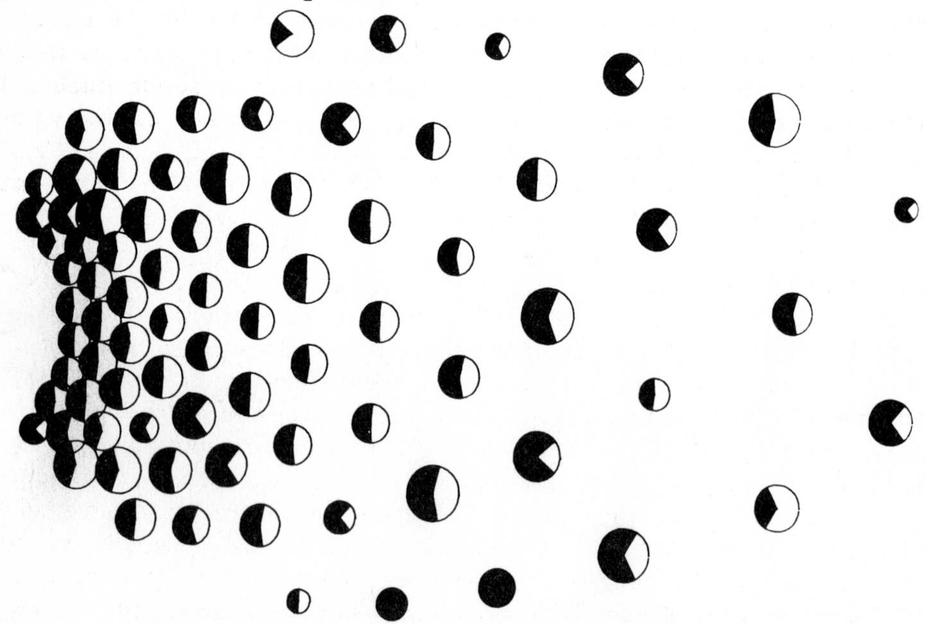

Fig. 2. Fitness of genetically variable and uniform progeny at each position in the density gradient. The center of each pie-chart indicates plant position in the hexagonal fan design (closest spacing is 1cm). Shaded segment is fitness of variable progeny, unshaded segment is fitness of uniform progeny, while area of each pie-chart is proportional to overall fitness at each position.

Results. Density had relatively little effect on reproductive output. A ten-fold difference in space available per plant produced only a two-fold difference in fitness (Fig. 2). The genetically variable progeny groups had a greater reproductive output than the genetically uniform groups. This effect was most apparent at the intermediate densities and at the edges of the fan designs, but was not seen at very high densities (Fig. 2).

LONG DISTANCE DISPERSAL EXPERIMENTS

Methods. Two experiments were carried out in successive years. In the first experiment single tillers from 22 genotypes were sampled from a limited area (c. 10 x 15 m) of the study site and cloned in the greenhouse for a year. One tiller of each genotype was then planted out along three transects spanning the field and intersecting the site from which the genotypes were originally sampled (see Fig. 3a). There were fifty transect positions and plants were arranged linearly within each transect position at 20 cm spacing. Tillers were planted in autumn 1978; survival and fecundity (inflorescence number) was measured over three years (for details, see Antonovics and Ellstrand, 1985). Since the position of "parental" individuals was known, the distances of cloned tillers from their parent could be determined.

The second experiment differed from the first in two ways. First, the genotypes were sampled from several diverse sub-habitats within the field rather than from a single area (Fig. 3b). Second, the individuals were sampled both as tillers and as seeds. The seeds were grown in the greenhouse to produce large individuals, from which tillers were cloned in a manner equivalent to that used in obtaining tillers from the parent individual. It was thus again possible to obtain seed-derived tillers, representing variable progeny (half-sib family) arrays, and adult-derived tillers representing uniform progeny (clonal) arrays. These tillers were planted in fall 1979 along transects parallel to the transects used in the first experiment, but separated from them by 20 cm. Again, genotypes were randomised within transect positions. Separate transects were used for the adult and seed derived transplants. About two weeks separated the time of planting of these transects. The experiment was followed through two flowering seasons.

Because reproductive output showed a highly skewed distribution, regular regression analysis of fitness on dispersal distance was not possible; instead contingency chi-square tests were carried out on five distance classes (10m intervals and distances >40m), and on 5 fitness classes (0, 1, 2-9, 10-49, and >49 inflorescences; the latter two were pooled in Experiment 2). Analysis was also done on the zero vs. other classes, and separately on classes with fitness >0.

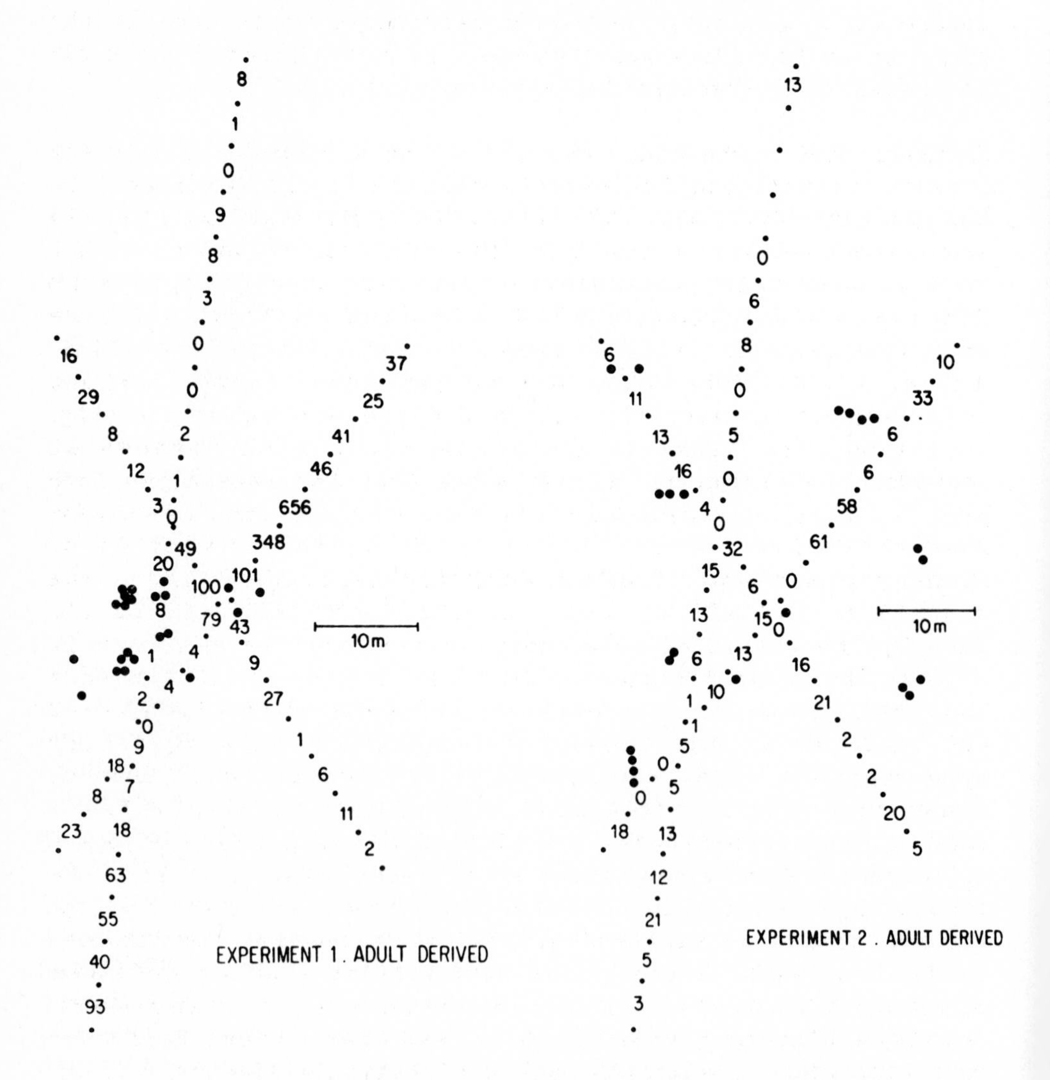

Fig. 3. Average fitness of 22 genotypes at each transect position for adult derived tillers of (a) Experiment 1, and (b) Experiment 2. The larger solid points indicate positions of the parent individuals used in the two experiments.

Results. In Experiment 1, with cloned progeny, reproductive output increased at moderate distances from each parent (to about 20m) and thereafter declined (Fig. 4a). The interaction of fitness class and distance class was highly significant (X^2 = 39.0, d.f.=16, $P<0.005$). Distance had a nearly significant effect on whether individuals flowered or not (X^2=8.8, d.f.=4, $P<0.10$), but the main effect was on individuals that flowered (X^2 = 30.1, d.f.=12, $P<0.005$). The higher fitness at intermediate distances could be attributed to an area in the field where fecundity was particularly high, located about 20m to the east of where adults were sampled (Fig. 3).

In Experiment 2 the adult-derived tillers showed a consistent decline in reproductive output with increasing distance from the parent (Fig. 4b). The effect of distance was nearly significant (X^2=18.3, d.f.=12, $P<0.10$). Distance had no effect on the proportion of individuals with greater than zero fitness (X^2=1.3, d.f.=4, n.s.), but had a significant negative effect on fitness of individuals that flowered (X^2 =17.0, d.f.=8, $P<0.05$). The seed derived tillers showed no clear relationship between distance and reproductive output (Fig. 4c); however, some individuals had a high fitness at considerable distances (40m or more) from the parent. No such high fecundity individuals were found that far from the parent when the tillers were adult-derived . The interaction of fitness class and distance for seed-derived tillers was significant (X^2=23.5, d.f.=12, $P<0.05$); this effect was largely the result of more individuals with greater than zero fitness at long distances (X^2=13.3, d.f.=4, $P=0.01$) rather than effects of distance on individuals that flowered (X^2 =10.2, d.f.=8, n.s.) . The three way interaction of tiller type (seed- or adult-derived), fitness class, and distance class was highly significant (X^2 =28.0, d.f.=12. $P<0.01$), showing that seed-derived and adult-derived tillers performed differently when transplanted at different distances from the parent. To examine this relationship more closely, the average fitness of adult- and seed-derived tillers was standardised to 1, and their relative fitness determined at each 5 m distance class (Fig. 5). With increasing distances the relative fitness of the seed-derived transplants increased significantly (Spearman's rank correlation = 0.54, $P<0.01$).

To examine if transplants had a higher fitness when near the parental genotype, the fitness of the eleven transplants nearest their respective parents and that of the eleven

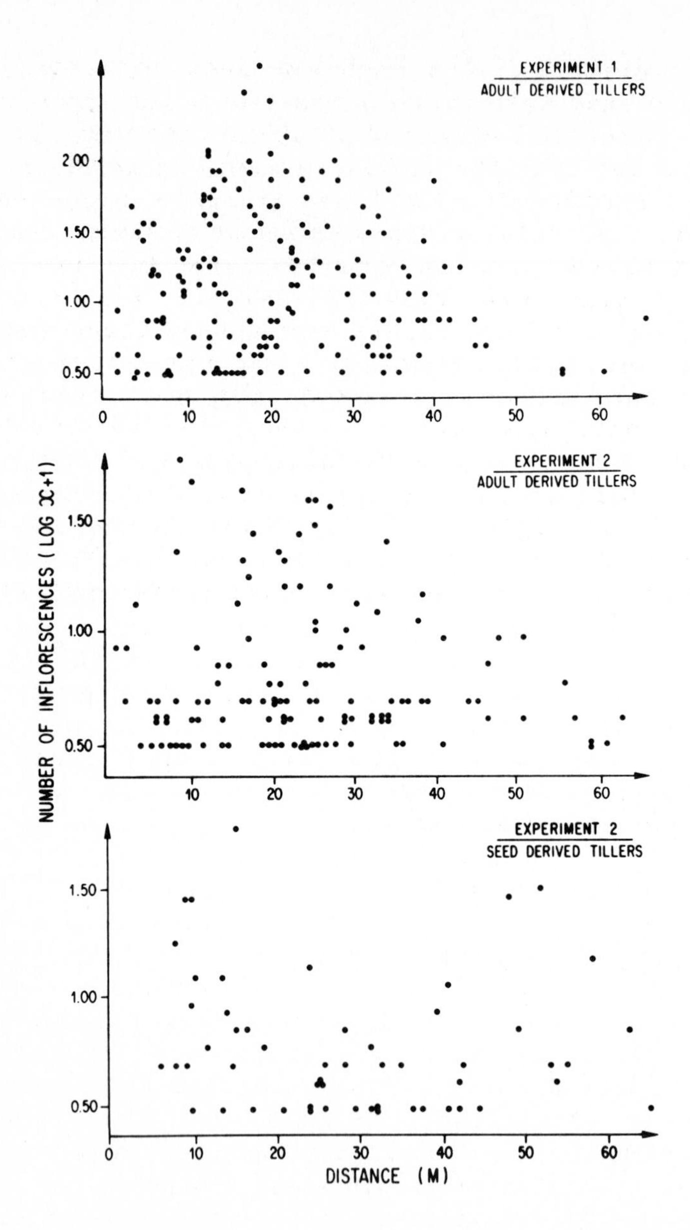

Fig. 4. Relationship between fitness and distance from the parent plant for (a) adult-derived tillers of Experiment 1, (b) adult-derived tillers of Experiment 2, and (c) seed-derived tillers of Experiment 2. Each point represents one individual. Individuals with zero inflorescences are not shown: numbers in each 10m distance class were as follows: (a) 174, 242, 207, 118, 51, 30, 2; (b) 127, 213, 233, 148, 77, 27, 12; (c) 133, 269, 279, 156, 96, 42, 14.

farthest was measured within each transect position. In Experiment 1, averaged over transect positions, genotypes near their parents surprisingly had a significantly lower reproductive output than those farther away (3.18 vs. 5.54; $P<0.027$). In Experiment 2, tillers near their parents performed better than tillers far away, nearly significantly so for adult-derived (1.15 vs. 0.94, $P<0.066$) but not for seed-derived tillers (0.46 vs. 0.34, $P<0.47$). Within each transect position, the individuals were ranked according to their fitness and their distance from the parent. A regression of rank fitness on rank distance, over all transects positions, was not significant in Experiment 1 (0.068x -0.0020x , $P<0.45$ & $P<0.80$ for linear and quadratic coefficients, respectively). It approached significance in the adult-derived transplants of Experiment 2, where there was a negative relationship (-0.164x+ 0.0059x , $P<0.07$ & $P<0.11$ for the coefficients) but was non-significant in the

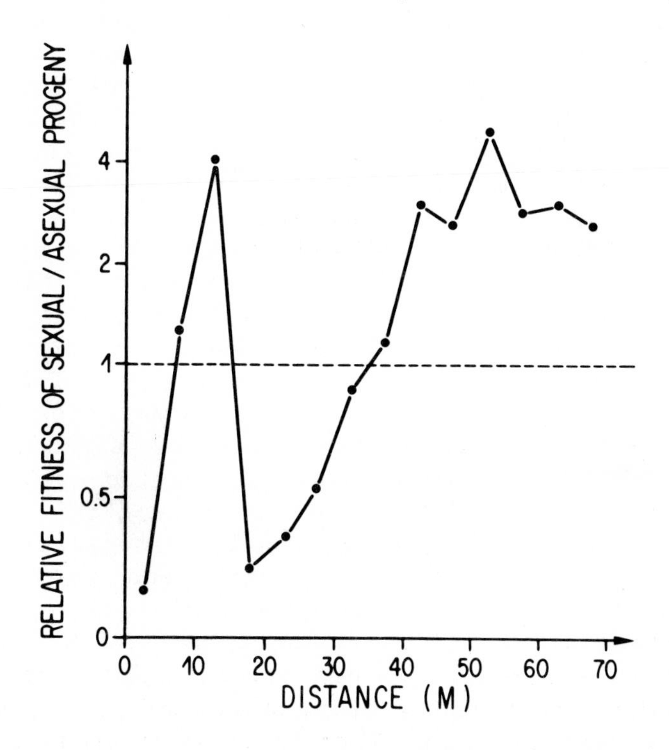

Fig. 5. Relationship between relative fitness of seed-derived vs. adult-derived tiller transplants and distance from the parent. The overall fitness of sexual and asexual transplants was standardised to 1; actual fitnesses were 0.412 and 1.086 inflorescences per plant, respectively.

seed-derived transplants ($-0.042x + 0.0023x$, $P<0.56$ & $P<0.44$, for the coefficients).

DISCUSSION

The present study shows that the fate of progeny dispersed to different distances from the parents depends both on the distance and on the genetic variance of these progeny. The short distance dispersal experiment tested for the presence of sib-competition among the descendants of one individual. Such competition is an obvious expectation given the skewed dispersal distributions typical of higher plants, where, as was the case with *Anthoxanthum*, most of the seeds fall around the maternal plant (Levin and Kerster, 1974). However, in spite of the large range of densities within the artificial 'seed shadow' of the present study, such competition was relatively weak. Because the community studied here is a dense sward of many different species, the dense arrays of *Anthoxanthum* may have been subjected to more inter- and intra-specific than intra-family competition. Moreover, the high rate of density-independent mortality precluded strong interactions among the transplants. Also, the experiment was started with tillers rather than seeds: tillers contain more initial resources, and probably grew relatively little prior to floral induction in the winter. These results therefore suggest that sib-competition in natural populations may be reduced by three major ecological factors, namely, density of associated species, level of density independent mortality , and initial size or starting capital. Genetic variance did not appear to alleviate sib-competition, since the effects of genetic varability were most evident at the lower densities and at the edges of the plots. A possible explanation (see Ellstrand and Antonovics, 1985) is that genetic variance is advantageous in the face of predators or pathogens, but that this advantage is not evident at high densities where the plants are resource-limited.

In the first of the long-distance dispersal experiments some unexpected results were obtained. When the parents came from a single small section of the field, there appeared to be an "optimum" dispersal distance. However, this optimum was not the result of any complex ecological process (cf Janzen, 1970), but the simple consequence of the most favourable habitat for *Anthoxanthum* in this field being at an intermediate distance from where the parents were sampled. These results indicate that plant populations are by no means uniform in their demography (see also Fowler and Antonovics, 1981; Antonovics and Primack, 1982), but contain regions of high population growth and reproduction ("fountainhead regions" or "hotspots" of Gill, 1978) as well as regions where population growth is negative ("reproductive sinks") and the numbers are maintained largely by immigrants from fountainhead areas. Such areas with intrinsically negative population growth can be considered demographically marginal, and the plants there may be subject to ecological and evolutionary forces which are quite different from those in the demographically central fountainhead areas (Antonovics,1978).

In the second experiment, where the parents were from different parts of the field, fitness of asexual progeny declined with increasing distance, partly because of a decline in the suitability of more distant habitats for *Anthoxanthum*, and partly because genotypes had been locally selected for high performance in their own region of the field; this latter explanation was supported by the observation that transplants performed better when they were nearer their own parental genotype, than when they were further away.

Only the seed-derived transplants gave individuals with high fitness at considerable distances from the parents. This observation is consistent with the classical view that sexual reproduction is favoured because it provides new and different genotypes that are adapted to a new environment (Williams, 1975). However, although this seems to be the first experimental demonstration of the commonplace assertion that genetically variable progeny are favoured in changing environments, whether such change provides a sufficient force for the maintenance of sexual reproduction is more problematical, especially since there was almost no natural dispersal beyond 5m. The effect of this on relative fitness of sexual and asexual types can be illustrated as follows. If there is a stable population where the average absolute fitness of each individual is one, and if 1% of a sexual progeny group are dispersed to a distance where they have a 5-fold advantage, then the relative fitness of this group (compared to an asexual progeny group) will be only $(0.99 \times 1) + (0.01 \times 5) = 1.04$. If however long distance dispersal involves a colonisation event, such that after dispersal the average absolute fitness is high (say 100) then the fitness of the sexuals will be (0.99 x 1) + (0.01 x 500) = 5.99, and that of the asexuals will be (0.99 x 1) + (0.01 x 100) = 1.99. The relative fitness of sexual vs. asexual in this second case is approximately 3:1. The plant community within which the present experiments were carried out is relatively stable (Fowler and Antonovics 1983), and probably sexual reproduction is not being maintained by the advantage of producing variable progeny in long distance dispersal events. However, if within the larger metapopulation (the interconnected populations of *Anthoxanthum* within its general area of distribution), colonisation was frequent, and if colonisation success depended on novel genotypes then sexuality could be maintained. We unfortunately have no data on colonisation and extinction rates of *Anthoxanthum* in the Durham area.

The experiments carried out here were not designed specifically to examine dispersal. They have been reinterpreted in this context as a stimulus to further experimental research on dispersal biology. The results demonstrate that studies of fitness at the individual level need to take dispersal into consideration, and that the ecological consequences of dispersal are often modified by genetic events.

Acknowledgements: The wind dispersal data were kindly provided by Steve Kelley. We wish to thank the National Science Foundation and the CNRS for financial support.

REFERENCES

ANTONOVICS J., 1976. - The nature of limits to natural selection. *Annals of the Missouri Botanical Gardens*, *63*,224-247.

ANTONOVICS J., 1984. - Genetic variation within populations. In: DIRZO R. & SARUKHAN J., eds., *Perspectives on Plant Population Ecology*, Sinauer Associates.

ANTONOVICS J. & ELLSTRAND N.C., 1984. - Experimental studies of the evolutionary significance of sexual reproduction. I. A test of the frequency dependent selection hypothesis. *Evolution*, *38*, 103-115.

ANTONOVICS J. & ELLSTRAND N. C., 1985. - Experimental studies of the evolutionary significance of sexual reproduction. III. A test of the spatially changing environment hypothesis.

Evolution (submitted).
ANTONOVICS J. & PRIMACK R. B., 1982. - Experimental ecological genetics in _Plantago_. VI. The demography of seedling transplants of _P. lanceolata_. _Journal of Ecology_, _70_,55-75.
ELLSTRAND N. C. & ANTONOVICS J., 1985. - Experimental studies of the evolutionary significance of sexual reproduction. III. A test of the density-dependent selection hypothesis. _Evolution_ (in press).
FOWLER N. L. & ANTONOVICS J., 1981. - Small scale variability in the demography of transplants of two herbaceous species. _Ecology_, _62_,1450-1457.
GILL D., 1978. - The metapopulation ecology of the red-spotted newt _Notophthalmus viridescens_ (Rafinesque). _Ecological Monographs_ _48_,145-166.
HOWE H. F. & SMALLWOOD J., 1982. - Ecology of seed dispersal. _Annual Review of Ecology and Systematics_, _13_, 201-228.
JANZEN D. H., 1970. - Herbivores and the number of tree species in tropical forests. _American Naturalist_ _104_,501-524.
JANZEN D. H., 1972. - Escape in space of _Stercularia apetala_ seeds from the bug _Dysdercus fasciatus_ in a Costa Rican deciduous forest. _Ecology_ 53:954-959.
LEVIN D. A. & KERSTER H. W., 1974. - Gene flow in seed plants. _Evolutionary Biology_, _7_,139-220.
WILSON D. E. & JANZEN D. H., 1972. - Predation on Scheelea palm seeds by Bruchid beetles: seed density and distance from the parent palm. _Ecology_ _53_,954-959.
WILLIAMS G. C., 1975. - _Sex and Evolution_. Princeton University Press.

A one species - one population plant : how does the common fig escape genetic diversification ?

G. Valdeyron, F. Kjellberg, M. Ibrahim, M. Raymond and M. Valizadeh
Unité de biologie des Populations et des Peuplements
Centre Louis Emberger (C.N.R.S.), BP 5051, 34033 Montpellier Cedex.

ABSTRACT

The caprifig tree, which is the male form of the common fig (*Ficus carica L.*), must produce the pollen but also permit the completion of the life cycle of the pollinator (the hymenopteran *Blastophaga psenes L.*).

In the South of France, this cycle has only two generations a year. Under the warmer climates of North Africa or Middle East, where the insect develops more rapidly, most authors have described three generations or more per year. However, the fructification phenology of the caprifig trees observed in the Montpellier region allows the tree to face the two situations as well, at the cost of a part of its pollinating potential. One is led to think that this adaptation to an extended geographic area is related to the efficiency of the long distance dissemination of seeds.

INTRODUCTION

Coevolution is commonly regarded as an important aspect of evolution. Field observations and experimentation on coevolutionary processes are, however, rather limited.

The interactions between fig trees and their specific mutualistic pollinators, which all belong to the family *Agaonidae* (*Hymenoptera*) may provide rather good material to study such processes. Cospeciation, an ultimate proof of coevolution, can readily be inferred from the fact that related *Ficus* species most often have related pollinators. Then the difference in generation

NATO ASI Series, Vol. G5
Genetic Differentiation and Dispersal in Plants
Edited by P. Jacquard et al.

time between the pollinator and the tree makes it inconceivable that the pollinator may benefit directly as an individual from fig offspring it would help to produce. We therefore expect that selective forces acting on the pollinator should tend to turn the pollinator into a parasite and ultimately lead to the disparition of the tree and its pollinator. As there are some 700 *Ficus*-pollinator couple of species, we may think, on the contrary, that we are faced with highly successful systems. The reasons of the success, however, are generally thought of as speculative.

The present work was initiated in an attempt to understand how individual selection on *Ficus carica* has led it to produce inflorescences each containing about a thousand flowers that give neither seeds nor pollen but which are necessary for the survival of the wasp. But let us first describe the biology of the system.

WHAT THE FIG TREE PAYS FOR THE SYMBIOSIS

While anatomically gynodioecious, the common fig (*Ficus carica* L.) is a functionally dioecious plant species: the female flowers of the hermaphrodite trees, which are short styled, are used by the pollinating wasp (*Blastophaga psenes* L.) for its reproduction (Fig. 1).

In the south of France, pollination takes place in July (all dates have been determined by following during several years the development of buds on marked shoots, monitoring the emergence of wasps in organdie bags and regularly opening syconia to ascertain whether they are visited or not).

On the female side, pollination involves the undelayed syconia of the female tree or domestic fig. These syconia form what is called (Condit, 1947) the second crop of the domestic fig but it is, in fact, the sole crop on most genotypes. They develop from buds at the axil of almost all leaves, at the beginning of June, so that they reach receptivity during July: the female flowers which line the inside of the syconium --the fig fructification-- are then ready to be fertilized. At the end of August these syconia become the edible figs, stuffed with seeds efficiently disseminated, especially by birds.

The pollen is produced by the male fig tree or caprifig in delayed syconia, called profichi. These syconia result from buds born, at the axil of the leaves, during the preceedings summer; they have stopped growing when the size of a pepper grain, before winter, and have resumed their development in spring, when vegetation growth started.

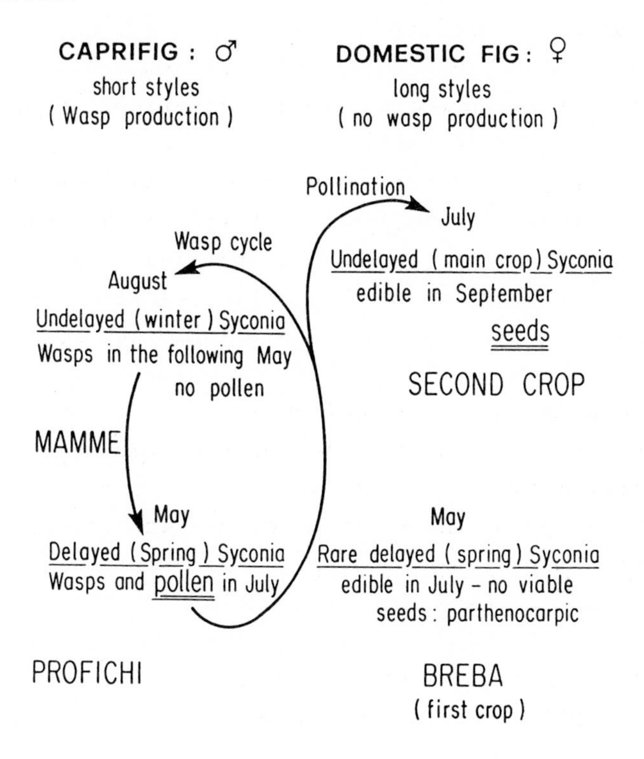

Fig. 1 - The common fig and its pollinating wasps: the symbiotic cycle in South France. Shown above are the types of syconia under the dates of receptivity of their female flowers to pollination and (if short styled) to egg-laying. Undelayed syconia develop during the same season as the shoot; delayed syconia develop in the following spring. The July-August gap in the wasp cycle must be closed by late emerging insects. Names of crops after Condit (1947).

The reproductive value of the female trees is related to the number of undelayed syconia, which contain all their seeds. Hence, one understands that the female trees produce mostly this sort of syconia (Valdeyron, 1967, Valdeyron and Lloyd, 1979).

Conversely, one would expect the male tree to produce almost only delayed polliniferous syconia --profichi--. Actually, no undelayed syconia are present on the caprifig at pollination time and, most generally the first buds do not appear before the end of July. However many buds develop later on, into undelayed syconia --at the expense of the future production of polliniferous profichi. This is necessary for the completion of the symbiotic cycle. The wasps issuing from the profichi which enter the domestic fig syconia cannot lay eggs in the long styled flowers: they are lost for the wasp reproduction. Some of these wasps must visit caprifig syconia, the female flowers of which are suitable for egg-laying. The profichi are receptive in May and there is no question for the fragile tiny wasp to spend nine months as an adult, waiting for the proper period. The short lifetime of the wasp can easily be inferred from the observation of one of us. When the Tunisian wind "sirocco" made the wasp emerge from all the winter syconia (or mamme) of isolated trees about one month before the receptivity of the profichi, all the profichi dropped. The undelayed caprifig syconia permit the wasps to complete their cycle: some of these, the mamme, remain on the tree during winter with the larvae inside.

Kjellberg (1983 and in prep.) has shown that any caprifig tree has an individual advantage in producing its own mamme, as this is an insurance against an insufficient intertree circulation of the wasps.

One point, however, remained to be explained.
In December 1982, Ibrahim (in prep.) examined a total of 392 year shoots on three caprifigs near the laboratory. These trees undoubtedly belong to the spontaneous fig population: domestic figs are frequently found, sometimes as planted trees, in the suburbs of Montpellier but planting caprifig trees is of no use

and never done in the region (Kjellberg and Valdeyron, 1984). The trees were re-examined in May 1983, when the delayed syconia were receptive, and in June when only the visited ones were remaining and in good condition for ripenning. A total of 2255 nodes were observed, that is 5,8 per shoot on the average. The repartition was as follows:

Undevelopped buds	1.8
Mamme	0.3
Undelayed fallen syconia scars	0.9
"Pepper grain" stage buds	2.8
of which, reached receptivity	2.1
of which arrived to maturity	1.4

Some hundred wasps escape from an average mamme. The existing mamme are then more than sufficient for the profichi production (the loss of profichi between receptivity and ripeness was mainly due to one of the three caprifigs, exceptionally low in mamme production).

Nevertheless, for the production of these mamme, four time as many buds (0.3+0.9) were used. This wastage of buds becomes still more intriguing when one looks at the ones which are wasted.

In 1982 and 1983 respectively 40 and 80 shoots were observed twice a week, from the beginning of August until the end of September, on the same trees. The developping undelayed syconia of these shoots became receptive on August 5th in 1982 and August 9th in 1983. Pooling the results of the two years --not significantly different-- we obtain 40 mamme out of the 183 observed undelayed syconia which became receptive, or one out of 4.6; this is quite similar to the ratio of the mamme present on the total undelayed syconia in the preceeding observation (97 out of 458 or one out of 4.7). Of the above total, the figures for the first week are 23 mamme out of 28 receptive syconia (82%), for the second week, 9 out of 24 (38%) and for the remaining of the period, 8 out of 131 (6%). It was verified that most, if not all of the syconia which fell did so because they had not been visited.

The question to be put is then: why are one third of the syconia to reach receptivity undelayed ones while only the earliest, representing at most one fourth of these undelayed syconia, are really efficient?

One answer might be that physiological constraint are such that the trees cannot produce undelayed syconia receptive in the beginning of August without producing later ones. An other explanation is that there is selection in other places in favour of syconia becoming receptive in late August and September and that gene flow is so important that this character cannot be supressed in southern France. Let us see what are the arguments in favour of this second hypothesis.

HAS THE CYCLE TWO GENERATIONS OR MORE?

The relation between the wasp and the fig was known in ancient times (Condit, 1947). A detailed description of the cycle, however, only became available by the turn of the nineteenth century, with the works of Solms-Laubach, (cited by Condit, 1947 and Condit and Enderud, 1956) and Leclerc du Sablon (1908) among others. After him, many authors such as Swingle (1899), Trabut (1901), Grandi (1924) Condit (1947) and recently, Galil and Neeman (1977) have contributed important discoveries. All of them agree on the point that some of the wasps emerging from the profichi lay their eggs in the earliest undelayed syconia of the caprifig to give adult wasps before the end of the season.

The female wasps of this generation, which is called the "second" by Grandi (1924) would lay eggs in later undelayed syconia in which the maggots would overwinter. Briefly, there would be, not one generation of wasps in the undelayed syconia, as we have found in Montpellier, but two or even more, more or less distinct. We have heard agronomists regarded as well informed, speak of such numbers of caprifig syconia "crops" in Greece and Tunisia.

Up to our observations, Leclerc du Sablon (1908, 1910) alone had found a cycle with a total of two generations a year --the wasps issuing from the profichi giving directly the larvae which spend the winter on the trees, inside the mamme. This is what we found by the observation of marked syconia, a procedure which does not seem to have been frequently followed and which permits one to be sure that if it is possible that the cycle has three generations or even more, it is certain that it can have only two. The discrepancy between these two sets of observations can easily be explained. Just as for Leclerc du Sablon, we observed spontaneous trees in southern France . Conversely, all the observations which led to a cycle with more than two generations a year have been done, as far as we know, to the South of the 40th parallel. Under these warmer climates, the vegetation lasts longer (in Greece, in Syria and in North Africa pollination takes place in June, that is at least one month before it does in Montpellier and the vegetation does certainly not stop earlier); above all, insect development is faster: the first undelayed caprifig syconia, if visited, cannot get through the summer without completing a generation of insects; these must then find other undelayed syconia for egg-laying to survive. The conclusion might be that the mamme (winter caprifig syconia) are undelayed syconia in which the egg-laying insects are those emerging from the profichi in the northern parts of the area (south France, for instance), from other --earlier-- undelayed syconia in more southern region.

The observation of marked syconia shows that it is the first situation which is met in the Montpellier region. Such observations have not been done elsewhere to our knowledge but there is a fact which confirms our conclusion.

The profichi alone produce pollen and the wasps emerging from them are therefore the only ones able to fertilize the flowers of the syconia they visit. While most of the druplets inside the mamme are, in fact, "galls", each one of them containing a larva, some seeds are sometimes found in these syconia in the Montpellier region. This had already been observed by Leclerc du Sablon (1908) on a caprifig not far from those observed by us (La Roque, in the Gard department)

Conversely, in mamme received from Syria at the end of 1982, we did not find any seeds. Leclerc du Sablon (1908) cites Solms-Laubach and Eisen, both of when have never found any seed on the mamme; as far as we know, both worked in the Southern part of the fig area.

GENE FLOW VERSUS SELECTION

It seems therefore logical that the caprifig trees of the warmer regions are selected to produce undelayed syconia up to the end of the vegetative season: the earliest ones are visited by the wasps emerging from the profichi, the last ones, which will harbor the larvae in winter, by wasps emerging from the preceeding ones or even from others, intermediate as to their date of receptivity.

There would appear to be a good case in favour of widespread selection for prolonged production of undelayed syconia. Let us now see whether gene flow might be sufficient to avoid differentiation of the fig tree in southern France into an ecotype having a short undelayed syconia production.

Rather little is known about the dissemination of fig seeds. Mustelines, foxes and hedge-hogs apparently eat quite a lot of figs; maybe more important, especially for long distance transportation, are the birds such as the Sylvia genus (one Sylvia species is named " becfigue" in southern France).

Electrophoretical studies of the genetic distance between groups of fig trees growing in different locations may give a better idea of gene flow. One of us (Valizadeh, 1978) analysed from 58 to 134 trees in 17 different locations scattered all over mediterranean France and in North eastern Spain. Four loci were used (acid phosphatase, esterase, G.O.T. and peroxidase). Using the Nei measure of genetical distance it is apparent (Fig. 2) that there is no correlation between that distance and geographical distance.

We therefore conclude that gene flow is very important in the fig tree and we may even say that tree growing 500 km apart belong to the same "population". Most authors call a population a group of individuals growing in one location; they do so because they believe that most seeds are not disseminated more than some meters. For the fig tree, most figs drop under the tree and the seeds do not go further; the electrophoretical study shows however that the biologically important seeds do go far. One should therefore always remember that the word population more often refers to a geographical entity rather than to a genetical entity).

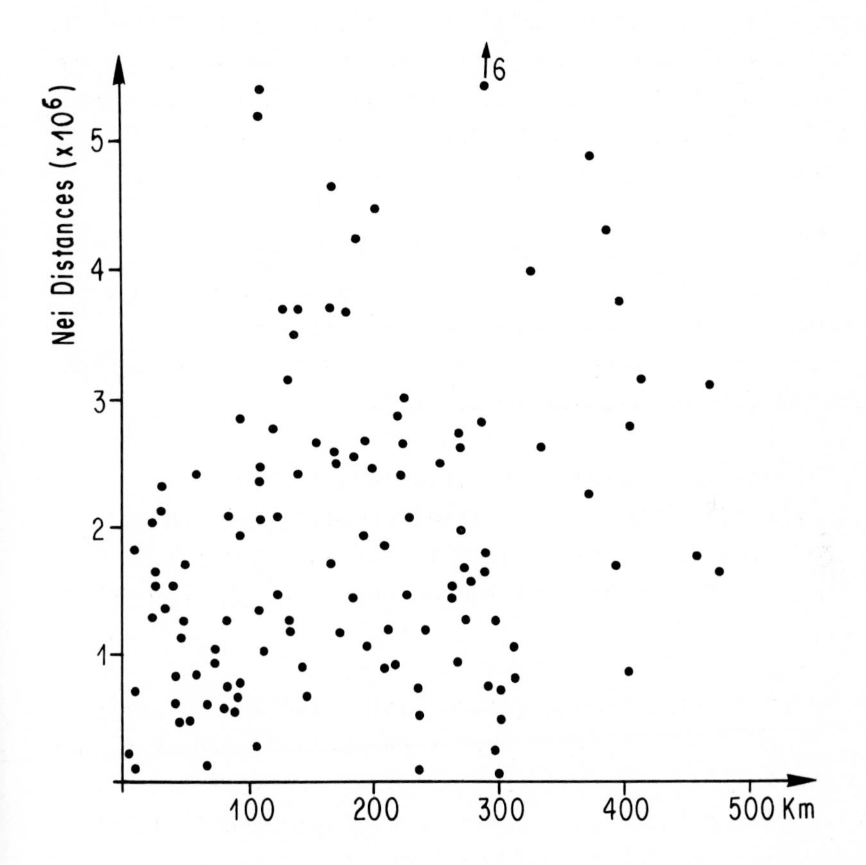

Fig. 2 - Nei distances between 17 populations scattered between North West of Spain and the Italian border of France plotted against geographical distances between the same populations (from Valizadeh, 1978)

We can therefore propose that the considerable wastage of syconia on male trees observed in southern France is the result of different selective pressures occuring in other parts of the Mediterranean basin. So even in this case of very strict coevolution a study limited to a 500 km long part of the distribution of the two species (Mediterranean France) may not be sufficient to understand the selective processes involved. One may wonder what insight into coevolutionnary processes one can get from punctual studies of the interactions between two species that happens to occur in the same location --as soon as at least one of the two species disseminates efficiently.

REFERENCES

CONDIT IJ (1947) The fig. Chronica Botanica, Waltham, Mass., 222 p.

CONDIT IJ, ENDERUD J (1956) A bibliography of the fig. Hilgardia, 25: 1-663.

GALIL J, NEEMAN G (1979) (The common fig). In Hebrew, 27 p.

GRANDI G (1920) Studio morfologico e biologico della Blastophaga psenes L. Portici R. Scuola super. di Agr. Lab. Zool. Gen. e Agr. Bol. 14: 63-204.

KJELLBERG F (1983) La stratégie reproductive du figuier (Ficus carica L.) et son pollinisateur (Blastophaga psenes L.). Un exemple de coévolution. Thèse INA-PG. Roneot. 78 p.

KJELLBERG F, VALDEYRON G (1984) The pollination of the fig tree (Ficus carica L.) and its control in horticulture. Oecol. Gener. 5: 421-426.

LECLERC DU SABLON M (1908) Observations sur les différentes formes du figuier. Congrès de l'Ass. Franç. Avanc. Sci. Toulouse, 39: 110-112.

SWINGLE WT (1899) The dioecism of the fig in its bearing upon caprification. Science (n.s.) New York, 10: 570-574.

TRABUT L (1901) La caprification en Algérie. Revue de viticulture, 16: 501-504 et 537-548.

VALDEYRON G (1967) Sur le système génétique du Figuier, Ficus carica L. Essai d'interprétation évolutive. Ann. Inst. Nat. Agro., V, 167 p.

VALDEYRON G, LLOYD DG (1979) Sex differences and flowering phenology in the common fig. Ficus carica L. Evolution,

33: 673-685.

VALIZADEH M (1978) Aspects génétiques, écologiques et agronomiques de l'étude de la variabilité des protéines chez les plantes supérieures. Cas de *Ficus carica* L. Thèse de Doctorat d'Etat. U.S.T.L. Montpellier. Roneot. 182 p.

Horizontal Structure of Populations : Migration, Adaptation and Chance. An experimental study on *Thymus vulgaris* L.

C. Mazzoni & P.H. Gouyon

Biologie des Populations et des Peuplements, Centre L. Emberger, C.N.R.S., B.P. 5051, F-34033 Montpellier cedex
& INAPG 16 rue Claude Bernard 75236 Paris cedex 05 France

ABSTRACT

The spatial structures of two thyme populations were studied using two transects. Under both transects the genetic composition of the population was recorded simultaneously with the environmental variations (as given by the vegetation composition).

The results provide insight into the fine scale variations and allow the authors to draw hypotheses about the respective roles of migration, selection and drift in the determination of the horizontal structure of populations. The results are in accord with the predictions from Wright's theoretical considerations about neighbourhoods.

INTRODUCTION

An important part of ecology deals with the patterns of distribution of individuals, populations and species and upon the mechanisms acting on them. These mechanisms can be described in different terms, using different concepts. We will try to use, in the following paper, the concepts from population genetics theory. The forces can be roughly classed into three main types (mutation, despite its major effect of introducing new genes in the species has probably little action at the scale of this study so that it

NATO ASI Series, Vol. G5
Genetic Differentiation and Dispersal in Plants
Edited by P. Jacquard et al.

will not be taken into account here).

1) First, a given individual, or genotype, must reach a site; thus, migration is obviously an important force acting on this pattern of distribution.

2) Once arrived, the individual, or genotype, is more or less fit (or adapted) that is to say able to live and produce an offspring in a given context. This results in what Darwin called natural selection. When observed at the level of a given trait, selection results in a change of the frequency of this trait. Note that the context acting on the fate of any trait must include both the external environment and the genetic composition of the individual since the different genes of a genome, which obviously interact, should be coadapted.

3) However, the presence or absence of a given genotype in a given site, or the presence of a given gene in a given individual cannot be entirely determined by these processes. Too many genes interacting at the genotypic and phenotypic levels are simultaneously involved in the existence of each individual and too many factors act simultaneously to determine whether or not the individual will establish (and, if so, where and when) so that part of the final pattern must be due to chance.

Being able to know in which way these forces interact to create the observed ecological distributions should be one of the main goals of evolutionary ecology if it was possible. Unfortunately, the action of chance cannot be directly proved, selection cannot be quantified in really natural conditions and only short distance migration events can be directly measured (long distance migration concerns such a low number of seeds per unit area that it is impossible to directly measure it). For these reasons, we have tried to understand the process of geographic and ecological distribution by an indirect approach. The pattern of distribution of a simple genetic marker has been studied at the finest possible scale (individual by individual) and the mechanisms determining this pattern have been sought.

I. MATERIAL AND METHODS

Thyme is a woody perennial species growing in northern mediterranean regions. It is polymorphic for the terpene composition of its essential oil (Granger & Passet 1971, 1973, Adzet et al. 1977). The genetic basis of this polymorphism was elucidated (Vernet 1976, Vernet & Gouyon in prep). Six chemotypes exist in the south of France, they will be referred to as G (geraniol), A (& terpineol), U (thuyanol), L (linalol), C (carvacrol) and T (Thymol). Their distribution in an area of approximately 8x10 km2 (Vernet et al. 1977b) has shown that adaptive effects had to be taken into account (Gouyon 1975, Vernet et al. 1977a) to explain this spatial pattern. Investigations on these adaptive effects have shown that predators were able to choose between these types (Gouyon et al. 1983) and that they had an effect on resistance to certain physiological constraints like heat and drought (Couvet, unpublished data). Migration of pollen is mainly achieved by bees and is thus probably over short distances (Schmitt 1980, Brabant et al. 1980). Direct observations show that seeds do not migrate and studies of seed migration on steep slopes of the garrigue confirm these observations (Bonnet et al. in prep.).

In order to investigate the finest scale at which a coherent ecological distribution could still be found, two transects of more than one hundred metres were studied. Each transect was a line under which the location of every plant of each species was noted. In addition, every individual of Thyme was analysed to determine its chemical composition. The data were treated by dividing the line into segments of one metre each. In each of these segments, the number of individuals of Thyme of each chemotype was known, as well as the presence or absence of the different species of the local flora. The floristic data enabled us to define as exactly as possible the direct ecological environment of the plants of Thyme; whether these species are indicators of edaphic differences or direct causes of ecological differences will not be discussed.

The two transects were located near Montpellier (France).

- Transect 1. was located on the Aumelas plateau, 20 km west of Montpellier. This transect measures 256m in length and passes through a slight depression of about 50m diametre and 1m depth. The vegetation here is very sparse and mainly composed of shrubs and grasses. The site is heavily grazed and regularly burned.

- Transect 2 is on the south side of the basin of Saint-Martin-de-Londres, 25km north of Montpellier. It is 112m long and is located on an outcrop of conglomerates surrounded by humid marls (a calcareous clay indicated by the presence of Schoenus nigricans) where Thyme is absent. This site has been undisturbed for at least 10 years, the vegetation is of the same type as in the previous transect but appears more homogeneous.

II. RESULTS

1) Transect 1. (Aumelas)

a) Floristic analysis.

Under each of the 256 1m long segments, the presence or absence of the 80 locally occurring species was determined. Each segment contained an average of 11 species. These data allowed us to build the 256x80 presence/absence matrix which was analyzed using correspondance analysis as discussed by Gounot (1969) and Dagnelie (1971). Note that, in order to allow this technique to provide an indication of changes in the spatial distribution of the vegetation, it is necessary that one of the axes be the geographical position of the segments themselves (Estève 1978). Consequently, in the graphic representation of the results, the horizontal axis will always give the location of the segment.

The distribution in space of the three principal components of the analysis (Fig. 1) allows us to define several different parts in the transect. As mentioned above, the variation of the values of these components will not be studied but considered as an indicator allowing us to define as generally and precisely as possible the environment of the plants.

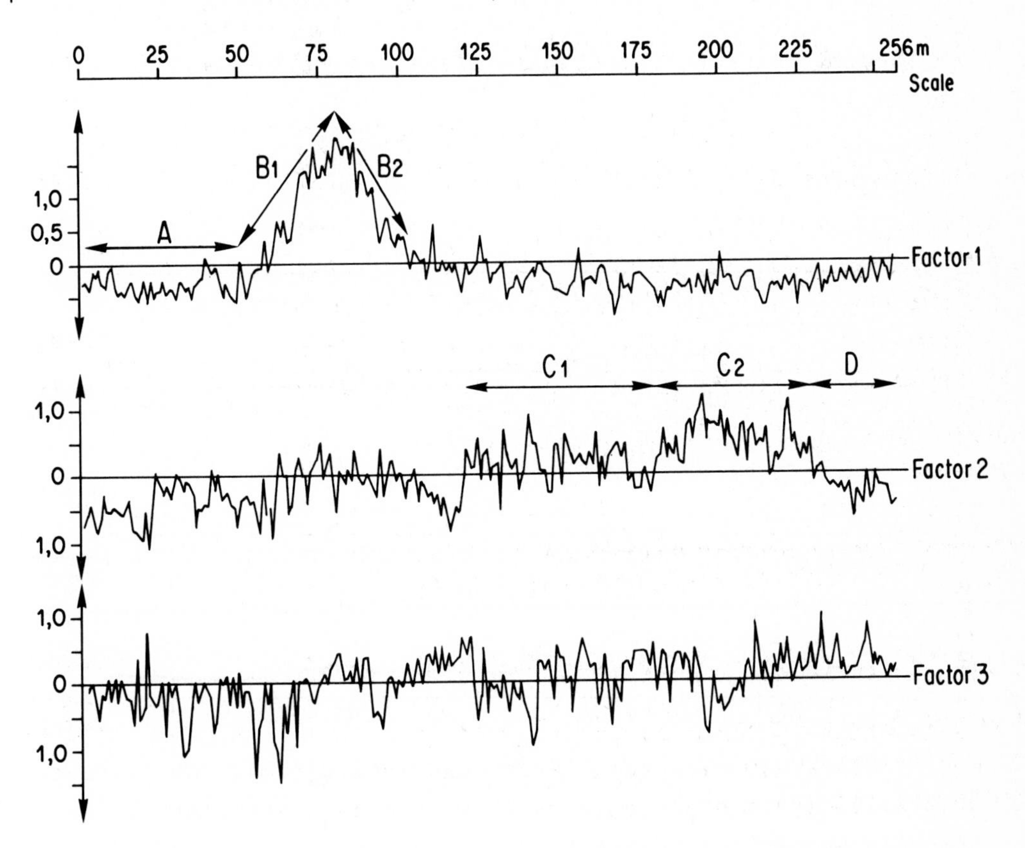

Fig. 1 Variations of the three principal factors of the correspondance analysis on the floristic composition in transect 1.

Component 1 allows us to divide the transect into three parts by distinguishing part B which begins at m60 and ends at m120. This plot corresponds to the slight depression mentioned above. This depression can be divided into two parts: B1 (m60 to m80) and B2 (m80 to m120).

Component 2 shows that parts C1 (m120 to m180) and C2 (m180 to m230) are different.

Component 3 is less clearly distributed but seems to show a difference between plots A (m0 to m60) and D (m230 to m256).

The species characterizing these plots are given in Table 1 (in appendix).

b) Chemotype distribution.

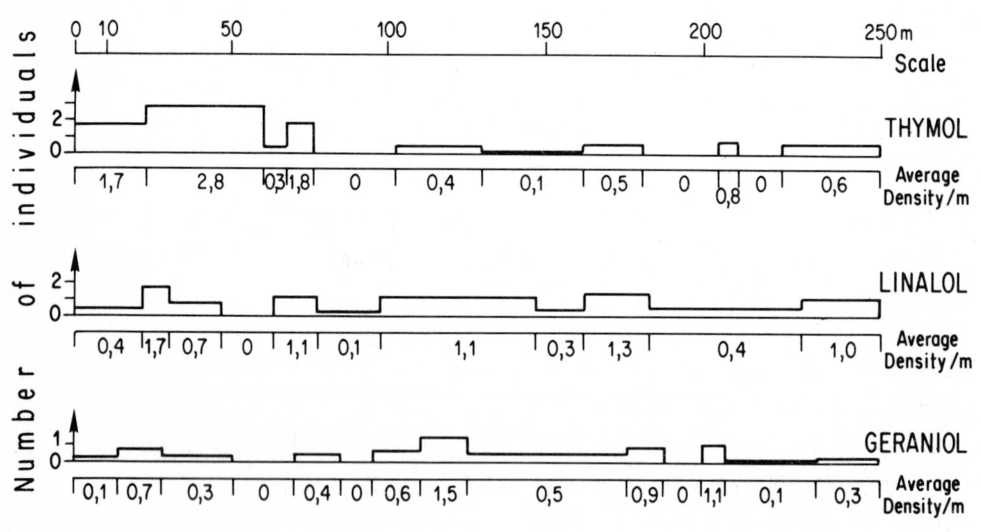

Fig. 2 Manhattan diagramme (from Ewan 1963) giving the average numbers of individuals of each chemotype per meter for each homogeneous part of transect 1.

The 505 Thyme individuals found under the transect belonged to 5 chemotypes (G, U, L, C and T) but only three were really frequent. The prevalent types are G (100 individuals), L (186 individuals) and T (211 individuals). The distribution of these types is given in Fig. 2 as a Manhattan diagram (Ewan 1963).

Since the selective effects have mainly been shown to influence the distribution of G+A+U+L (non phenolic types) versus C+T (phenolic types), the observation of the proportions of these two different types must also be considered.

These results (Fig. 3) clearly show that the floristic composition and the chemotypes found in the different parts of the transect are related (p(X2_138,5df)=0).

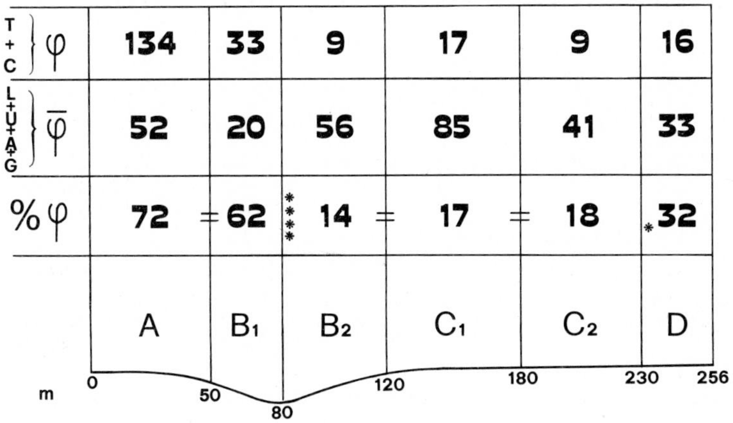

	A		B1		B2		C1		C2		D
T+C φ	134		33		9		17		9		16
L+U+A+G $\bar{\varphi}$	52		20		56		85		41		33
%φ	72	=	62	****	14	=	17	=	18	*	32
m	0		50		80		120		180		230 / 256

Fig. 3. Distribution of phenolic (T+C) and non phenolic types (G+A+U+L) in transect 1. Differences in the composition of the plots are indicated as follows.

= p _ 15%

* p _ 5%

****p _ .00001% .

- Plot A, which is characterized by the most xeric species contains a majority of phenolic types, a result which fits with the preceeding results found in another geographic region and on another scale (Vernet et al. 1977a).

- Plot B1 corresponds to a progressive variation of the floristic composition and also corresponds to a transition area for the chemical types. Note that B is a depression so that the genetic composition of B1 can be strongly influenced by the seeds produced in part A which is located just above. As a result, the composition of B1 does not differ significantly from A (p(X2_1.9,1df)=.17). On the contrary, the part B2, which represents the other side of the depression is totally protected against seeds carried in runoff from part A and contains very few phenolic types. Plot B2 can be strongly influenced by plots C1 and C2 and contains a similar chemical composition (p(X2_.4,2df)=.8). These results indicate that B1 and B2 are both influenced by the surrounding population and are well isolated so that differences between B1 and B2 are highly significant (p(X2_30,1df)=0).
- Plot D seems slightly different from plots B2, C1 and C2 (p(X2_7,3df)=.04), the proportion of individuals belonging to the phenolic types seems to increase in this plot.

2) Transect 2. (St Martin de Londres)

a) Floristic analysis.

Transect 2 was analyzed using a correspondance analysis on the presence/absence matrix (112 segments x 62 species). Although this area was chosen to be relatively homogeneous, the results (Fig.4) show the existence of an important heterogeneity at the two ends of the transect. The characteristic species of the different parts are given in Table 2 (in appendix).

- Plot A (from m0 to m25), particularly marked by factors 2 and 3, corresponds to a flora similar to that which is found on the clay next to it (and where Thyme is generally absent). A few individuals of Thyme, probably produced in the central part of the transect, were established at this interface zone.
- Plot B (m25 to m80) contains the characteristic species of xeric environments.

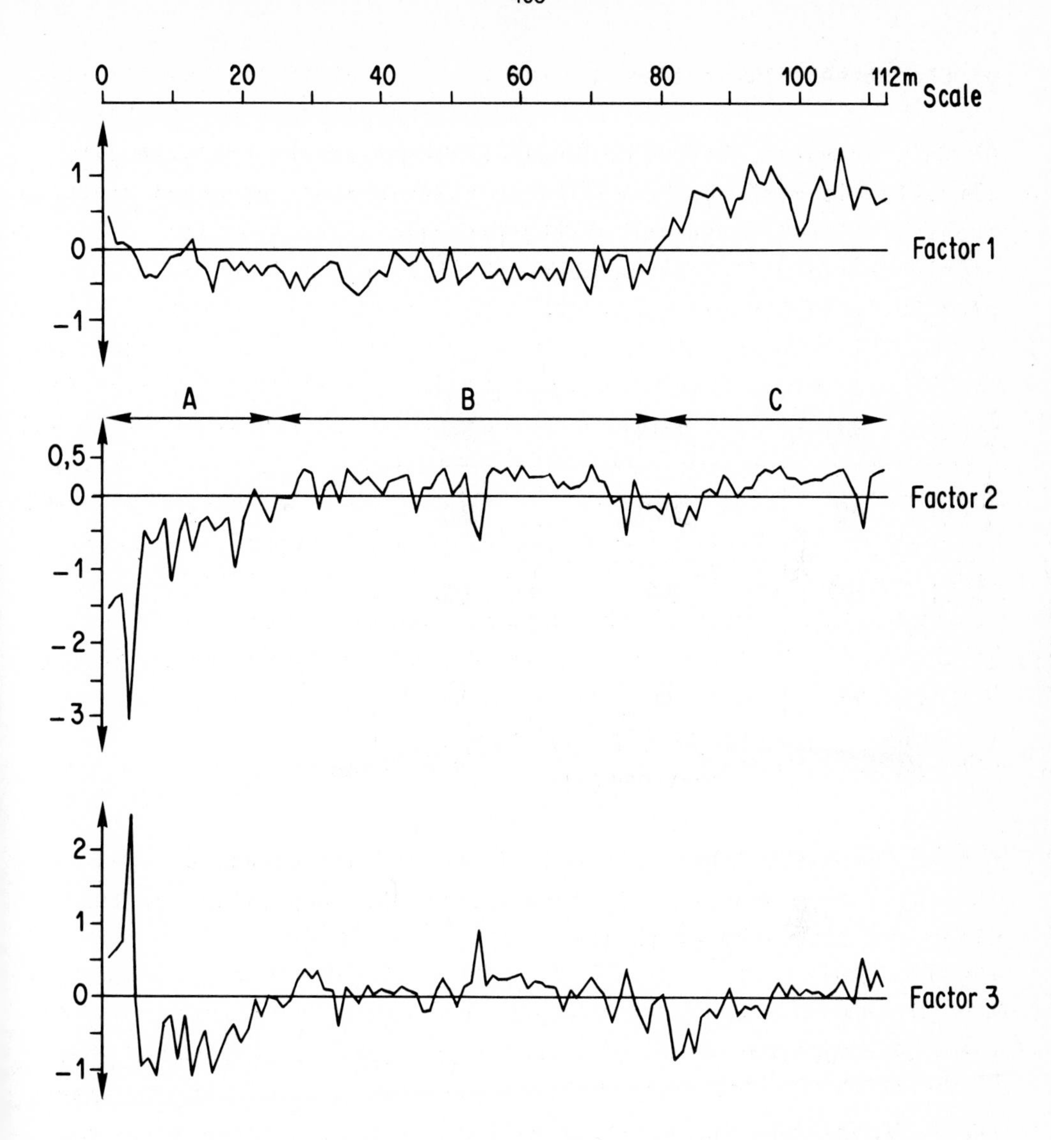

Fig. 4 Variations of the three principal factors of the correspondance analysis on the floristic composition in transect 2.

- The last part of the transect (plot C, from m80 to m112) is discriminated by factor 1 and corresponds to a higher proportion of herbaceous species (Carex spp. and Gramineae). Like plot A, the flora in plot C is related the typical flora found on the clay outside of the transect.

b) Chemotype distribution.

On this transect, 717 individuals were marked and analyzed. The situation is much more complicated than in the preceeding transect. The frequencies of the phenolic types are also significantly different between the three plots (Fig. 5, p(X2_33,2df)=.000001).

	A		B		C
T+C φ	36		294		36
L+U+A+G $\bar{\varphi}$	16		247		88
% φ	69	*	54	***	29

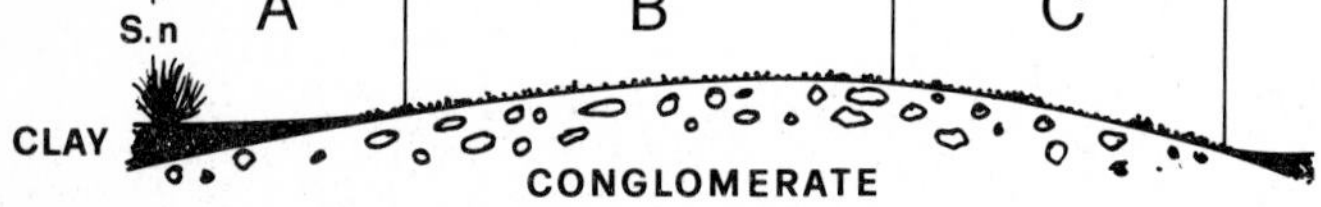

S.n
CLAY
CONGLOMERATE

Fig. 5 Distribution of phenolic (T+C) and non phenolic types (G+A+U+L) in transect 2. Differences in the composition of the plots are indicated as follows.

= p _ 15%

* p _ 5%

****p _ .00001% .

The relative frequencies show the existence of a cline (Fig. 6a). However, these data, which are the usual data treated in population genetics, do not take into account the actual frequency distribution of the individuals. The cumulative absolute frequencies should contain more information. Unfortunately these results are very difficult to interpret (Fig. 6b). A possible interesting insight can be provided us by the observation of absolute frequencies (Fig. 6c). These frequencies seem to show a pattern which can be more clearly shown

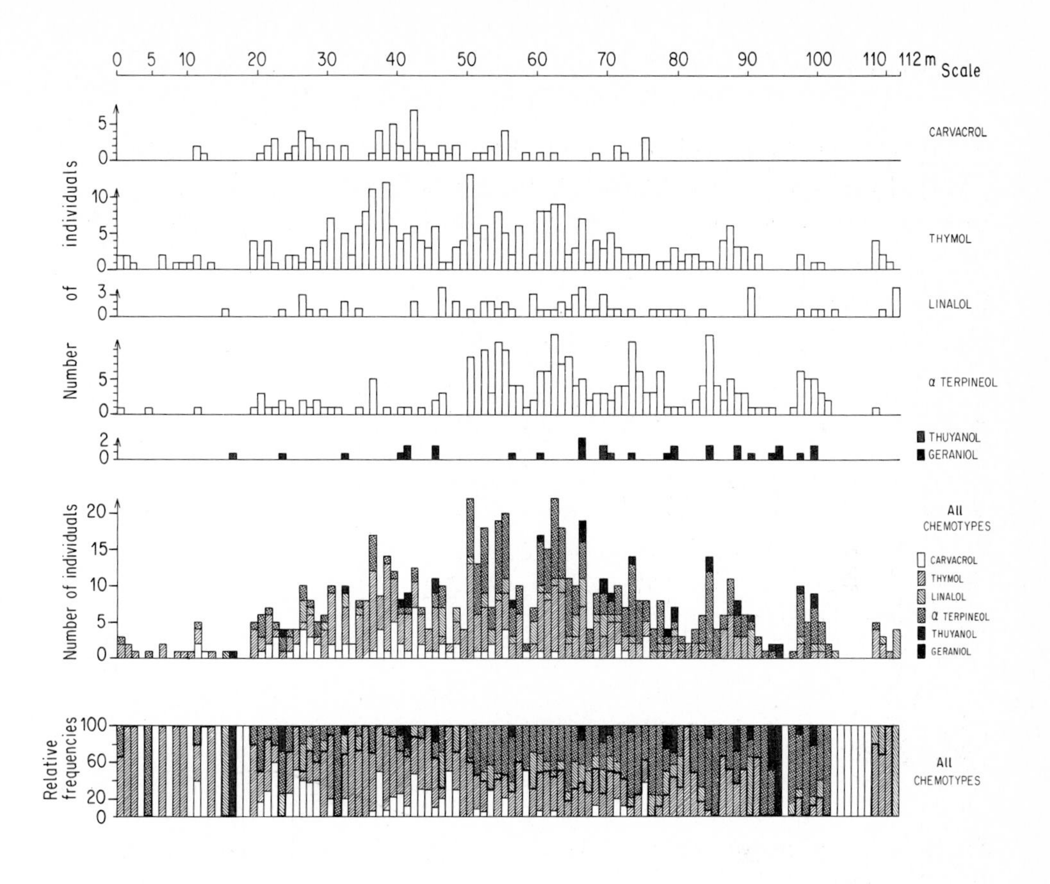

Fig. 6 Observed numbers of individuals and proportions of the different chemotypes for each meter of transect 2.

using the technique of Ewan (1963): the Manhattan diagram (Fig. 7) shows that the distribution of the chemical types on the line is not homogeneous at all. The distribution of the A type is particularly remarkable; it exhibits a relatively regular pattern of six patches of approximately 10m diametre.

Although there is still a relation between the phenolic types and the most xeric environments the distribution of the chemical types shows no clear relation with the environment as described by the floristic analysis. Even in very homogeneous parts of the transect, important differentiation of the genetic composition of Thyme population(s ?) can be found.

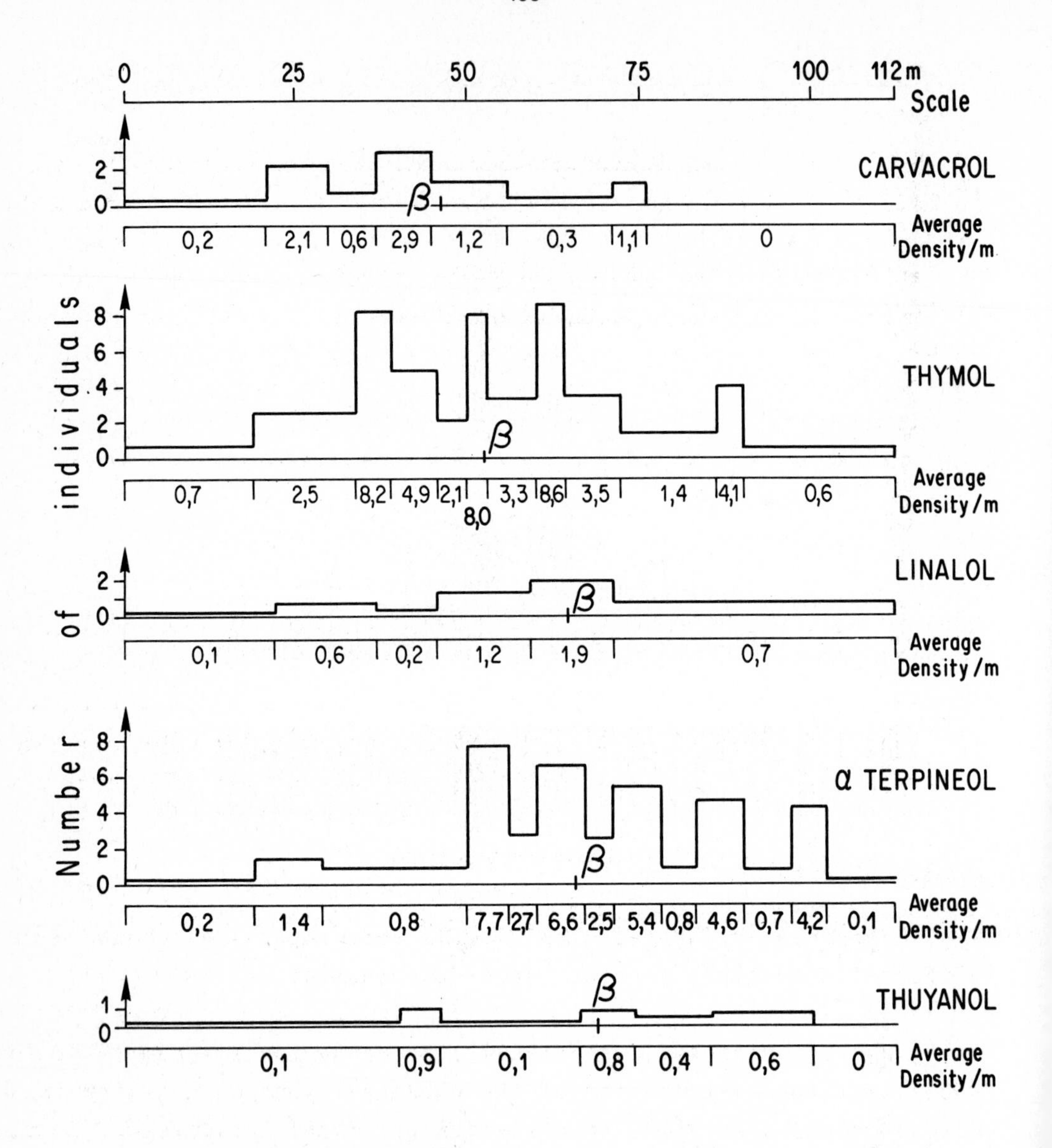

Fig. 7 Manhattan diagramme (from Ewan 1963) giving the average numbers of individuals of each chemotype per meter for each homogeneous part of transect 2.

III. DISCUSSION AND CONCLUSION

The two transects presented here exhibit both similarities and differences which can allow us to form precise ideas on the mechanisms acting on the horizontal structure of plant populations. In both cases, a definite pattern could be found but interpretation demands more detailed information and analysis. Indeed, it would be easy to conclude that, since the chemical types have been found to be of adaptive significance, the pattern found here reflects an environmental heterogeneity. However, even with environmental heterogeneity, the population is not necessarily structured. Several models (Levin 1978) have shown that the structure of the environment would create a structure in the population only if the migration distances are small enough compared to the grain of the environment. Selection (of reasonable strength) alone is thus not enough to create a definite pattern in the population, migration must also be low. When migration distances are small, theoretical models (see for instance Kimura & Maruyama 1971) show that, even without selection, spatial heterogeneity of genotypes occur. From these two types of approach, it appears that the above results must first be considered in terms of migration and not in terms of selection and/or chance. The observed pattern indicates mainly that, among the populations studied here, migration is small enough to allow and probably to produce precise differentiation even over a few metres. In that case, from an ecological and evolutionary point of view, the population level is not relevant anymore. The neighborhood (Wright 1943, 1946, 1978, Crawford 1984 and van Dijk, same issue) analysis corresponds much better to what is observed. Experimental studies published concerning this problem were designed to measure the dispersal distances of pollen and seeds (Cahalan & Gliddon in press). In some species, like *Trifolium*, the neighborhood size calculated from these results allows the authors to predict the existence of local differentiation under the effect of any pressure (i.e. selection, drift...). The present study starting from the other end of the scale, allows us to predict that dispersal distances must be very low in Thyme.

Even if low migration is both necessary and sufficient to cause local differentiation of neighborhoods, the question of the possible role of chance and necessity in the resulting pattern is still addressed.

The action of chance cannot be demonstrated. We can only infer it from the absence of any other force. The only technique which we can use is to look as carefully as possible for any selective event and conclude random determination when absolutely no selective process has been detected. The action of selection is likely to produce a simultaneous change in the genetic composition of the population when the environment changes.

This selective effect is clearly found in the study transects and particularly in transect 1 where the phenolic types, which are prevalent in the first metres of the transect are replaced within a few metres at the exact point where the floristic composition changes. However, precise insight into the structure of the population in the transition area shows that, at the scale of a metre, the composition of the population can be influenced by local migration effects (such as directionnal seed dispersal by water runoff from A to B1 and from C to B2). In this place, the environment seems to select different genotypes in the different parts of the population. This selection can result in local differentiation because migration does not mix the genotypes at each generation.

Genotypic changes without any relation to the environment, such as could be produced by random events associated with low migration, seem to appear in transect 2. In this transect, important heterogeneity is found within each plot and it is absolutely impossible to find any relationship between the patchy distribution of the chemical types (particularly A) and any environmental parametre.

We propose the following hypothesis as being the most plausible.

The actual structure of the population reflects a complex sum of events starting from the colonization of the site. If the founders exhibited a polymorphism, the very low migration of both seeds and pollen would have maintained the heterogeneity generated by these founder effects, even though the population appears approximately continuous. The spatial heterogeneity is probably not maintained by low migration alone but also by non uniform distribution of seeds and pollen (Handel 1982, 1983a, 1983b).

The distribution of the genes determining the chemical type in Thyme has been studied at different scales (Gouyon et al. 1979). The present study presents the finest possible scale and shows two main results. (1) The pattern of distribution of the genotypes can be found even at a very fine scale (an order of 10m) because of the effect of very low migration rates. (2) The differentiation of the different patches is influenced by the environment which results in directional selection.

However, even without any selective pressure, differentiation can be found which can be attributed to the effect of random events like founder effects, heterogeneous gene flow... These results indicate that, in a species like Thyme, the concept of population is not very useful since it is impossible to define what is the population in such a context. The concept of neighborhood, although it is not stable in time (the neighborhood of a given plant is very large when it colonizes a new environment, very small when it is invading this new habitat by creating a patch, and perhaps larger when the patches have covered the whole surface of the site (see also van Dijk as well as Gliddon & Saleem, same issue) describes, for such species, the level at which the evolutionary and ecological forces may be acting.

AKNOWLEDGEMENTS

The authors are grateful to M. Godron and G. Valdeyron for their scientific help during this work and to C. Gliddon and G. Williams for their scientific and linguistic comments on this text. This research was supported by a DGRST grant.

REFERENCES

Adzet T, Granger R, Passet J & San Martin R (1977) Le polymorphisme chimique dans le genre Thymus: sa signification taxonomique. Bioch. Syst. Ecol. 5: 269-272.

Brabant P, Gouyon PH, Lefort G, Valdeyron G and Vernet P (1980) Pollination studies in Thymus vulgaris L. (Labiatae). Acta Oecol. Oecol. Plant. 15: 37-44.

Crawford TJ (1984) The estimation of neighbourhood parameters for plant populations. Heredity 52 (2): 273-283.

Dagnelie P (1971) Some ideas on the use of multivariate statistical model in ecology. In:Statistical ecology, Patil, Pielou & Waters eds. 3: 167-180.

Esteve J (1978) Les méthodes d'ordination: éléments pour une discussion. In Biométrie et écologie 1: 223-250.

Ewan WE (1963) When and how to use Cu-Sum charts. Technometrics 5 (1): 1-22.

Gounot M (1969) Méthodes d'étude quantitative de la végétation. Masson, Paris 314p..

Gouyon PH (1975) Note sur la carte provisoire de la répartition des différentes formes chimiques de Thymus vulgaris. Oecol. Plant. 10 (2): 187-194.

Gouyon PH, Valdeyron G & Vernet Ph (1979) Sélection naturelle et niche écologique chez les végétaux supérieurs. Bull. Soc. Bot. Fr. Act. Bot. (2): 87-95.

Gouyon PH, Fort Ph & Caraux G (1983) Selection in seedlings of Thymus vulgaris L. and grazing behaviour of a slug. J. of Ecol. 71: 299-306.

Granger R, Passet J & Girard JP (1972) Methyl-2 methylene-6 octadiene-2,7 ol isole de Thymus vulgaris. Phytochemistry 11: 2301-2305.

Granger R & Passet J (1973) Thymus vulgaris spontané de France: races chimiques et chémotaxonomie. Phytochemistry 12: 1683-1691.

Kimura M & Maruyama T (1971) Pattern of neutral polymorphism in a

geographically structured population.Genet. Res. Camb. 18: 125-131
Handel SN (1982) Dynamics of gene flow in an experimental population of cucumis melo (cucurbitaceae). Amer. J. Bot. 69 (10): 1538-1546.
Handel SN (1983) Contrasting gene flow patterns and genetic subdivision in adjacent populations of Cucumis sativus (Cucurbitaceae). Evolution 37(4): 760-771.
Handel SN (1983) Pollination ecology, plant population structure, and gene flow. in Pollination Biology, L. Real ed. Academic press: 163-209.
Levin DA & Wilson JB (1978) The genetic implications of ecological adaptations in plants. In:Structure and Functionning of Plant Populations. Freyssen & Woldendorp eds. North Holland Publishing Company.
Schmitt JM (1980) Pollinator foraging behaviour and pollen dispersal in Senecio. In:Abstracts of ICSEB II p. 338.
Vernet Ph, Guillerm JL & Gouyon PH (1977) Le polymorphisme chimique de Thymus vulgaris ; I. Répartition des formes en relation avec certains facteurs écologiques. Oecol. Plant. 12 (2): 159-179.
Vernet Ph, Guillerm JL & Gouyon PH (1977) Le polymorphisme chimique de Thymus vulgaris L. ; II. Carte à l'échelle de 1/25000 des formes chimiques de la région de St Martin de Londres. Oecol. Plant. 12 (2): 181-194.
Wright S (1943) Isolation by distance. Genetics 28: 114-138.
Wright S (1946) Isolation by distance under diverse systems of mating. Genetics 31: 39-59.
Wright S (1978) Evolution and the genetics of populations. 4. Variability within and among Natural Populations. University of Chicago Press.

Appendix.

Table 1. Characteristic species for each plot of transect 1. (d = dominant)

Table 2. Characteristic species for each plot of transect 2. (d = dominant)

Appendix

Table 1. Characteristic species for each plot of transect 1.

Zone	Species
Zone A	*Brachypodium ramosum (d)*
	Koeleria vallesiana (d)
	Bromus erectus (d)
	Avena bromoides
	Aphyllantes monspeliensis (d)
	Anthyllis vulneraria
	Euphorbia exigua
Zone B	*Brachypodium phoenicoides (d)*
	Trifolium campestre (d)
	Medicago minima
	Bellis perennis (d)
	Potentilla reptans
	Phleum nodosum
Zone C	*Genista scorpius (d)*
	Brachypodium ramosum (d)
	Aphyllantes monspeliensis (d)
	Filago spathulata
	Linaria striata
Zone D	*Brachypodium ramosum (d)*
	Bromus erectus (d)
	Hieracium pilosella
	Koeleria vallesiana

Table 2. Characteristic species for each plot of transect 2.

Zone	Species
Zone A	*Juniperus communis (d)*
	Lavandula latifolia (d)
	Avena bromoides (d)
	Carex humilis
	Aphyllantes monspeliensis
Zone B	*Brachypodium ramosum (d)*
	Bromus erectus (d)
	Helianthemum canum
	Thymus vulgaris (d)
	Carex halleriana
Zone C	*Carex glauca (d)*
	Globularia vulgaris (d)
	Coris monspeliensis (d)
	Brachypodium phoenicoides
	Thymus serpyllum

Symbol (d) refers to dominant species

Seed Dimorphism for Dispersal : Physiological, Genetic and Demographical Aspects.

I. Olivieri and A. Berger
Biologie des Populations et des Peuplements, Centre L. Emberger, C.N.R.S., B.P. 5051, 34033 Montpellier cedex.

ABSTRACT

Seed dimorphism for dispersal characteristics is widespread and essentially found in annual plant species. Each individual of such species is able to produce two types of seeds: some adapted for dispersal and others with no obvious dispersal properties. The morphological position of these dimorphic seeds is usually constant within the plant architecture.

Using a few examples, it is shown that seed dimorphism for dispersal is often associated with physiological dimorphism with regard to dormancy and growth characteristics : dispersed seeds have no dormancy while non dispersed seeds may be dormant. Plants grown from dispersed seeds are more vigourous and produce more seeds than plants grown from non dispersed seeds, when environmental conditions are favorable. On the other hand, plants grown from non dispersed seeds are non stress-tolerant : they can produce some seeds in extreme conditions while the others cannot.

The proportion of dispersed seeds produced by a plant is variable, depending on the species, the population, and the individual. For some species, it is the product of both environmental conditions and genetic background. Moreover, for self-compatible species, seed dimorphism may be associated with different selfing rates : the non dispersed seeds being more likely to be inbred.

The evolutionary significance of dimorphism for dispersal, and its possible association with other aspects of dimorphism, are discussed.

NATO ASI Series, Vol. G5
Genetic Differentiation and Dispersal in Plants
Edited by P. Jacquard et al.

INTRODUCTION

Seed dimorphism, especially with regard to dispersal, is found in many plant species, mainly of the *Asteraceae*, *Cruciferae*, *Gramineaceae* and *Chenopodiaceae* families (Harper 1977). In these species, each plant can produce both seeds which are normally not dispersed and seeds with dispersal mechanisms. These seeds are usually produced on well-defined parts of the plant or of the inflorescence (e.g., ray and disk achenes of some *Asteraceae*). According to Harper (1977, p.71), "somatic seed polymorphism appears to be largely restricted to relatively short-lived, fugitive species, particularly weeds".

The present paper deals with some of the different aspects which may be associated with dispersal dimorphism, based on some well-studied cases. Follows a discussion on the evolutionary significance of this dimorphism.

MORPHOLOGY OF THE TWO TYPES OF SEEDS.

Seed dimorphism can usually be easily appreciated by through features other than the presence or absence of dispersal mechanisms. Table 1 summarizes the main morphological differences (including dispersal characters) observed in eight species with seed dimorphism (achenes in the case of *Asteraceae*).

The most important differences concern the size: non-dispersed seeds are usually much smaller, except for *Plantago coronopus* f. *pygmeae* (Schat 1982). The non-dispersed seeds then appear as low-cost products, whereas the dispersed seeds are more costly. This idea is emphasized by the observation of "cleistogamous species", i.e. species where each plant may produce both open, possibly outcrossing flowers (chasmogamous flowers, CH) and closed, obligate selfing flowers (cleistogamous flowers, CL): CL flowers are known to be of low cost when compared with CH flowers (Lord 1981). When dimorphism for dispersal is observed in such species, as for example in *Danthonia spicata* or *Gymnarrhena micrantha*, the dispersed seeds are produced by CH flowers and the non-dispersed by CL flowers.

Table 1. Morphological differences between the two types of seeds.

Species	Dispersed	Non-dispersed	References
Carduus pycnocephalus and Carduus tenuiflorus	Inner achenes pappus striated sticky straight cream coloured larger	Outer achenes no pappus not striated not sticky curved dark brown smaller	Bendall 1973 Parsons 1977 Olivieri 1983
Centaurea solstitialis	Inner achenes pappus cream coloured larger	Outer achenes no pappus darker smaller	Olivieri (unpubl.)
Heterotheca latifolia	Disk achenes pappus two-sided sericeous hairs no waxy coats larger embryo thinner pericarp	Ray achenes no pappus three-sided glabrous smooth waxy coats smaller embryo thicker pericarp	Venable 1979
Salicornia patula	central seeds perianth kept (high buoyancy) lightly colored larger	lateral seeds perianth lost (low buyoancy) darker smaller	Berger 1983
Gymnarrhena micrantha	aerial chasmogamous large seeds	subterranean cleistogamous small seeds	Koller & Roth 1964
Danthonia spicata	aerial chasmogamous large seeds	along stem cleistogamous small seeds	Clay 1982
Plantago coronopus	small seeds	large seeds	Schat 1982

GERMINATION REQUIREMENTS OF THE TWO TYPES OF SEEDS.

The most frequently described aspect of somatic seed polymorphism associated with dispersal dimorphism is related to the germination requirements of the two types of seeds (Table 2).

Table 2. Germination requirements and germinating behavior of the two types of seeds.

Species	Dispersed	Non-dispersed	References
Carduus pycnocephalus and C. tenuiflorus	Quick germination at a high rate	Light Slow germination at a low rate Secundary induced dormancy	Parsons 1977 Olivieri 1983
Heterotheca latifolia	No dormancy	Dormancy	Venable 1979
Salicornia patula	No dormancy Low longevity	Light,low salinity Dormancy High longevity	Berger 1983
Gymnarrhena micrantha	No dormancy	Dormancy	Koller and Roth 1964
Centaurea solstitialis	No dormancy	No dormancy	Maddox (pers. comm.)
Plantago coronopus	Slow germination at a low rate	Quick germination at a high rate	Schat 1982

The germination of smaller seeds is often light-sensitive, and indeed it was found that in the case of _Carduus pycnocephalus_, _C. tenuiflorus_, and _Salicornia patula_, non-dispersed seeds or achenes were more light sensitive (Parsons 1977; Berger 1983). Light response may determine the maximal depth from which germination may take place (Koller 1964). The larger seeds may be able to germinate in a wider range of depths than smaller seeds. The small seed size may also increase the probability of it being buried in the soil (Silvertown 1981).

While a high percentage of the dispersed seeds is usually observed to germinate quickly, the non-dispersed seeds possess a variable dormancy or have precise germination requirements (Table 2). For example, non-dispersed seeds of *Salicornia patula* require a low salinity to germinate, while dispersed seeds can germinate over a wide range of salinities (Berger 1983, Philipupillai and Ungar 1984)). In *Carduus*, as well as in *Gymnarrhena micrantha* (Koller and Roth 1964) and *Heterotheca latifolia* (Venable 1979), non-dispersed achenes are more or less dormant, while dispersed seeds have no dormancy. In *Carduus* (Parsons 1977), a secondary dormancy is induced by imbibition in the dark, and this is likely to occur for non-dispersed seeds, which are not released from the capitula.

However, Maddox (pers. com.) found no difference between dispersed and non-dispersed achenes of *Centaurea solstitialis*. Venable and Lawlor (1980) cite some cases where dispersed seeds have a higher degree of dormancy, but these examples are in a minority. In *Plantago coronopus* f. *pygmeae*, dispersed seeds also have a slower germination rate than non-dispersed seeds (Schat 1982). Clay's data (1982) on *Danthonia spicata* are not very clear: under controlled conditions, non-dispersed seeds from cleistogamous flowers, germinated better than dispersed seeds from chasmogamous flowers. However, under natural conditions, the latter showed a higher germination rate.

The rapid germination of dispersed seeds will allow an early establishment in new sites; they then appear as colonizers (escape in space). Delayed germination of non-dispersed seeds allows the maintenance of a seed-bank (escape in time). This is emphasized by the low longevity observed by Berger (1983) in dispersed seeds of *Salicornia patula*, while its non-dispersed seeds have a high longevity. The delayed germination of non-dispersed seeds also provides these seeds with an immediate selective advantage: being non-dispersed, they would be in sib-competition if they were to germinate all at the same time.

FITNESS, PLASTICITY AND STRESS-TOLERANCE OF PLANTS GROWN FROM THE TWO TYPES OF SEEDS.

Few studies have been conducted on the growth characteristics of seeds grown from the two types of seeds. Seed size may have consequences on seedling vigour. The biochemistry of dimorphic seeds has been studied to a limited extent. According to Flint (pers. com.), non-dispersed seeds of some species seem to be less protein-rich than dispersed seeds. However, the differences in energy invested per seed are essentially due to weight differences: the non-dormant morphs generally are energetically more expensive on an absolute basis (Flint, pers. com.). The higher percentages and absolute amounts of protein found in the non-dormant, rapidly germinating morphs, suggest a role of high protein content in germination performance. Indeed, it has been shown that a higher protein level increases the rate of germination (Brockelhurst and Dearman 1980). High amounts of protein apparently result in a greater protein synthesis for seedling growth (Ching and Rynd 1978) and an increased seedling photosynthesis due to more fraction 1 Protein and a large first-leaf area (Metivier and Dale 1977). In *Salicornia patula* for instance, seedlings grown from dispersed, larger seeds are much bigger than seedlings grown from non-dispersed, smaller seeds (Berger 1983).

In *Carduus pycnocephalus* (Bartolome and McLeod, in prep.) and *Heterotheca latifolia* (Venable 1979) (Table 3), plants grown from dispersed achenes produced more seeds under average conditions than plants grown from non-dispersed achenes.

These authors also showed that plants grown from the dispersed achenes of these species were more plastic, in the sense that they were able to considerably increase their seed production under good conditions, while plants grown from non-dispersed achenes showed a much lower variability in seed production.

On the other hand, plants grown from non-dispersed achenes were shown to be more tolerant to water stress, in *Carduus* pycnocephalus, Heterotheca latifolia, and Gymnarrhena micrantha. In particular, Bendall (1973) showed that in C. pycnocephalus, seedlings originating from non-dispersed morphs produced 75 per cent more roots under drought conditions than seedlings grown from dispersed achenes.

Table 3. Growth differences between plants grown from the two types of seeds.

Species	Dispersed	Non-dispersed	References
Carduus pycnocephalus	Higher fitness under good conditions More plastic	Lower fitness under good conditions Less plastic	Bartolome & McLeod (in prep)
	Intolerant against water-stress	Tolerant against water-stress	Bendall 1973
Heterotheca latifolia	Same characteristics as above	Same characteristics as above	Venable 1979
Gymnarrhena micrantha	Low water-stress tolerance	High water-stress tolerance	Koller & Roth 1964
Salicornia patula	Larger seedlings	Smaller seedlings	Berger 1983

It is commonly observed that mediterranean species, adapted to drought, have a lower performance (vigor, seed set) than their non-mediterranean equivalent species or subspecies, even if grown under high moisture conditions. This is probably a consequence of a generally slower growth rate. The same pattern occurs when comparing plants grown from dispersed seeds, which are high performing but in tolerant against water-stress, and plants grown from non-dispersed seeds, which are low performing but tolerant against water-stress.

GENETIC ASPECTS ASSOCIATED WITH DISPERSAL DIMORPHISM

The species used as examples in the preceding sections are mainly self-pollinating. Carduus pycnocephalus and C. tenuiflorus do not present any inbreeding depression of their selfed offspring (Olivieri et al. 1983); Clay (1982) estimated the outcrossing rate of Danthonia spicata to be of an average of 8.5 per cent for chasmogamous flowers; cleistogamous flowers are obligate selfers; data on other species of Salicornia indicate that they might be highly autogamous (Jefferies and Gottlieb, 1982). This can be related to Harper's observation that seed dimorphism is mainly observed in weeds (see introduction), which are well-known for their usual autogamy. More generally, colonizing species are often self-pollinated (Allard 1965).

However, when selfing was forced upon Centaurea solstitialis, a decrease of 98 per cent of the usual seed production was observed (Olivieri, unpub.). This suggests that this species is mainly cross-pollinating.

Dispersed seeds may have a higher probability of being outbred than non-dispersed seeds (Table 4).

Table 4. Inbreeding rate of the two types of seeds.

Species	Dispersed	Non-dispersed	References
Danthonia spicata	Chasmogamous outbred 8%	Cleistogamous inbred	Clay 1979
Gymnarrhena micrantha	Chasmogamous ?	Cleistogamous Obligate inbreeders	
Carduus pycnocephalus, C.tenuiflorus	Inner seeds more likely to be outbred	Outer seeds more likely to be inbred	Olivieri et al. 1983

This is obvious in cleistogamous species such as Danthonia spicata or Gymnarrhena micrantha. In studying the Asteraceae in which the two types of achenes are produced in each capitulum, we found that, because of the centripetal development inside the capitulum, and because of protandry of florets, the non-dispersed achenes, produced first , have a higher probability of being inbred than the dispersed achenes (Olivieri et al. 1983).

However, in species like *Heterotheca latifolia*, in which ray florets are male-sterile, the outcrossing rate of ray, non-dispersed achenes can be either lower or higher than of disc, dispersed achenes, depending on the flowering phenology (the male-sterile, ray florets can still be selfed if they can receive pollen from disc flowers of the same capitulum or from other capitula of the same plant) . In *Senecio vulgaris* for instance, Marshall and Abbott (1984) found a higher outcrossing rate for plants with male-sterile ray florets than for rayless plants.

ENVIRONMENTAL AND GENETIC DETERMINANTS OF THE PROPORTIONS OF THE TWO TYPES OF SEEDS PRODUCED PER PLANT

According to Harper (1977, p. 72) "somatic seed polymorphism permits a degree of sensitivity in adjustment of the proportions of morphs which is lacking in genetic polymorphism". Indeed, a variation in allocation of resources to the two types of seeds, i.e. the production of relatively more or less of one type, may change the colonizing ability, the maintenance of a seed bank, and the outcrossing rate (see above).

The range and average values of the percentages of non-dispersed seeds produced per plant of seven species (Table 5) are very variable within and between species.

In *Salicornia patula* (Berger 1983), the growing of mother plants at very low salinity (0.1 g/l NaCl) leads to the disappearance of the seed dimorphism, and only the central, dispersed seed type is produced, with no dormancy but high longevity. Seeds harvested during the autumn of 1983 in various natural populations, represented only the lateral, non-dispersedd type, and this might have been due to the dry and saline conditions during the growing period (table 6).

In *Danthonia spicata*, plants which are attacked by a fungus produce only cleistogamous flowers, i.e. only non-dispersed seeds (Kelley, this issue).

Table 5. Range and average proportion of non-dispersed seeds (data from Olivieri and Gouyon; 1985) (C. pycnocephalus and C. tenuiflorus); Olivieri (unpub.) (C. solsticialis); Venable (1979) (H. latifolia); Clay (1982) (D. spicata); Berger (1983) (S. patula); Koller and Roth (1964) (G. micrantha); Schat (1982) (P. coronopus).

Species	Percent age of non-dispersed seeds	Regulating system
C. pycnocephalus C. tenuiflorus	8 to 40% Average value: 14%	Numbers of both types per capitulum
Centaurea solstitialis	0 to 30% Average value:20%	Numbers of both types per capitulum
Heterotheca latifolia	36 to 66% Average value: 50%	Number of dispersed seeds per capitulum
Danthonia spicata	0 to 60% Average value: 30%	Numbers of cleistogamous and chasmogamous flowers
Salicornia patula	0 to 100% Average value: 66%	Resources allocation to the two types of seeds
Gymnarrhena micrantha	up to 100%	Numbers of both types of flowers
Plantago coronopus	Average value: 80%	Resources allocation to the two types of seeds

In Heterotheca latifolia (Venable 1979), as well as Carduus pycnocephalus and C. tenuiflorus (Olivieri unpub.), (table 6) the percentage of non-dispersed achenes per capitulum increases with the age of the plant. In H. latifolia, it also increases with a size decrease of the plant. This is related to the capitulum size:

when resources are limited, as at the end of the flowering period, or when plants are of a small size, capitula are smaller and bear less seeds. Geometrical constraints lead to the production of about the same number of ray, non-dispersed seeds, but less disc, dispersed seeds. However, in *Carduus*, more outer seeds per capitulum are produced, on an absolute basis. This suggests a regulating mechanism for the percentage of non-dispersed seeds produced per plant. In some cases, this percentage might therefore be determined by other factors than geometrical constraints. Clay (1982) observed the opposite in *Danthonia spicata*, bigger plants producing less dispersed (chasmogamous) seeds (table 6).

The most instructive case of environmental factors acting upon the ratio was described in Gymnarrhena micrantha by Koller and Roth (1964). Plants of this species produce only cleistogamous flowers in years when rainfall is low (table 6). It should be noticed that plants grown from seeds of these flowers are more tolerant to water stress.

The only case of demonstrated and calculated heritability of the ratio of non-dispersed vs. dispersed seeds concerns cleistogamous species. In particular, Clay (1982) estimated that for *Danthonia spicata*, the heritability was of about 50%, under greenhouse controlled conditions (table 6).

In *Carduus pycnocephalus* and *C. tenuiflorus* (Olivieri and Gouyon 1984), we observed a decrease in the percentage of pappus bearing achenes produced per plant with increasing age of the population and its position in ecological successions. Plants of old populations produced a higher absolute number of outer seeds per capitulum. Environmental factors may play a role, but a genetic component is also likely to be present, if we consider the heritability of the ratio in cleistogamous species. In this case new sites are colonized by individuals grown from dispersed seeds, i.e. seeds likely to issue from plants producing a high percentage of dispersed seeds. In these new populations, non dispersers are selected, so that old populations consist of plants producing a high percentage of non-dispersed seeds.

Table 6. Environmental and genetic determinants of the proportion of the two types of seeds (same references as before). The symbols / and -- mean "increase of" and "corresponds to", respectively.

Species	Environmental factors	Genetic factors
Danthonia spicata	/ % Non-dispersed -- / Plant size -- Low plant density	50% Heritability of the percentage
Gymnarrhena micrantha	/ % Non-dispersed -- Bad climatic conditions for water availability	?
C.pycnocephalus and tenuiflorus	/ % Non-dispersed -- / Plant age	/ % Non-dispersed with population age and degree of stability
H. latifolia	/ % Dispersed -- / Plant size / % Non-dispersed -- / Plant age	?
Salicornia patula	/ % Dispersed -- Very low salinity conditions	?

CONCLUSION

A possible origin of somatic seed diversification in relation to dimorphism for dispersal is described (Fig 1).

Figure 1. Possible origin of seed dimorphism for dispersal,

associated with physiological and genetic aspects.

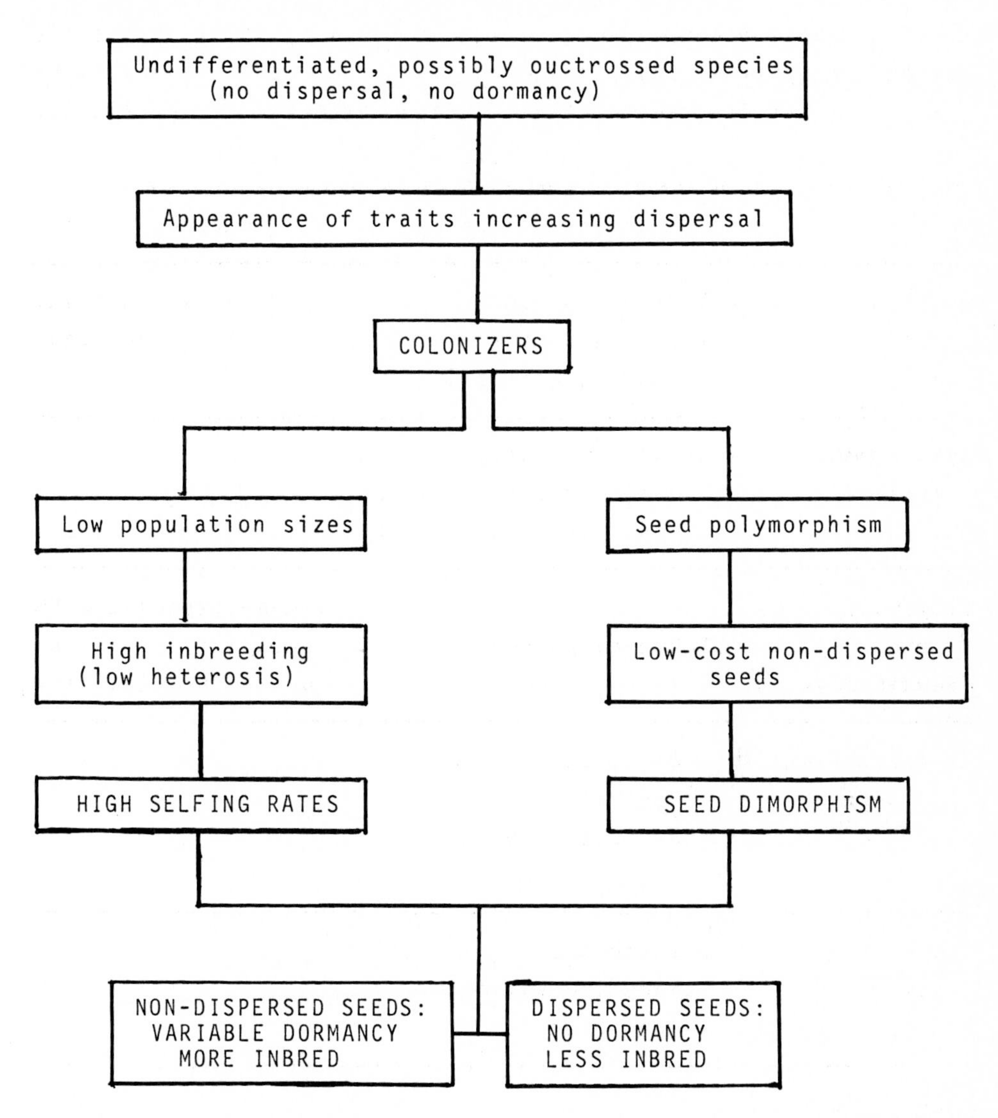

We assume an originally undifferentiated, probably outcrossing species, with no dispersal and no dormancy. In such species, some traits favoring dispersal, such as a pappus, appear. For these new colonizers, the funding of new populations will be associated with small size populations, and therefore with a high degree of inbreeding, so that some of these species might have rapidly become selfers. Seed polymorphism for dispersal, which can simply have arisen because of morphological constraints, could have been selected for if the non-dispersed seeds have the advantage of being produced at low costs. This seed polymorphism would in that case have led to complete dimorphism for dispersal. Since species had become selfers, selection probably acted in favor of variable dormancy in non-dispersed seeds, in order to escape sib-competition. It also allowed for the maintenance of a seed-bank. In most cases dormancy was probably not selected for in dispersed seeds as they first had to establish in new sites; they might however have been selected as being more outcrossed. Indeed, the wider their genetic basis, the higher their chances of funding new populations in new sites. Also, since dispersal features are often not very efficient, most of these seeds do not fall very far from the mother-plants, and are therefore likely to be in situations of sib-competition, as they have no dormancy. It is then advantageous for them to be outbred. A final reason for outcrossing dispersed seeds is given by Antonovics'results (this issue): he found that the advantage of being outcrossed decreased with the distance to the mother-plant, on a short scale.

The lower fitness of non-dispersed seeds is compensated for by their low-cost (they are usually smaller), their higher inbreeding rate (more genes of the mother plant are transmitted), their delayed germination (escape in time). Dispersed seeds, with no special germination requirements in most cases, will germinate as soon as they are in suitable habitats. The high plasticity of plants grown from these seeds will allow a quick response to environmental variation.

The problem here is to determine what can and what cannot have been selected. For instance, can plants really "choose" to outcross the seeds they disperse, can they "choose" to produce more or less of one type or another, and so on. Could selection have lead to the production of bigger non-dispersed seeds in most cases? This has yet to be determined, before entering a fully "adaptationist programme" (as defined by Gould and Lewontin 1979).

A trait may have an adaptive advantage although it has been selected for another purpose in the past (Gould and Vrba 1982) or at another level (Gouyon and Couvet 1984). This problem of discriminating between the "possible" and the "obligate", that is, the study of genetic and physiological correlations between characters and their consequences upon fitness, might constitute an important interface between physiological and genetic studies. response to environmental variation.

REFERENCES

Allard RW (1965) Genetic system associated with colonizing ability in predominantly self-pollinated species. In: Baker HB, Stebbins GL (eds) The genetics of colonizing species, Academic Press, NY, London, pp 50-78

Bartolome JW, McLeod SA (in prep) Population characteristics and control of Italian thistle

Bendall GM (1973) Some aspects of the biology, ecology and control of slender thistles, *Carduus pycnocephalus* L. and *Carduus tenuiflorus* Curt. (*compositae*) in Tasmania. M Agric Sc Thesis, Univ of Tasmania

Berger A (1983) Dimorphic germination behaviour in *Salicornia patula* Duv-Jouve. Int symp "Towards a synthesis of vegetation science, population dynamics and ecophysiology in coastal vegetation", 22-25 March 1983, Haampstede

Bocklehurst PA, Dearman J (1980) The germination of carrot (*Daucus carota* L.) seed harvested on two dates: a physiological and biochemical study. J Exp Bot 31: 1719-1725

Ching TM, Rynd L (1978) Developmental differences in embryos of high and low protein wheat seeds during germination. Plant Physiol 62: 866-870

Clay K (1982) Environmental and genetic determinants of cleistogamy in a natural population of the grass Danthonia spicata. Evolution 36 (4): 734-741

Gouyon PH, Couvet D (1985) Selfish cytoplasm and adaptation: variations in the reproductive system of thyme. 2nd Int Symp on Struct and Funct of Plant Popul, 7-11 May 1984, Wageningen, Woldendorp (Ed) (in press)

Gould SJ, Lewontin RC (1979) The sprandels of San Marco and the panglossian paradigm: a critique of the adaptationnist programme In "The evolution of adaptation by natural selection", J Maynard Smith, R Holliday (Ed.) London

Gould SJ, Vrba ES (1982) Exaptation -a missing term in the Science of form- Paleobiology 8(1) : 4-15.

Harper JL (1977) Population biology of plants London. Academic Press

Jefferies RL, Gottlieb LD (1982) Genetic differentiation of the microspecies Salicornia Europeae L. (sensu stricto) and S. ramosissima J Woods. New Phytol 92: 123-129

Koller D (1964) The survival value of germinating-regulating mechanisms in the field. Herbage Abstracts 34: 1-7

Koller D, Roth N (1964) Studies on the ecological and physiological significance of amphicarpy in Gymnarrhena micrantha (Compositae). Amer J Bot 51: 26-35

Lord EM (1981) Cleistogamy : a tool for the study of floral morphogenesis, function and evolution. Bot Rev 47: 421-450

Marshall DF, Abbott RJ (1984) Polymorphism for oucrossing frequency at the ray floret locus in Senecio vulgaris L. III.Causes. Heredity 53(1) (in press)

Metivier JR, Dale JE (1977) The utilization of endosperms reserves during early growth of barley cultivars and the effect of time of application of nitrogen. Ann Bot (Lond) 41: 715-728

Olivieri I, Swan MC, Gouyon PH (1983) Reproductive system and colonizing strategy of two species of Carduus (Compositae). Oecologia 67: 114-117

Olivieri I, Gouyon PH (1985) Seed dimorphism for dispersal: theory and observations. 2nd Int Symp on Struct and Funct of Plant

Popul, 7-11 May 1984, Wageningen, Woldendorp (Ed) (in press)

Parsons WT (1977) The ecology and physiology of two species of Carduus in Victoria. PhD thesis, Univ of Melbourne

Philipupillai J, Ungar IA (1984) The effect of seed dimorphism on the germination and survival of Salicornia europaea L. populations. Amer J Bot 71(4):542-549

Schat H (1982) On the ecology of some dutch sclack plants. PhD Thesis, Univ of Amerstam

Silvertown J (1981) Seed size, life span, and germination date as coadapted features of plant life history. Am Nat 118: 860-864

Venable DL (1979) The demographic consequences of achene dimorphism in Heterotheca latifolia Buckl. (Compositae): germination, survivorship, fecundity and dispersal. PhD Thesis, Univ of Texas, Austin

Venable DL, Lawlor L (1980) Delayed germination and dispersal in desert annuals: escape in space and time. Oecologia 46: 272-282.

Heterocarpy, Fruit Polymorphism and Discriminating Dissemination in the Genus *Fedia* (*Valerianaceae*)

J. MATHEZ and N. XENA de ENRECH*

Laboratoire de Systématique et Ecologie Méditerranéennes
Université des Sciences et Techniques du Languedoc
Institut de Botanique, 163 rue Auguste Broussonet
34000 Montpellier, France

ABSTRACT

The genus *Fedia* (*Valerianaceae*) is constituted by annual entomogamic plants which are found in the southern part of the Mediterranean Basin. Each species of *Fedia* shows both heterocarpy and genetic polymorphism in the fruits (monospermal achenes). Heterocarpy associates persistent and deciduous fruits on each individual. Each species exhibits two morphs differing in the shape of their deciduous fruits. Both morphs are present in each population. Deciduous fruits of *F. cornucopiae* are inflated in the [G] morph and narrow in the [E] one. The [G'] morph of *F. graciliflora* has inflated deciduous fruits, while the [C'] morph presents horny ones. In a similar manner, *F. pallescens* has a [G''] morph and a [C''] morph with, respectively, inflated and horny deciduous fruits. A first experimental study of *F. cornucopiae* and *F. graciliflora* leads to the conclusion that each species is dimorphic for a single locus with the allele for inflated deciduous fruits dominant upon the alleles for narrow (*F. cornucopiae*) or horny (*F. graciliflora*) ones.

The object of this paper is to report the first results related to the determination of the fruit shapes as well as the observations and hypothesis of the dispersal patterns which are peculiar to each type of the fruits. Fruits G and E, which are the only ones provided with elaiosom, are abundantly found with their embryos intact in the refuse piles gathered around the entrances of the *Messor* nests. The hypothesis of an ecological mechanism based upon differential dissemination, which would maintain the polymorphism and determine its evolution in the genus, deserves a close examination, particularly since other very similar cases of fruit polymorphism exist within american species of other genera of *Valerianaceae*, *Valerianella* and *Plectritis*.

* Universidad Central de Venezuela, Facultad de Ciencias, Escuela de Biologia, Departamento de Botanica, Apartado 21201, Caracas 1020, Venezuela.

NATO ASI Series, Vol. G5
Genetic Differentiation and Dispersal in Plants
Edited by P. Jacquard et al.

INTRODUCTION

The genus *Fedia* consists of annual plants occuring in the south of the Western Mediterranean Basin (from the southern part of the Iberian peninsula to Macedonia, including North Africa, the Mediterranean Islands and Southern Italy). A preliminary revision of this genus (Mathez 1984) shows a taxonomical situation complicated by morphological particularities appearing at fructification. *Fedia*'s terminal and lateral inflorescences are cymose, dichasial at their base and monochasial at their distal parts. Fruits, dry and indehiscent, with 3 locules (1 monospermic and fertile and 2 sterile ones), have their shape determined both by their position in the inflorescence and by the genotype of the individual that carries them. Thus, heterocarpy and polymorphism operate simultaneously.

Presently, we are carrying out the inventory of the species and their variations using traditional biosystematic methods : chorology, floristics, numerical taxonomy, karyology, fine morphology (palynology, anatomy), biochemistry (flavonoids) and experimental hybridization ; we are alloting a special interest to the polymorphism analysis of these species (Mathez and Xena de Enrech, in preparation).

The object of this paper is to report the first results related to the determination of fruit shapes as well as the observations and hypothesis on the dispersal patterns peculiar to different fruit types of the three species already described (Mathez 1984) : *Fedia cornucopiae* (L.) Gaertn., *F. pallescens* (Maire) Mathez, and *F. graciliflora* Fisch. and Meyer. Finally, we present research prospects raised by this genus' peculiar morphology and biology.

FRUIT SHAPE TYPES AND THEIR DETERMINATION

Heterocarpy

A morphological and functional gradient determines the shape of fruits from the infructescence bases upto their tips : fruits located at the dichasial lower zone are generally prismatic, showing poorly developed and almost solid sterile locules, and are often provided with a calyx reduced to a few irregular teeth (Fig. 1 : P). Furthermore these fruits have the peculiarity of remaining firmly bound to their bearing axis after maturity is reached. Thus,

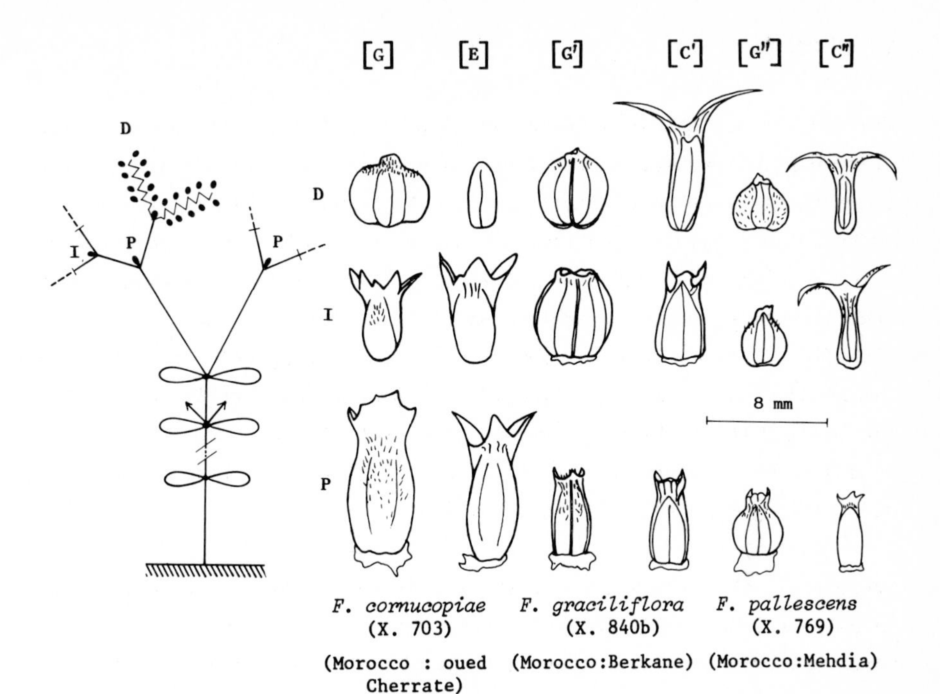

Fig. 1 - Schematic representation of an individual of the genus *Fedia* which shows, on the left, the position, within the inflorescence, of persistent (P), intermediates (I) and deciduous (D) fruits. On the right : the morphology of each fruit category (P, I and D) for every morph of every studied species. In parenthesis : sample's collection of number and sampling site location. Morphs based on different deciduous fruits are labeled as follows : [G], [G'], [G''] for inflated, [E] for narrow, [C'] and [C''] for horny. Additional information in the text.

they are persistent (P), unlike those of monochasial ends, which are easily (or very easily) deciduous (D) at maturity. However, at the tip of the dichasial zones some intermediate fruits may be found assuring by their shape and their solid attachment the transition between P and D fruit types (Fig. 1: I). In *Fedia cornucopiae* and *F. graciliflora*, the infructescences' axis thicken at dichasial zones, becoming fistular and easily detachable, thus breaking down into diaspores that associate low density axis fragments with P fruits. These diaspores can easily roll on the ground surface in windswept open regions (van der Pijl 1972 : 59), or be carried away by running waters (Ernet 1978 : 105). Very likely, stronger winds (particularly ascending whirlwinds) occasionnally allow an anemochoric dispersal of a greater range.

Polymorphism

As was long ago suspected (Sernander 1906 ; Ernet 1978 ; Pignatti 1982 ; Mathez 1984) polymorphism in *Fedia*'s fruits is now a certainty

(Mathez and Xena de Enrech in preparation). It affects almost exclusively the deciduous fruits of the monochasial zones (Fig.1: D). In all sampled sites, the three studied species appear normally dimorphic and in every one of them, one of the two morphs is distinguished by an important development of sterile locules in the deciduous fruits, which results in a blown up, inflated shape (type G fruits, from french "gonflé") that defines morphs [G], [G'] and [G"]. On the contrary the other morph's deciduous fruits in every species have in common poorly developed sterile locules which remain narrowly cylindrical. However, depending on the species, these fruits appear under two different shapes. In *F. cornucopiae* (Iberian peninsula and Western Morocco) these narrowly ovoid and calyxless deciduous fruits constitute type E (from french "étroit") and characterize morph [E]. In *F. pallescens* (Central and Western Morocco) and *F. graciliflora* (from Eastern Morocco to Macedonia as well as in the Mediterranean Islands) these deciduous fruits exhibit a calyx outgrowth consisting in 2 (seldom 1 or 3) appendages or horns and constitute type C (from french "cornu" : horny) defining morphs [C'] and [C"] in every species. These two species, both associating similar morphs with G and C deciduous fruits, are distinguished by other morphological characters (corolla's colours, dimensions and symmetry, Mathez 1984).

These three types of deciduous fruits, G, E and C are so different that, until very recently, they were considered as belonging to three distinct species. The right-hand side of figure 1 sums up the distribution of the different fruit shapes according to their position in the inflorescence and to any of the morphs of the three species described here.

Raised in the field, the polymorphism presumption has been reinforced by first experimental cultures. Thus, the cultivation of field collected fruits yields individuals with unforeseeable fruit types. In *F. cornucopiae*, for instance, a plant born of an E fruit may as well yield E as G type fruits, and reciprocally a plant born of a G fruit can yield either G or E fruits. In *F. graciliflora* or in *F. pallescens* we have found the same situation where a C fruit yields a plant carrying either C or G fruits, and a G fruit yields a plant with either G or C fruits. On the other hand we have never been able to obtain an E fruit plant in *F. graciliflora* nor in *F. pallescens*, or a C fruit in *F. cornucopiae*.

In *F. cornucopiae*, self-fertilization of E plants has yielded a 108 [E] plants progeny, without any [G] plant, while in *F. graciliflora*, self-fertilization of [C'] plants has yielded 23 [C'] ones and not a single [G'] plant : these are expected results for recessive homozygotes. On the other hand, nothing similar has been obtained with G fruit individuals of these two species : in *F. cornucopiae*, self-fertilization of a [G] plant has yielded a 9 [G] and 2 [E] progeny ; in *F. graciliflora*, self-fertilization of a [G] plant has yielded an 8 [G'] and 2 [C'] progeny. Finally, the complete results of those cultures, which included cross-fertilizations between morphs (Mathez and Xena de Enrech in preparation) confirm the hypothesis of one single locus with two alleles and also the one of dominancy of G fruit's allele over the other one, this in both species. As for *F. pallescens*, the control analysis is not finished yet, but up to now, it seems to lead to similar results. Whichever these precise modalities of genetic control may be, the polymorphism shown in the genus *Fedia* makes these plants a convenient and practical-experimental material since concerned characteristics are particularly evident.

In sampling more than 150 collecting sites, across most of the genus' area, we have not found any population of any taxon which is not at least dimorphic in its fruits' shape. In april 1983 and april 1984 we collected large quantities of *F. cornucopiae* individuals in several sites without taking their morphs into account. For 7 samples of 3 collecting sites from Southern Spain and one from Western Morocco, summing up 351 individuals, morph [E] is always preponderant with a mean frequency of 73% ([G]/[E]: 1/25 ; 5/15 ; 3/22 ; 33/94 ; 45/56 ; 7/29 ; 2/14).

In *F. graciliflora*, morph frequencies are closer to each other : an average of 52% for [C'] and 48% for [G'] for a total of 239 individuals collected in 8 sites in Eastern Morocco and Western Algeria ([G]/[C]: 11/6 ; 7/4 ; 32/44 ; 3/9 ; 2/8 ; 11/16 ; 48/38). Finally, in *F. pallescens*, three samples summing up 53 individuals gave an average of 88% [C''] and 12% [G''] ([G]/[C]: 1/28 ; 2/6 ; 1/6 ; 0/7 ; 2/24 ; 6/21). In this species, morph [C''] seems the most frequent within a majority of sites.

These observations suggest that, in spite of fluctuations, selective pressures maintain polymorphism within *Fedia*'s species with comparable morph frequencies.

The *Valerianaceae* family presents other cases similar to this one, in North-American species belonging to two other genera, *Plectritis* (Ganders *et al.* 1977a, b) and *Valerianella* (Eggers-Ware 1969). Hence fruit polymorphism could very well be an ancestral character of the family. Such fidelity may not be fortuitous.

We therefore dare to postulate a working hypothesis which assumes that selective pressures are exerted directly and in a very differentiated manner upon the different fruit types ; thus maintaining some kind of balance between the two morphs of each species. This is why we concentrate our interest on the dissemination patterns proper to the different fruit types observed.

DISSEMINATION PATTERNS OF DIFFERENT FRUIT TYPES OF *FEDIA*

We will not mention the persistant fruits' dissemination, discussed above. The deciduous fruits dissemination, poorly known, is presently still a speculation subject (Ernet 1978 : 103). Yet, already in 1902, by the means of simple and convincing experiments, Sernander proved the myrmecochory proper to certain fruits of *F. graciliflora* from Sicily. Obviously, this is the most remarkable of all dissemination types of *Fedia*.

Myrmecochory (Dissemination by Ants)

"Harvesting" ants (genus *Messor* and other related genera) pick up *Fedia*'s fruits fallen on the ground, take them into their nest galleries and, later on, take them out along with other refuses that they pile up around certain holes and are sometimes found in great quantities. Embryos of fruits that have stayed in the galleries are intact and germinate easily. On the contrary, the pericarp of G and E type fruits, during this transit, has suffered the ablation of a zone which is located at the junction of sterile locules. Specific colorants (Sudan III) show in this zone the existence of epidermic (and eventually hypodermic) cells, rich in fats, that constitute an elaiosom (Bresinsky 1963). Figure 2 shows these elaiosoms as they appear in each of the three species, after being stained by Sudan III, in front and cross sections. In *F. cornucopiae*, deciduous fruits of both morphs are provided with an elaiosom and are myrmecochorous, while in *F. graciliflora* and *F. pallescens*, only deciduous G fruits have an elaiosom. The cross section shows that the elaiosom is diametrically opposed to the seed ; when ants eat it they never hurt the embryo.

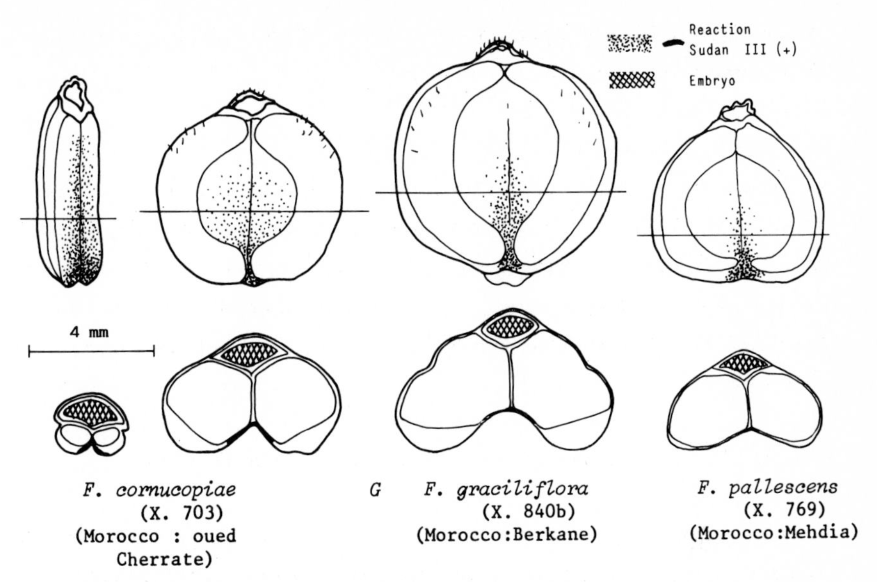

Fig. 2 - Semi-schematic representation of myrmecochorus fruits in three species of the genus *Fedia*. Elaiosom position is revealed by its reaction to Sudan III. On the top : a frontal view (sterile locule side) ; bottom : cross section of the fruit. In parenthesis, sample's collection number and sampling site location.

The study of some refuse piles gathered at different seasons around *Messor*'s ant-nests shows that G and E fruits of *F. cornucopiae* are abundantly harvested, but *F. pallescens*' C fruits (devoid of elaiosom) are practically absent. In *F. graciliflora* our observations are yet insufficient but Sernander (1906) reports the carrying of only G fruits (and possibly not all of them). Finally, P fruits are never found in the refuse piles. Hence, it looks as *Messor* showed a peculiar selectivity while harvesting. However, when the whole range of fruits are manually proposed on the ground to active harvesting ants, all of them are carried away, including C and P fruits. Our recent observations are not acute enough to reveal a preferendum. However, the harvest of P fruits is excluded in natural conditions and that of C fruits is probably very limited ; indeed either in natural or in culture controlled conditions, C fruits break off much less easily than G or E ones. In september and october, the dead and dry *F. pallescens* plants still have C fruits attached while all G are gone long ago. Furthermore, C fruits can yet be found on the ground, which allows us to believe that ants have lost their interest after the harvest's maximal phase. Likewise, it can be supposed that C fruits and elaiosom fruits do not have the same transit pattern through ant-nests, and that their calycinous outgrowth represents an obstacle to their transportation within and along the galleries. However, all the above demands a study programm with methodical observations and experimentations.

Other Disseminating Modes

Ernet (1978 : 104) reports the possibility of elastic projection (ballochory). This procedure actually works for G and E fruits, which fall very easily far before the plant's axis dries up and becomes fragile. It is immediately followed by myrmecochory, hence reaching an extraordinary efficiency : thus it is almost impossible to find G or E fruits on the ground under the plants that have produced them. Regarding C fruits, they do not detach easily enough to be projected by the axis which quickly turns fragile when becoming dry.

On the contrary, the morphology and the presentation of C fruits seem very well adapted to a dissemination by Mammal's fur (epizoochory). An indirect argument suggesting the assumption of the efficiency of this procedure is supplied by the finding of *F. graciliflora*, as a casual alien, in diverse sites of Southern France : at Pont-Juvénal of Montpellier and Toulon (Cosson 1860 : 612), at Bédarieux (herbier Sennen in Barcelone, BC) and at Hérépian (Coste et Sennen 1894 : 106) : in all these sites the richness in adventitious species at that time is well known and related to an imported wool industry then prosperous.

Limits and Possible Effects of Selective Dissemination

In spite of the preliminary character of the available data, we believe that the dissemination of deciduous fruits of different individuals from a single population of *F. graciliflora* or *F. pallescens* is very different for each morph : horny C fruits are disseminated at the foot of the mother plant (late barochory) or at a long range in open environments visited by furry mammals (presently sheep) ; while inflated fruits are efficiently concentrated at medium range in the areas of implantation of *Messor* ant-nests.

In *F. graciliflora* myrmecochorous fruits are born of individuals with (G-) genotype. With every generation they perform a preferential flow of dominant alleles towards ant-nests. Hence, a spatial diversification of the population genetic structure could be expected in relation with the ant-nests distribution : periodical exportation of myrmecochorous fruits should progressively purify some part of the population of their G alleles and create recessive homozygote sub-populations when sufficient isolating conditions are given. As a matter of fact, such a segregation, unfavorable as a rule to polymorphism persistence, encounters quite a number of obstacles : gene flows following upon pollen and other fruit types (P and C) dispersal, eventual ant-nests' or their refuse holes' displacement from one year to the next, high ant-nests density, are the first obstacles that come to our mind. However, we plan to study in a bigger scale the biocenosis of *Fedia* and harvesting ants to test for eventual correlations between morphs and ant-nests distributions.

Regarding *F. cornucopiae*, both types of deciduous fruits are efficiently dispersed and only a much more acute study will confirm whether ants show a preference for any of them under certain circumstances (abundant fruit production year, for instance). Actually, our first observations have revealed that G type fruits are much more abundant than those of type E in a few refuse piles even though they are in a much lower number in the living surrounding population.

SOME THOUGHTS ON FUTURE RESEARCH ORIENTATIONS

Many species show, as these of the genus *Fedia*, different dispersal patterns : it may be easily concluded that this ability can be an acquired selective advantage. But to prove this act of faith we do not yet have sufficient information and data.

In the first place, myrmecochory is an example of mutualism where the advantage obtained by the vegetal species dispersed is not always obvious. Regarding *Fedia* it seems that *Messor* are frequently found in open environments (cultures, fallow fields, orchards, sparse woody formations) which are convenient to *Fedia* and shield it against elimination by the pressure of progressive vegetation dynamics. But many works (Davidson and Morton 1981 ; Heithaus 1981 ; Milewski and Bound 1982) report more precise advantages for other species : protection against predators (particularly rodents) during certain seasons, higher concentrations of some nutrients (K, N) in the soil around ant-nests, diaspore protection against fires... A modification of the germinating capability of diaspores after elaiosom ablation can also be conceived. Our first observations do not show any confirmation of this hypothesis in *Fedia*. On the other hand, the modalities of introducing diaspores into the protection of ant galleries could interfere with the germination phenology inducing ecological consequences : this seems to us one of the more promising questions to further explore.

Additionally, once all respective characters of different types of dispersal under different ecological conditions are known, as well as the advantages obtained by one given species in holding that plurality of types, a reflexion arises. Many are the species that, like *Fedia*, are morphologically and functionally heterocarp (or heterosperm, which recalls, cf. Van der Pijl 1972 : 84). But in *Fedia*, heterocarpy is reinforced by a real genetic polymorphism acting upon the diaspores. This phenomenon seems rare outside the *Valerianaceae* family : which peculiar advantage does it add to the sole heterocarpy ?

Finally, supposing that the advantages of all these complex patterns are understood, there remains to explain the mechanism of their stability ; is the sole differentiated action of selective pressures upon the different fruit types able to maintain polymorphism ? Or will other less obviously operating environmental pressures be needed, as in heterozygote advantage for example ?

The answer to all these questions requires first, as was shown earlier, a thorough field study, for which the cooperation of ant behavior specialists is absolutely necessary.

ACKNOWLEDGEMENTS

We are especially grateful to Professor G. Valdeyron for his comments on the manuscript.

REFERENCES

Bresinsky A (1963) Bau, Entwicklungsgeschichte und Inhaltsstoffe der Elaiosomen. Biblioth Bot 126: 1-54

Cosson E (1860) Appendix florulae juvenalis. Bull Soc Bot Fr 6: 605-615

Coste H, Sennen F (1894) Plantes adventices observées dans la vallée de l'Orb à Bédarieux et à Hérépian. Bull Soc Bot Fr 41: 98-113

Davidson DW, Morton SR (1981) Myrmecochory in some plants (F. *Chenopodiaceae*) of the Australian Arid Zone. Oecologia 50: 357-366

Eggers-Ware DM (1969) A revision of *Valerianella* in North America. PhD These Vanderbilt University, Nashville

Ernet G (1978) Fruit morphology and dispersal biology of *Valerianella* and *Fedia* (*Valerianaceae*). Pl Syst Evol 130: 85-126

Ganders FR, Carey K, Griffiths AJF (1977a) Outcrossing rates in natural populations of *Plectritis brachystemon* (*Valerianaceae*). Can J of Bot 55: 2070-2074

Ganders FR, Carey K, Griffiths AJF (1977b) Natural selection for a fruit dimorphism in *Plectritis congesta* (*Valerianaceae*). Evolution 31: 873-881

Heithaus ER (1981) Seed predation by rodents on three ant-dispersed plants. Ecology 62 (1): 136-145

Mathez J (1984) Introduction à une révision du genre *Fedia* Gaertn. Mem Soc Brot 27: 129-175

Mathez J, Xena-Enrech N (in preparation) Le polymorphisme génétique de la morphologie des fruits du genre *Fedia* Gaertn (*Valerianaceae*). I. Détermination du mécanisme de contrôle génétique chez les espèces *F. cornucopiae* (L.) Gaertn. et *F. graciliflora* Fisch. et Meyer

Milewski AV, Bond WJ (1982) Convergence of myrmecochory in mediterranean Australia and South Africa. In: Buckley RC (ed) Antplant interactions in Australia. Junk, The Hague, p 89-98

Pignatti S (1982) Flora d'Italia, vol 3. Edagricole, Bologne, p 2324

Van der Pijl L (1972) Principles of dispersal in higher Plants, 2nd edn. Springer-Verlag, Berlin

Conference Commentary

Janis Antonovics
Botany Department
Duke University
Durham, North Carolina 27706 U.S.A.

The past few years have seen a burst of conferences on plant population biology. Many of these conferences, like the present one, have been directly concerned with the interactions between ecology, genetics and physiology. There is clearly a widely felt urgency to find interfaces, bridge them, and thereby to integrate different disciplines and to exchange ideas. The expected products of this integration are creativity and new directions. There is no question that plant population biology is now a vigorous area of inquiry. So where does this need for new directions come from? Is it simply a natural manifestation of the vigour of a growing area, or, as some cynics might say, has the firework fizzled and is it in need of new spark and inspiration? As I sit here a few months after the conference, reading the transcript of the final discussion and reflecting back on the papers, I think the answer is all and more of the above. Indeed, since many of the comments made in the concluding discussion reflected both the history and the current crises in the subject, it seems appropriate to summarize some of that discussion here.

A historical perspective was provided by Pierre-Henri Gouyon. As a student ten years ago he felt excited by the arrival of plant population ecology. Its arrival in France was as in other countries of Europe, and as in the United States. It came (it must be said, its incarnate form was John Harper!), it was seen, and it certainly conquered an ecology rather moribund with the excesses of descriptive phytosociology and "multi-factor ecology." The latter was an autecology based on the assumption that if only we could measure all the environmental factors impinging on every species we would then understand their ecology and distribution. Plant population ecology promised a more dynamic view dealing with process, it was placed in a Darwinian context and integrated with evolution, and its Harperian origins gave it a thoroughly experimental tradition. The success of this invasion was alluded to by a number of speakers during the final discussion.

NATO ASI Series, Vol. G5
Genetic Differentiation and Dispersal in Plants
Edited by P. Jacquard et al.

Pierre-Henri Gouyon told us that in France "now Population Biology exists; even in the same corridor you can find geneticists, physiologists, ecologists, together, even not fighting, even discussing!". Mike Hayward also told us we should be encouraged that ecologists and physiologists are now getting together and looking at common problems. Other speakers were more pessimistic, more yearning for direction and unity. Pierre Jacquard felt that we had been given a tool, demography, but not a science, and that our pride in being able to measure more things than zoologists may be leading us into a new descriptive phase. He feared continued progress might be in the direction of measuring smaller and smaller plant parts! George Williams also remained perplexed: "One of the things that struck me first at the Population Biology Conference in Waageningen and continues to puzzle me now is what is the central theme of population biology, what is the question that is being asked by population biologists"..." there seems to be no unification or even any way to get at unification."

When Pierre Jacquard surmised that "perhaps some of our defects are the result of past successes" I think he went a long way towards explaining the source of the disquiet expressed by many speakers. Plant population biology conquered precisely because John Harper showed us how it could be done, and how it could be done in an exciting way. The subject was not born because there was some pre-ordained issue to be solved (cf. the development of population genetics as a means of reconciling Mendelian processes with evolutionary change), nor did it have a strong theoretical base. Its self-thinning laws were empirically derived; the competition theory of De Wit was a gem, but borrowed more from the physical chemist than from the plant physiologist; and while a vast theoretical literature existed in animal ecology and population genetics, these were two separate strange and alien ponds, in which few botanists had the tools or the courage to fish. Could it have been otherwise? During the discussion Jim Cullen, confessing to being a zoologist watching from the sidelines of the meeting, pointed out that a unifying force in animal population biology had historically been the question of what regulates populations. This question had its source in a powerful theory of population regulation and in real-world contrasts such as the subtly synchronized cycles of Lynx and Hare versus the wildly chaotic cycles of Australian insects. The question of population regulation has persisted in animal ecology, perhaps no longer with the full force of the density-dependent/independent battle, but nevertheless with sufficient intensity to generate sophisticated approaches to population dynamics, and to theoretical and empirical determination of community structure. In contrast we have almost no theories of plant population dynamics. This is true at both the conceptual and at the mathematical model building level. Thus not only do we hardly know or even discuss, beyond decrying Lotka-Volterra models, which if any population models are valid for plants, but it seems not to be an issue whether density dependence occurs, whether metapopulation concepts are useful, or how component dynamics affect community process. In the discussion Steve Kelley reminded us that "What you see is dependent on your theory and if the theory is not developed to any degree of sophistication, you end up doing a lot of unproductive research." While I strongly support an eclectic approach to all science and therefore would not exhort everyone to go out and do theory, there is a need for such theory to balance the historically empirical basis of the discipline.

The difficulties of integrating physiological, ecological, and genetic approaches do not simply stem from the absence of common questions. Often, highly conflicting paradigms form the fabric of the different disciplines. For example, I have argued that explaining the distribution of a species is not an ecological but a genetic question. This is perhaps an extreme view, but it stems from the notion that any explanation should account for evolutionary failure and for why a population does not evolve beyond its existing limits. So many current explanations are purely ecological and wedded to the assumption that to explain distribution one only has to find physiological tolerance limits. Evolutionary issues are therefore by-passed and integrative approaches delayed.

Another somewhat related problem came up in discussion. It was pointed out by Pierre-Henri Gouyon that evolutionary biology is in a crisis. "One of the best signs of a crisis in any field is when a scientist begins to question its fundamental concepts: thus adaptation has been questioned as a useful concept simultaneously by Gould and

Lewontin, and by Harper"..."We are in an active destructive phase." As I mentioned during discussion, this is a serious point of inhibition between physiological ecology and evolutionary ecology. Physiological ecologists assume the paradigm of adaptation, assume populations are different in a way that is adaptive. In contrast, evolutionists are seriously trying to destroy the concept of adaptation or at least get by without it. It is no wonder communication is at times difficult!

During the discussion I outlined a hierarchy of types of interactions that can occur between disciplines. We can interact socially. We can use each other's techniques. We can work on the same species. Or we can ask shared questions. (Appropriately, at this point during the conference, a clap of thunder was heard as a Mediterranean storm passed overhead.) Asking shared questions is the hardest but also potentially the most valuable way to interact. One thinks of a chemist and biologist (with a little borrowing from an X-ray crystallographer) combining to resolve the structure of DNA. Nearer to home, one thinks of geneticists combining with ecologists in the early seventies to merge models of selection with models of population growth and so formally intergrate theoretical population ecology and theoretical population genetics. However such instances are in my opinion very rare. Certainly this conference produced no agreed upon question. I personally shared beer, shared ideas on techniques, and enjoyed listening to others describe studies with some of my past, present, and perhaps future research organisms! Indeed, much of science proceeds more by insights that come from these modest hybridizations, rather than by some saltation, some interdisciplinary allopolyploid, that sends us back to our laboratories clutching a golden question.

The second focus of the conference was to examine the role of gene flow and dispersal in the population biology of plants. One outcome of the papers and the ensuing discussions was that, by the end of the conference, we were all very hesitant to use the word "population." Although this surprised me at the time, I now think that this was to be expected. The term "population" is not only used differently by ecologists and geneticists, but the use of summary population statistics (such as neighborhood area, or gene frequency) obscures the complexity of pollen and seed dispersal events. These events may have consequences for the fitness of individuals, they may therefore themselves be the objects of selection, and they will affect genetic structure and propensity for differentiation. A focus on dispersal events also takes us out of the conventional view of populations as isolated units having only an internal dynamic and leads us to consider the broader scale of inter-population dynamics.

Another clear impression that I was left with after the second part of the conference was that the documentation of genetic variability should no longer occupy the center stage when we consider factors determining the rate and direction of evolution. Demographic structure could be all important. For example, a few large established individuals in a perennial population could be seen as a gerontocracy making disproportionate genetic contribution and so preventing or slowing evolutionary change. We are all aware of the homogenizing effects of gene flow, but evidence was presented of such effects among fig "populations" over distances that we think of as being geographic in dimension. And genetic correlations were shown to play a crucial role in selection response for life-history traits in natural populations. Certainly many factors, and not just simply the "raw amount" of variation, influence evolutionary rates and directions. Moreover, in their eagerness to measure genetic variation, population geneticists have perhaps forgotten that most of the traits of importance to the physiologist and ecologist show continuous variation. Variation in such traits cannot be easily quantified in terms of specific allelic frequencies at specific loci. These characters are therefore of little use for measuring levels of variation in a way that permits "character-free," unbiased comparisons across taxa. This concern with genes that are clearly identifiable electrophoretically or as overt polymorphisms could explain in large measure the lack of interactions between ecological physiological and genetic approaches that were alluded to earlier. Recent theories of evolution of quantitative traits will give population geneticists tools for studying ecologically and physiologically relevant traits. Conversely, the ecologist and physiologist will see genetic techniques as accessible, and if still not easy to master, at least not leading them down a seemingly endless avenue of population statistics. Thus a view of evolution more firmly based in ecology and quantitative genetics may well bring about the integration we seek.

It remains for me to again thank the organizers of this symposium. Georges Valdeyron and Pierre Jacquard were largely responsible for much of the initial planning and conceptualization. Georges Valdeyron also planned a memorable field trip into the countryside of the Montpellier area, which included field sites, castles, a meal (or should I say feast), and an organ concern in the Romanesque Abbey of St. Guilhelm-le-Desert. Rosalyne Lumaret impeccably handled the registration and general organization of the conference. Georges Heim did much of the editorial leg-work, and with Pierre Jacquard, coped with author-reviewer interactions. We thank them, the authors, and the participants, for a memorable meeting.

SPECIES INDEX

SUBJECT INDEX

NATO ASI Series G

Vol. 1: Numerical Taxonomy. Edited by J. Felsenstein. X, 644 pages. 1983.

Vol. 2: Immunotoxicology. Edited by P.W. Mullen. VII, 161 pages. 1984.

Vol. 3: In Vitro Effects of Mineral Dusts. Edited by E.G. Beck and J. Bignon. XIII, 548 pages. 1985.

Vol. 4: Environmental Impact Assessment, Technology Assessment, and Risk Analysis. Edited by V.T. Covello, J.L. Mumpower, P.J.M. Stallen, and V.R.R. Uppuluri. X, 1068 pages. 1985.

Vol. 5: Genetic Differentiation and Dispersal in Plants. Edited by P. Jacquard, G. Heim, and J. Antonovics. XVII, 452 pages. 1985.